VOLUME 1 ARCHAEOLOGY OF THE MAYA

Maya Ceramic Technology and Ceramic Socio-economy

A multifaceted analysis of Late Postclassic ceramic production and distribution in Northern Yucatán, México

CARMEN GIOMAR SÁNCHEZ FORTOUL

BAR INTERNATIONAL SERIES 2899

Published in 2018 by
BAR Publishing, Oxford

BAR International Series 2899

Archaeology of the Maya, 1
Maya Ceramic Technology and Ceramic Socio-economy

ISBN 978 1 4073 1640 6

COVER IMAGE *Small test vessels used for experimental pottery making and firing.*

Printed in England

BAR
PUBLISHING

BAR titles are available from:

BAR Publishing
122 Banbury Rd, Oxford, OX2 7BP, UK
EMAIL info@barpublishing.com
PHONE +44 (0)1865 310431
FAX +44 (0)1865 316916
www.barpublishing.com

ARCHAEOLOGY OF THE MAYA

Series Editors: Elizabeth Graham (UCL), David Pendergast (UCL) and Genner Llanes Ortiz (Leiden)

The archaeology, epigraphy, art and history of the Maya of Mesoamerica are increasingly studied throughout the world. As discoveries multiply, new venues of publication are a critical means of spreading knowledge both within and outside the field. The BAR sub-series publishes monographs and excavation or survey reports on the Maya area in a format that permits extensive illustration, a feature not found in the common American Royal (229 mm × 152 mm) book format. Conference proceedings and revised doctoral theses/dissertations are also considered.

If you are interested in publishing in the *Archaeology of the Maya* series, please contact editor@barpublishing.com.

ACKNOWLEDGEMENTS

I would like to express my deepest appreciation to the people and institutions that made this research possible. I would like to thank my family, Armando, Cristina, Verónica, and Mercedes for their support and understanding during all the years of this project, and to my father and mother who taught me the value of perseverance and hard work.

This book is based on my Ph.D. project. My deepest appreciation goes to Dr. Ian Whitbread, my Ph.D. supervisor, for his constant guidance and suggestions that profoundly shaped and benefited this study. His expert knowledge in a wide range of practical and theoretical areas greatly influence me and this project. I am also grateful to Dr. Clifford Brown, who supervised my master project about Mayapán, from which this study emerged. I appreciate his ongoing help, guidance, and expert knowledge about Mayapán and all things that are Maya. He greatly benefited this project.

I am grateful to Professor Elizabeth Graham (UCL Institute of Archaeology) and Dr. Richard Thomas, my Ph.D. examiners, for their insightful observations. Professor Graham commented profusely on this study, and this work and I benefited greatly from her comments.

I would like to extend my gratitude to the Mexican institutions and people for allowing me to analyse archaeological ceramics of México. In particular, I would like to thank Dra. María de los Angeles Olay Barrientos, Consejo de Arqueología, Coordinación Nacional de Arqueología, Instituto Nacional de Antropología e Historia (INAH), México and the Arqueólogo (Arqlg.) Eduardo López Calzada, Delegado Centro INAH Yucatán, for providing the necessary permits for this research (Oficio 401.B(4)19.2013/36/0926).

I am very grateful to all the archaeologists who provided the ceramic materials for this study, including the Arqlg. Alfredo Barrera Rubio, Arqlg. María José Con, Arqlg. Enrique Terrones González, Arqlg. Carlos Peraza Lope, and Dr. Clifford Brown. I am particularly grateful to Arqlg. Carlos Peraza Lope for his help throughout the years of this project.

I am also thankful to the Department of Geosciences, Florida Atlantic University (FAU), Boca Raton, Florida for providing access to the petrographic microscopes. I would like to thank Pittsburgh Mineral & Environmental Technology, Inc. (PMET), and, in particular Mr. Randolph Shannon, who expertly performed x-ray diffraction of a clay and coarser portions. In addition, my gratitude goes to Dr. Daniel Pierce and Dr. Michael D. Glascock of the Archaeometry Laboratory at the University of Missouri for performing the chemical analysis (supported in part by NSF grant BCS-1415403 to the Archaeometry Laboratory at the University of Missouri Research Reactor). I want to thank them for going the extra mile in the analysis of the chemical data. I also want to thank Dr. Leslie Cecil for providing her Belize data and for her assistance during chemical analysis, without which some of the results of this study would not be reached.

I would like to thank the individuals who contributed to this project providing their assistance. Thanks to Luis Flores Cobá for some of the pottery illustrations and his assistance throughout the years. I am also thankful to the Arqlg. Wilberth A. Cruz Alvarado for his time and help of many years. My gratitude goes to Don Fernando Cobá for his help during the search for raw materials and for sharing with me his deep knowledge of the area. I am also very grateful to the four anonymous readers of this book. I deeply appreciate their knowledgeable comments and suggestions, which greatly benefited this work. Finally, I thank Chris Myers of the BAR publishing team for his patience and contribution during the editing process.

Madrid Codex (from Léon de Rosny, "Codex Cortesianus", Libraires de la Société d'Ethnographie, París, 1883, p. 40)

TABLE OF CONTENTS

LIST OF FIGURES

LIST OF TABLES

ABSTRACT

This book is about the Maya who inhabited northern Yucatán, México, during the centuries around the European arrival, the Late Postclassic (AD c. 1100-1500). The Europeans arrived only decades after the abandonment of Mayapán, the last Maya capital in this region. The nature of Late Postclassic (LP) society is not well understood, primarily because its socio-economic nature eludes researchers. To gain a better understanding of the Late Postclassic social and economic environment in northern Yucatán, this study focuses on ceramic production, a crucial socio-economic aspect of the ancient Maya society.

Through a combination of analytical methods that include petrographic, chemical, and surface features analyses, this research reconstructs ceramic production technology and examines regional patterns in this technology for vessels from Mayapán, minor north-central sites, and eastern sites. The research questions can be summarised as follows. Were there patterns in raw material selection and ceramic technology reflecting zones of production, groups of potters, technological traditions, or other social divisions? What might such patterns tell us about the organisation of production and the nature of interactions, including networks of ceramic exchange and technological traditions that may reflect social divisions or integration?

The results of this study have provided new insights into the technology involved in the manufacture of Late Postclassic ceramics. The potters who produced these ceramics had a deep working knowledge of the very challenging raw materials available for pottery making in northern Yucatán.

This study shows that raw material selection is not random and follows patterns determined by overarching technological traditions dictating the appropriate materials for different types of vessels at different geographical locations. These traditions can be traced back to, at least, the northern Terminal Classic, and continue up to the present. Furthermore, cultural divisions between Mayapán and minor north-central sites may be reflected in the data. Divisions are also observed between eastern fabrics in utilitarian vessels and fabrics in north-central vessels. At least two orientations to production are emerging because the association between raw materials and types of vessels at Mayapán differs from minor north-central sites. This study found that many potters' groups may have supplied the minor north-central centres, while the compositionally and technologically homogeneity of Mayapán pottery points to few raw materials sources.

New perspectives have been gained on long-standing questions about the location of zones of productions, the extent of the export and import of Mayapán vessels, and the mechanisms of exchange that may have been involved in this movement of vessels. Mayapán imported few vessels and exported many. However, vessels with Mayapán fabric style are found only at sites less than two days' march from this major site. A ritual context or a limited sub-regional market context may explain this movement. These findings have informed current views about Late Postclassic ceramic production, exchange, and advanced our understanding of the socio-economic nature of this period.

Chapter 1
INTRODUCTION

This book is about the inhabitants of northern Yucatán, México, during the last centuries before the European arrival (1511, the shipwreck of Aguilar). They are part of a people known as the Maya, and the period of study is the Late Postclassic (AD c. 1100-1500). The Europeans arrived only decades after the abandonment of Mayapán, the last Maya capital in this region. The Late Postclassic is then crucial to gain a better understanding of the state of northern Yucatán (Figure 1-1) society at the time of contact and the impact of this contact on the Maya people. Despite extensive ethnohistorical records and current research, the nature of the Late Postclassic Maya society is not well understood, primarily because details of its socio-economic nature still eludes researchers, prompting the proposal of multiple and many times conflicting perspectives about this period.

To gain a better understanding of the social and economic environment in northern Yucatán during the Late Postclassic, this study focuses on ceramic production, a socio-economic aspect of the Maya society. Through a combination of analytical methods that include petrographic, chemical, and surface features analyses, this research reconstructs ceramic production technology and examines regional patterns in this technology for vessels from Mayapán, secondary north-central sites, and eastern sites. These patterns are examined in term of the wider context, including ethnohistorical, ethnographic, and geological to gain a better understanding of ceramic manufacture, production organisation, and networks of interactions, such as ceramic technological traditions and mechanisms of exchange. These are crucial aspects for understanding the socio-economic environment of this society and informing current perspectives about this critical period.

The questions about the socio-economic environment that this study is addressing can be summarised as follows. Were there patterns in raw material selection and ceramic technology that might reflect zones of production, groups of potters, technological traditions, or other social divisions? What might such patterns tell us about the organisation of production and the nature of interactions, such as networks of ceramic exchange and technological traditions that may reflect social divisions or site integration?

New insights have been gained on the manufacture of Late Postclassic ceramics. The ceramics from Mayapán have usually been regarded as of less quality than ceramics from previous periods. However, the results show that the potters who produced these ceramics had a deep working knowledge of the very challenging raw materials available in northern Yucatán.

This study showed that raw material selection is not random and follows patterns determined by overarching technological traditions for different types of vessels at different geographical locations. These traditions can be traced back to, at least, the northern Terminal Classic, and continue up to the present. Furthermore, cultural divisions between Mayapán and minor north-central sites are reflected in the data.

At least two orientations to production are emerging. The association between raw materials and types of vessels at Mayapán differs from minor north-central sites. In addition, many potters' groups may have supplied the minor north-central centres, contrasting with the compositional and technological homogeneity of Mayapán pottery. Divisions in ceramic technology are also observed between vessels from north-central sites and vessels from sites further to the east.

New perspectives have been gained on long-standing questions about the location of zones of productions, the extent of the export and import of Mayapán vessels, and the mechanisms of distribution that may have been involved in this movement of vessels. Mayapán imported few vessels and exported many. However, vessels with Mayapán fabric style were found only at studied sites located less than two days' march from this major site. A ritual context or a limited sub-regional market context may explain this movement. These findings have informed current views about Late Postclassic ceramic production and exchange and advanced our understanding of the socio-economic nature of this period.

This chapter introduces the Late Postclassic and the challenges and problems it raises for researchers. It summarises overarching views that have been proposed about the nature of the Late Postclassic. The objectives of this project are presented in detail, and the methodology selected is outlined.

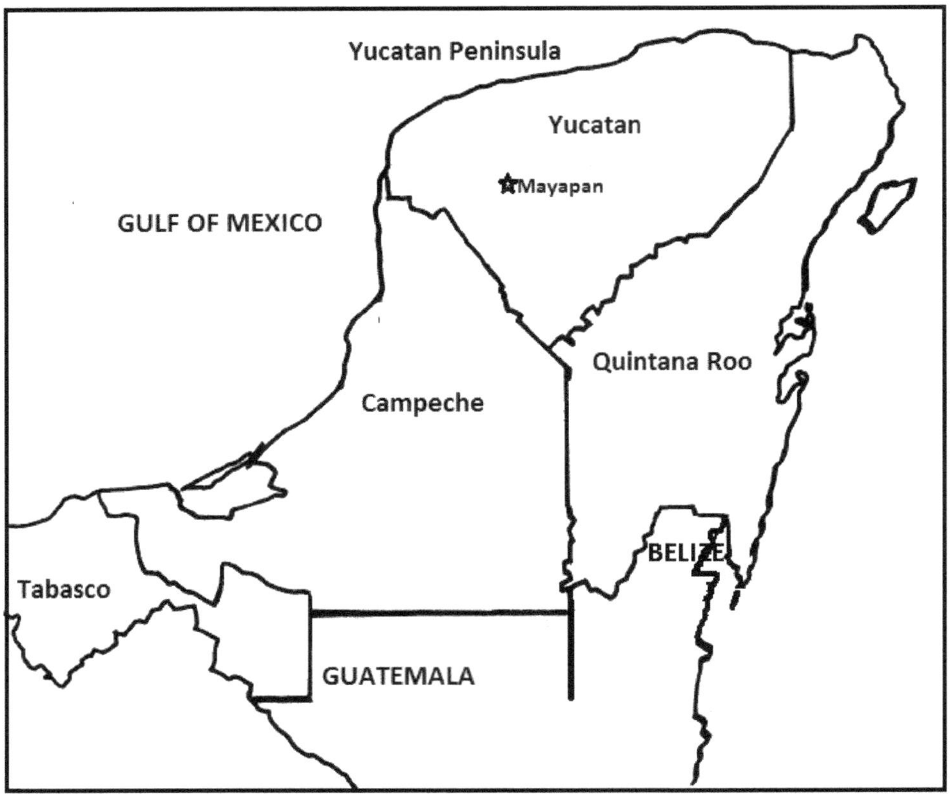

Figure 1-1. Yucatán Peninsula and three main sites (map based on INEGI 2005)

1.1 Brief Introduction to the Late Postclassic in Northern Yucatán

Mayapán (Figure 1-1, Figure 1-2, Figure 1-3) was the largest Maya capital and urban centre during its peak of 250 years (Masson and Peraza 2014, p.1) from AD 1200 to around 1450. During that time, Mayapán was a centre with a political centralisation without rivals in the Maya lowlands (Masson and Peraza Lope 2014). It was a capital uniting much of northern Yucatán in confederacy of territories under its dominion (Roys 1962). It was a walled city, and its size and population were unrivalled reaching a population of 15,000-17,000 and an enclosed area of 450 hectares (Brown 1999; Masson and Peraza Lope 2014, p.266).

Construction at Mayapán may have started as early as AD 1050/1100 (Milbrath and Peraza Lope 2003, p.24), coinciding with the decline of the Terminal Classic (AD 850-1000, from Masson and Peraza Lope 2014, p.63) centre of Chichén Itzá (Peraza Lope et al. 2006) and new political arrangements in northern Yucatán. The founding of Mayapán may have been related to waves of migrations of the Itza (Roys 1962), a Toltec-influenced or hybridised Maya originally from Tabasco (Figure 1-1) that, according to the interpretation of the chronicles (Roys 1962), had entered the peninsula and retreated probably to the Petén after their defeat at Chichén. In a second wave of migration and conquest, they entered the peninsula from the east coast. According to the informants in Landa (1986), the founding of Mayapán resulted from an agreement between the Itza and the different lords of the country to found a new city to which the lords moved and to which all matters of the country were brought. The north-central Yucatán area was governed from Mayapán by this league of lords (Landa 1986; Roys 1962).

Social developments were associated with changes in material culture including changes in architectural design and differences in building construction techniques. Most important for this research, a distinctive red-slipped ceramics appeared fully developed at Mayapán (Smith 1971a) with forms that, up to the current state of research, bear little relation to previous ceramics of the area (Brown 1999); Smith 1971a)These ceramics dominated in all assemblages from inland sites of northern Yucatán (Peraza Lope et al. 2006). They are also found in the eastern coastal area where they are somewhat different in texture and compactness (Brown 1999, p.316; Peraza Lope 1993, p.308; Smith 1971a, p.241).

Mayapán was abandoned around AD 1441-1461 following a rebellion in which the chiefs, thinking that the ruling lineage, Cocom, was getting too close to the Mexicans killed all the members of the Cocom house, except for the now famous son who, according to Landa's (1986) informants who traced their ancestors to Mayapán, was trading in Honduras. After the massacre, the inhabitants of Mayapán abandoned this centre, and the chiefs returned to their towns (Landa 1986).

Figure 1-2. Mayapán (photograph by the author)

Figure 1-3. The observatory of Mayapán (photograph by the author)

Mayapán was the last Maya capital in the lowlands. Until the conquest, no other centre rose to become the new Mayan political centre. The Europeans arrived a few decades after Mayapán was abandoned, marking the start of the conquest and colonisation period.

1.2 Late Postclassic Debates

An eclectic array of perspectives exists aiming to explain the nature of Late Postclassic society. The overarching debate that encompasses these views is whether the socio-economic institutions and organisation of this period represent, paraphrasing Robles Castellanos and Andrews (1985, p.56), a slow disintegration of Maya culture or a refinement of its development.

In the south, during the Classic, the rulers were the living reincarnation of the Maya supernaturals, and political power was transferred through pure blood lines (Friedel 1985, p. 424). In later periods, the north shared the basic cosmology with the Classic south, such as the twin heroes (Friedel 1985, p. 425), but religion affirmed the individual rulers rather than dynasties. There was a change in what Friedel (1985) calls political religion. As a result, royal tombs and temples dedicated to deceased kings were no longer constructed during the Postclassic (Masson and Peraza Lope 2014, p.49), and the institution of priesthood probably took more relevance with the decline of divine kingship (Masson and Peraza lope 2014, p. 53).

During the northern Yucatán Postclassic, the list of new material remains traits at Mayapán is long. Regarding ceramics, there was little continuity with the Terminal Classic ceramics (Brown 1999; Smith 1971a; Willey 1985), except for one ware, Peto Cream Ware, which has been found in Terminal Classic contexts at Chichén (Smith 1971a). Pottery composition was coarser and with fewer components (mostly limestone) when compared with Terminal Classic ceramics (which include volcanic ash, grog, limestone, and clay lumps) at Chichén Itzá or the Puuc area (Pollock 1952, Smith 1971b). Another break is the discontinuity in some architectural aspects between Chichén and Mayapán (Willey 1985, p.38). Masonry is different. Small shrines appeared at Mayapán, while ball courts and bath houses, commonly found at Chichén, disappeared (Willey 1985, p.38). A particular house plan forming a straight-legged C that differed from previous periods is commonly found at Mayapán (Pollock 1952; Willey 1985, p.38). There was also a change in the way of approaching architectural monumentality and the construction of elaborate objects, for the need for monumental or large-scale construction and elaborate objects decreased.

The rest of this section outlines some of the most influential, and sometimes overlapping, views about the nature of the Late Postclassic. Some of them, like the decadent view, have been fading away but remnants exist. Most of these views argue for a society overwhelmingly dominated by one socio-economic aspect over the others.

1.2.1 The Decadent View

This view of the Postclassic prevailed over most of the 20th century. The alleged decadence had started with the political and demographic collapse that occurred in the southern lowlands around A.D. 700-900; collapse that continued until the conquest. Society had fallen in a state of cultural decay, despondency and carelessness (Pollock 1952, p.24; Roys 1962, p.44-45; Shepard 1964; Willey 1985, p.51). The main symptoms convincing researchers of a 'collapsed' or 'decadent' Maya Late Postclassic included the lack of centralised government at the time of contact (Roys 1957), the perceived low quality of the construction and materials involved, and the small number of sizable ceremonial structures. For instance, Pollock (1952, p.239) commented on the poor structural design of buildings, which might lead to their faster destruction. The smaller size and number of ceremonial structures were attributed to cultural decay (Pollock 1952, p.239-240). In northern Yucatán, the change to a ceramic composition using almost exclusively the most common rock in the region – limestone – has been attributed to the simplification of techniques resulting from cultural decline, carelessness, and lower resourcefulness (Shepard 1964, p.518 - 519).

1.2.2 The Revivalist Views

According to this view, during the northern Late Postclassic, rulers and priests reintroduced some of the Classic and Terminal Classic traditions and monumental structures, particularly those present at Chichén Itzá. Pollock (1952, p.237) suggested that, with this, rulers asserted their political power and reasserted themselves over the foreign practices of the Toltecs (Pollock 1952, p.237). However, there was a selective revival of structures from Chichén Itzá in which, for instance, temples honouring the mythical Cuculcán hero, colonnaded halls, and round temples where adopted while ballcourts and gallery-patio structures were rejected (Masson et al. 2006, Figure 12.3; Willey 1985, p.38). The appropriation of the mythical Cuculcán by the Itza group may have been a legitimisation effort from their part (Masson 2005, p. 260) by associating their lineage to Mayapán's foundational myth. Similarly, the revival of the Classic period stela tradition may have been intended to rewrite history participating in ancient local traditions.

There was also a reversion or revival to some features of the Classic, such as block masonry and corbelled vaults, that were missing during the Late Classic (Andrews IV 1975); features, in particular, observed in buildings (Pendergast 1985, p.240; Pollock 1962, p.240; Roys 1957, p.44). Because of this, the term 'Maya Renaissance' has also been used for the northern Late Postclassic (Andrews IV 1975).

Mayapán is the type site for the Late Postclassic. Elsewhere in Yucatán, and in particular on the east coast, pottery and architectural remains have been found with stylistic and construction methods with affinities to Mayapán (Masson 2015; Pollock 1951, p.239; Sanders 1960, p.228; Smith 1971a). Some researchers consider the Late Postclassic changes so transforming that they have attributed them to a conscious effort to restructure the society as a whole (Robles Castellanos and Andrews 1985, p.90). Others see a gradual change, a process of cultural transformation rather than a break with the past (Milbrath and Peraza Lope 2009), with Willey (1985, p.43) considering Mayapán culture to have been a new synthesis that was essentially Maya.

1.2.3 The Political Economy Views

Political economy commonly refers to studies drawing on economics and political science in explaining how the political environment and the economic system influence each other (Weingast 2008). In Mesoamerican research, political economy is usually associated with the principle of rational choices and maximisation, most often from the part of the elite, but increasingly also from the part of the commoners. Rational choice in political science stands for the application of the economics approach in the study of political phenomena (Lohmann 2013).

The control of the environment through technology and the belief that ancient people used technology in a quest for efficiency or for material's performance are the basis for views known, usually, as the managerial (Wells 2006) or adaptational (Brumfiel and Earle 1987) models. In these models, the elite organise and mobilise labour and resources to take advantage of the environment or to avoid famine and other societal failures.

The financial (Wells 2006), also known as political (Brumfiel and Earle 1987), framework also emphasises the elite and rulers as organisational agents. Elites manipulated ideological, political, economic, or other aspects of society to create, maintain, or increase the inequality that defines the existence of an elite class (Brumfiel and Earle 1987; Wells 2006). The elite and rulers control economic activities, such as long-distance trade and the flow of goods, artefacts needed for ritual activities, luxury and prestige items, or food to extend their power.

Mercantilist or commercial development models (Brumfiel and Earle 1987) of ancient Mesoamerica economies emphasise economically rational choices to maximise elite and/or individual gains. These models are usually intertwined with ideas of the rationalist approach and individual agency in which the individuals strive for efficiency, personal gain, and competitiveness (Wells 2006, p.278). These models propose a Maya society engaged in market exchange and 'World-Systems' dynamics, as discussed by Brumfiel and Earle (1987), Kepecs (1999), Rathje (1975) and others.

1.2.4 The Collective Reciprocity View

In this view, society was organised around reciprocal obligations and perceived as a collective enterprise directed by the elites involving the gods, nature, and humanity (Farriss 1992, p.6). Society was organised in interdependent widening networks that incorporated the extended family, through the *cah* (community), *chibal* (patronym group) to the polity and the sacred (Farris 1992, p.6; Restall 1997, p.90; Fernández Tejedo 1990, p.34). Through the extended family, individuals were incorporated into this network and collective enterprise based on reciprocal obligations (Farriss 1992; Fernández Tejedo 1990, p.34; Restall 1997, p.185), such as gift giving.

That many views exist about the organising principles of Late Postclassic society makes evident that current research is recognising the complexity of Late Postclassic society and that several types and levels of socio-economic organisations, exchange, or institutions may have co-existed. The goal has become to determine the relative role that the different political and socio-economic aspects played in the community, while, usually, emphasising a set of factors. The following four models exemplify the variety of perspectives that have been proposed to explain the organising principles and foundations of the Postclassic political economy.

For instance, the power of a polity might have been based not on the domination of territory (Graham 2011, p.34-44). It was based on the control of the rights of what was produced, administered through spheres of tribute obligations, interpersonal relationships, or administrative control centred on a person or group of people (Graham 2011). In this model, spheres of tribute obligations and interpersonal relationships are emphasised, while acknowledging the importance of other economic aspects such as inter and intra-regional exchange. In research at Mayapán, Masson and Peraza Lope (2014) consider that market exchange and craft production with a surplus take a prominent role in the Late Postclassic economy while acknowledging the significance of other aspects such as tribute and gift giving.

Contrasting with perspectives that emphasise the control of tangible wealth and the display of valuables as the source of power, Rice (2009a) proposes that power could have been based on the control of knowledge, in particular, astro-calendrical knowledge from part of the elite. Political power was achieved by the privileged access to production, exchange, and consumption of knowledge (Rice 2009a). Lastly, Friedel (1981) proposed that Maya political economy was based on the control of scheduled, regulated pilgrimages to focal centres and on the control and taxing of the distribution of goods. At the focal centres, periodical festivals and markets were organised and structured by the calendrical nature of rituals. Markets gave elites the opportunity to control and tax the distribution of local and regional goods (Friedel 1981).

1.3 The Problem

Chase (2004, p.118) has said that any economy can be characterised in terms of its production and distribution. The study of craft production and the interdependencies between production, exchange, and distribution are crucial to characterise the socio-economy of a society. Several models have been proposed to explain the organisation of Maya ceramic production; in particular, for the Classic period for which more studies over larger areas exist. In chapter two, current research and models proposed to explain ceramic production, organisation and exchange are summarised and discussed. One of the objectives of the current research is to evaluate how the results emerging from this study fit these models. Despite the extent of research and multiple proposed models, the study of the ceramic production, organisation, and exchange in Late Postclassic Yucatán has had many challenges, and it is still unclear.

One of these challenges results from the stylistic similarity of ceramics throughout northern Yucatán. When compared to the previous period, the Late Postclassic presented a reduced palette of ceramic types (Masson 2001). A ceramic type, in the type:variety ceramic classification commonly used in Mesoamerica, comprises ceramics sharing diagnostic features, including decoration, form, and paste (Sabloff and Smith 1997), while a ceramic group is a broader category comprising types of the same ware with a consistency in the range of form and colour (Sabloff and Smith 1997, p.279; Smith 1971a, p.7). For instance, summaries in Masson (2001, p.172-173) of the slipped sherds at Mayapán show that 92 per cent of them are classified in one of the ceramic types, the Mama Red type. Contrastingly, at Altar de Sacrificios during the Late Classic, the slipped ceramic type with the highest frequency represents 40 per cent of the total.

In addition, within a ceramic type, each of the vessel forms and its surface finish, decoration, style, and macroscopic paste composition are highly similar from site to site, in particular from north-central Yucatán to Quintana Roo (Masson 2001, p.161), with affinities to

Mayapán. Similarities to those at Mayapán spread over large areas of the Yucatán Peninsula including current-day Belize, Yucatán, and Quintana Roo states (Pendergast 1985, p.240; West 2002, p.178; Rathje 1975; Sabloff and Freidel 1975). However, only a few studies have attempted to measure the observed uniformity of Late Postclassic ceramics (Brown 1999, 2015; Howie, Aimers, and Graham 2014; Masson 2001; Masson and Rosenswig 2005).

Composition, as seen in hand specimens, also appears very homogeneous (Smith 1971a). For instance, red-slipped ceramics commonly found at any of the north-central sites (chapter three) cannot be distinguished from site to site. In light of this uniformity, it is not known whether Mama pots (described in chapter three) were produced at one or several production centres. Therefore, it is not known if paste similarities result from local production under common choices of materials, from a high degree of economic integration between centres with production and distribution from a few of them, or from a lack of geological diversity. The same can be said about the unslipped ceramics commonly found in this area. There is variation, however, on the texture of the paste of most sherds from north-central sites, when compared to sherds from eastern coastal sites. At the coast, red-slipped ceramics with a finer and compacted paste dominate, that nevertheless present stylistic similarities to the north-central red-slipped ceramics. This similarity in the ceramics, combined with the challenges of a mostly geologically homogeneous northern Yucatán, has hindered research on ceramic production, organisation, and exchange.

Some researchers (Howie 2012; López Varela et al. 2001) have argued that in Maya archaeology, this situation will not change and little will be known about production organisation until ceramic production is studied in term of technological choices and techniques applied in the construction of pottery. The way ceramic products are made results from choices and actions made by potters during the selection and processing of raw materials, and during forming, finishing, and firing of the pots.

How can this help the understanding of ceramic production and organisation, and Late Postclassic socio-economic organisation? Pottery reflects the choices and actions made by artisans with regard to materials, the techniques applied, and the sequence of steps followed for the construction of the product. There is a growing body of literature supporting the notion that these choices are modulated by the wider cultural, symbolic, economic, and political environment surrounding the artisans (Dobres 2010; Lemonnier 1993, p.3; Gosselain 2000, Jordan 2015, p.67; Sillar and Tite 2000). Therefore, differences in the wider environmental and social context are reflected in differences in the choices made by potters. These, in turn, may result in similarities and dissimilarities in the way pottery is made, or technological variations in the pottery produced.

The study of how ceramics were made, the choices made from the raw materials that may have been available to potters (chapter four), and variations in technology have, as a result, an archaeological value for they may point to divisions within the potters' environment including socio-economic context, technological traditions, or production groups and zones (Roux and Courty 2005, p.202). These groups and zones of production may have implications for understanding ceramic manufacture, organisation, and exchange. Chapter five lays out the methods designed to gather the data needed to test the hypotheses and achieve the objectives of this research. The chapter addresses the *chaine opératoire*, the theoretical basis for the methodology, and the reconstruction of pottery production. The sampling strategy and samples taken are discussed as well as the analytical methods used, mainly hand-specimen and thin-section petrographic analyses, and chemical analysis. The study of technology is then used to investigate the Late Postclassic socio-economic environment.

More scientific studies are needed in which, for one, the variability in the ceramic composition and manufacturing techniques over large areas of northern Yucatán are compared, and second, these variations are seen from the context of potters' choices for the construction of the pots. Current ceramic research for the Late Postclassic of northern Yucatán, with few but significant exceptions (Kepecs 1999; Cecil 2012), has been overwhelmingly typological or stylistic in nature based on macroscopic examination (Brown 1999, Cruz Alvarado 2012, Ochoa Rodríguez 2004, Peraza Lope 1993), or at the site level (Masson and Peraza 2014, Sánchez Fortoul 2009, 2013). It has therefore not been possible to assess ceramic variability and address questions of ceramic distribution and exchange. As a result of this gap, research has not been able to address basic questions of intra- and inter-site variations in composition, number and location of ceramic work groups, zones of raw material procurement, manufacturing technologies, and common choices or technological traditions. Basic scientific research through a methodology that combines analytical techniques such as petrographic (chapters six and eight), chemical (chapter seven), and surface feature analyses (chapter nine) to examine the variability in ceramic composition and manufacturing choices and technology (chapter nine) can overcome many of the challenges that studies of the Late Postclassic have traditionally faced. This will pave the way to addressing questions dealing with the wider Late Postclassic context such as the existence and scope of technological traditions (chapter 10), organisation of production (chapter 10), and distribution or exchange (chapter 10), as well as the role of Mayapán in the overall Late Postclassic Yucatecan society (chapter 11).

1.4 The Objectives of the Research

In light of the significant role of pottery in the ancient Maya socio-economy and the many questions that remain about ceramic production traditions and networks of interaction and exchange, this research is aimed at investigating the following research questions:

1. Are there sufficiently varied patterns in raw material selection and ceramic attributes or characteristics to allow the characterisation of pottery fabrics into fabric classes and distinct technological classes? This first and most basic question needs to be addressed.
2. What might such technological patterns tell us about how the different classes of pottery (such as with different surface finish or forms) were made?
3. Are the observed homogenisation of style and macroscopic composition maintained through different levels of analysis, e.g. microscopic composition and chemical analysis? Do the observed homogenisation of style and macroscopic composition reflect a shared technological tradition? For instance, a technological tradition defined by a shared understanding of how red-slipped Mama vessels should be made.
4. What might such technological patterns and traditions say about the organisation of pottery production, such as number and location of potters groups, or associations with and between geographical areas or specific sites, and its social significance?
5. What might such patterns and organisation of pottery production tell us about the distribution of utilitarian ceramics and the types of exchange that may have taken place? In particular, were ceramics produced at each centre? Alternatively, was production centralised and, if so, where?
6. How are the results of this research informing current models for the organisation of Late Postclassic ceramic production, distribution, and exchange?

Crucial to addressing the social and economic nature of the Late Postclassic and informing current models is determining the role of Mayapán, the northern capital, in the regional socio-economic sphere. Equally important is the understanding of the role of the minor or peripheral communities in the socio-economic environment, including the nature of their interactions amongst themselves and with Mayapán. For instance, did Mayapán provide ceramics to the lesser centres? Different models explaining aspects of ceramic production, organisation, and exchange for the Late Postclassic have been proposed. They are presented in chapter two. In the concluding chapter of this book, the patterns observed in this study concerning the organisation, traditions, distribution of ceramics, and types of exchange are examined in relation to these models.

Based on observations from a preliminary survey of a fraction of the hand-specimen samples, which indicated that patterned differences in the compositions of slipped and unslipped samples may exist in the data, three hypotheses were proposed as a guide to addressing the research questions. The patterns observed in the preliminary survey (section 5.5 and chapter six) are:

1. In the north-central area, the macroscopic composition of red-slipped samples appears more uniform throughout the sites than for the unslipped ones.
2. Samples classified as Payil, the eastern group of red-slipped ceramics commonly found at eastern sites, and also found at Mayapán, have similarities to a few other samples from the north-central sites and could have been locally made.

With these hypotheses, answers are proposed to particular cases of the research questions and, therefore, in the process of collecting the data to test them, data addressing the questions are gathered. The hypotheses proposed are presented below. They are examined in greater depth with alternative interpretations in chapter five.

Hypothesis A: Unslipped jars found at the different sites were locally produced at minor centres and Mayapán.

This hypothesis is based on a variety of fabrics observed within the unslipped samples.

Hypothesis B: In the north-central area, red-slipped pots (Mama) were produced at one production locality that supplied the rest of the north-central centres.

The basis for this hypothesis is two compositional groups, one with white limestone and one with dark inclusions, observed within the red-slipped samples.

Hypothesis C: Payil, the fine-grained red-slipped ceramic type commonly found at eastern coastal sites, when found at north-central sites, was locally made.

This hypothesis is based on paste characteristics of samples categorised as Payil and found at Mayapán that have fabrics similar to unslipped samples considered probably to be local to the north-central area.

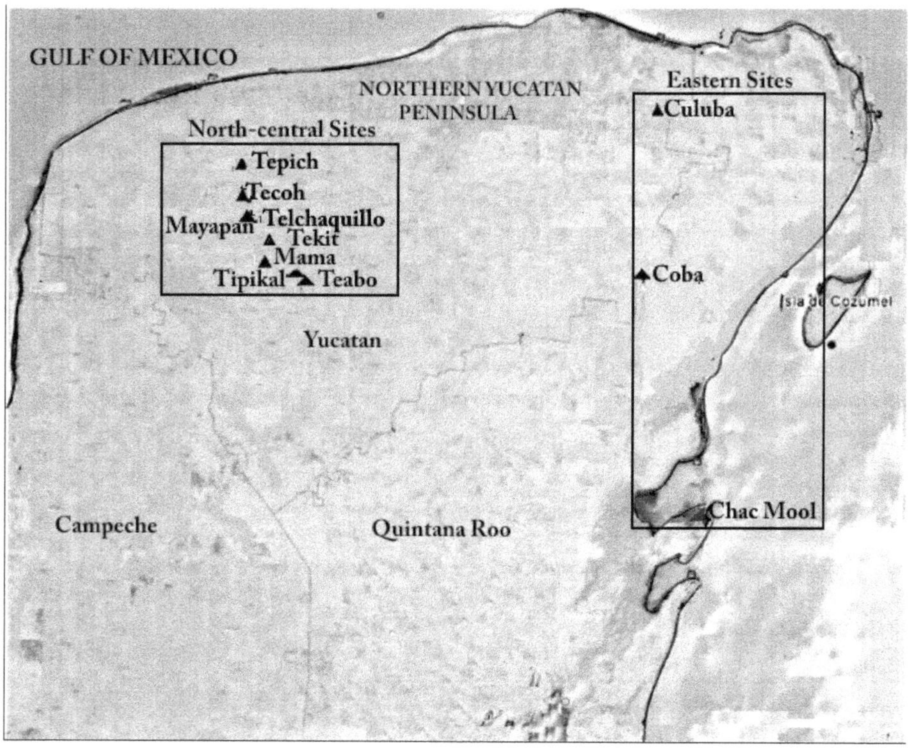

Figure 1-4. Map of studied region with sites in this research

1.5 Outline of the Methodology

The methodology selected to test the hypotheses listed above and to achieve the objectives of this research uses a combination of chemical, petrographic (in hand-specimen and thin-section) and surface feature analyses. These analytical techniques will be used to examine the variability in composition and ceramic manufacturing methods based on pottery samples from Late Postclassic contexts.

Local clays were collected with the objective of evaluating the available raw material resources and, in the absence of direct evidence for ceramic production, seeking clays compositionally comparable to the pottery. The technical attributes are examined as indications of potters' actions and choices embedded in the wider socio-economic environment. Conversely, interpretations given to the raw materials analysis will be based on current knowledge of the environment in which the potters lived. Ethnohistorical accounts complement integral parts of this study, as well as the presentation of the wider environment including the geological setting and current research on the socio-economic environment of the Maya people during the Late Postclassic in northern Yucatán.

The study is structured using a regional approach with two test cases that represent sites from the north-central and eastern areas of the Yucatán Peninsula (Figure 1-4), an area presenting a Mayapán-style ceramic assemblage. Both cases seek to study the characteristics and technological patterns of the local ceramics. The location of the north-central sites, adjacent to the main Late Postclassic centre of Mayapán, provides the opportunity of studying pottery production, distribution, and socio-economic integration between closely located major and minor centres, while the eastern sites address the research questions concerning interaction over longer distances.

Petrographic analysis of samples as found when collected, known as hand specimens (as opposed to, for instance, samples cut into thin sections), was used to divide the samples into broad groups based on characteristics observed at relatively low magnification (chapter six). From these groups, subsets of samples were selected for further chemical and thin-section petrographic analyses.

The chemical analysis used neutron activation analysis (NAA) and multivariate statistical analysis of the chemical compositions of the samples with the aim of characterising the samples into distinct chemical groups. The analysis and resulting chemical groups are presented in chapter seven.

Using petrographic analysis of thin-sections, the samples were grouped by their petrographic characteristics into technological fabric classes. These fabrics are described in chapter eight together with a description of the main inclusions found.

The objectives of this research call for an examination of the analytical results to find patterns in ceramic production. To this end, chapter nine evaluates results from the above analyses, complemented with geological and ethnographic data, to seek compositional and technological patterns. The objective is to draw inferences about the ceramic techniques that might have been used in the construction of the pottery vessels and about the number of different potters' groups. Chapter 10 further develops the interpretation of technological patterns to determine the existence and scope of technological traditions, the organisation of production, and movement of pots. This study ends (chapter 11) with a discussion of the findings and directions for future research.

Chapter 2

THE LATE POSTCLASSIC: A VIEW FROM THE NORTH

This chapter aims to present current research on the ceramic economy within the study region, as well as models that have been proposed explaining the organisation of ceramic production, distribution, and exchange during the northern Yucatán Late Postclassic period. At the completion of this monograph, the results of this study are examined to assess how well they fit these models.

The current models are linked to perceptions of what the organising principles of Maya society were and to interpretations of ancient Maya socio-economic institutions. This chapter starts reviewing current interpretations of the socio-political and economic institutions during the Late Postclassic, including political, tribute, and market institutions. The chapter continues with a review and discussion of current research on ceramic economy and models that have been proposed.

2.1 The Late Postclassic in Northern Yucatán

This section presents current interpretations and research informing about some of the main Maya social institutions. Aspects of the political and social environments surrounding the potters of the Late Postclassic are also discussed.

2.1.1 Political Organisation of Northern Yucatán

When compared to the northern Terminal Classic (AD 850-1000), the Late Postclassic shows changes in the political arena. The approaches taken to the political organisation by the inland and eastern areas of the peninsula differed (Robles Castellanos and Andrews 1985).

In the inland areas, Robles Castellanos Castellano and Andrews (1985, p.91) see the rise of Mayapán as representing an attempt to bring back the power and grandeur of Chichén Itzá during the Terminal Classic, but with a different political organisation. In this area, there was a shift from the increased political and economic centralisation observed since the Classic period up to the Terminal Classic at Chichén Itzá (Chase and Chase 1985, p.2; Kepecs 1999, Robles Castellano and Andrews 1985, p.54). This reversal is evident in the league of lords who, for a time, shared rulership at Mayapán (Landa 1986).

In the north-eastern areas of the peninsula (mainly Chikinchel and Ecab areas, see map in Roys 1957, p.2), no Late Postclassic site seems to be larger or more prominent than the others, suggesting political decentralisation. Based on ceramics, settlement pattern information, and ethnohistorical records, it is believed that the eastern area was outside the direct political dominion of Mayapán and without a main political centre (Roys 1962, p.32). Robles Castellano and Andrews (1985) suggest that two political experiments were running concurrently: in the west, an attempt to bring a large area under centralised but shared government, and, in the east, decentralisation.

According to Landa (1986, based on informants), the lords, the heads of the different lineages or houses, of northern Yucatán were persuaded by the leader of the foreign groups to agree to found a new city to which the lords moved and to which all things (*cosas*) and matters (*negocios*) of the country were brought. The nature of this agreement and whether it was voluntary or enforced under duress are not clear. The Cocom, Xiu, and Chel and many heads of local main lineages and their entourages moved to Mayapán and administered their domains from there (Landa 1986). It is thought that most of the north-central Yucatán area was governed by this league (Roys 1957). It was apparently a new type of shared government, a government by many. After the Itza leader had left (who called himself C'uc'ul can, the mythological Toltec leader), the house of the Cocom became the head of the coalition (Landa 1986). As noted by Landa in the *Relations*, it was the duty of all the lords to honour, gift, visit, and entertain the Cocom.

As mentioned earlier, Graham (2011, p.34-44) had proposed that, right before the Conquest, the power of a polity was not based on the dominion of a territory. It was based on the control of the rights of what was produced. A lord could acquire rights over resources by appropriation through warfare of part of the tribute historically owed to another lord. Applying this model to Mayapán, forced extraction of the resources in the form of tribute owed to each local lord may have been behind the rounding up of lords at Mayapán, first by the Itza and continued by the Cocom. This interpretation gives new meaning to the passage in the *Relations* (Land 1986) recalling that the Itza leader and the heads of the north-central linages had come to an agreement to move to Mayapán and that, as a result, all kind of products were brought to this centre. Blanton et al. (1993) call the lords at Mayapán 'invited hostages.'

Based on Landa (1986) and his informants, it can be gathered that northern Yucatán was already divided into multiple polities before Mayapán was founded (Robles Castellano and Andrews 1985, p.91). After the abandonment of Mayapán, the lords and their entourages returned to their provinces (Landa 1986).

Our understanding of the political order at the time the Spanish arrived owes an enormous debt to the work of Roys (1957). He compiled Spanish and Native documents interpreting that, at the time of contact, Yucatán was divided into sixteen provinces (Roys 1957, Map 3.1), or *cuchcabalob*, most of which were territorial in nature. Most north-central provinces are considered included in the Mayapán confederation (Roys 1957), while north-eastern and eastern areas of the peninsula are thought to have been independent (Roys 1957). Since Roys's fundamental work, new analyses have amended the limits proposed by Roys (Andrew 1984) or suggested that the model is too rigid to represent the dynamics that may have existed (Fox et al. 1996b; Graham 2011; Okoshi Harada 2012, 2015; Restall 1997) while recognising the extraordinary contribution of his work to Maya research.

A model by Okoshi Harada (2012) puts forward a fluid political organisation. In this model, the basic political unit was not the province but the community or *cah* (2012, p.5). In Roys' model, the emphasis is on the territorial dominion of the chief (Graham 2011), while in the new model, the focus is on the network of relationships and obligations to the rulers.

Maya people appear to identify themselves as part of the extended family, the patrilineage *(ch'ibal),* and the community *(cah)* (Restall 1997:13). The name of the *ch'ibal* and the *cah* to which a person belonged were the only specific terms used to identify a person as part of a distinctive social group (Restall 1997). A significant proportion of the population lived in detached residential compounds or residential clusters spread around in the *montes* (bushes), and sometimes far from a larger concentration or town. The *cah* also included farming land (*milpa*) that could lie miles from the *cah* residences (Restall 1997). Therefore, usually, a *cah* could not be linearly demarcated (Okoshi Harada 2012, p.5).

A group of dispersed *cahob* (*-ob* for plural) comprised the *batabil* or *señorio*, under the authority of a *batab*. Both, the *cah* and the batabil were political units. Each *cah* had an *ah cuch cab* selected by the *batab* that ensured that tribute and services were promptly provided, organised the community festivities, or prepared it for war (Okoshi Harada 2012, p.6). However if a *batab* could not provide needed benefits to the *cah*, the *ah cuch cab* could decide recognising a different *batab*. Therefore, the *batabil* usually did not occupy a continuous space (Okoshi Harada 2012, p.5).

According to Okoshi Harada (2015, p.6), many *batabilob* were independent, but most depended upon a *batab* with the title of *halach uinic.* The relationships between the different levels were more political than hierarchical, immersed in a web of alliances and rivalries (Fernandez Tejedo 1990, p.32). This political environment bears relation to the potters and the result of their work, such as when a vessel or the content of a vessel is destined to be part of the tribute. Therefore, the distribution of a vessel from its place of production can reflect a political area. Such is the case at Palenque. The range of distribution of most of the pottery from Palenque coincided with the spatial range of Palenque emblem (Rands and Bishop 1980).

2.1.2 The Institution of Tribute

The power of the rulers was rooted in complex relationships that allowed them to collect offerings or tribute, a crucial part of the socio-economic arrangements (Fernandez Tejedo 1990, p.34; Graham 2011; 2012). Benefits such as divine protection, defence, and administrative services were provided by nobles to the population, while commoners were obligated to provide recognition to their authority and compensate the lords for the benefits showered on them (Fernandez Tejedo 1990, p.34). During the northern Late Postclassic, Mayapán was at the centre of the tributary system (Landa 1986), and many centres were linked to Mayapán by their tribute obligations. It is possible that in some of these centres, people considered the lords at Mayapán their natural lords, and in others, tribute could have been extracted based on rights obtained by coercive means such as warfare (Graham 2011).

Tribute was paid at the family, most probably extended family, and the community levels. It consisted of a fraction of the production of agricultural goods, such as cotton, corn, salt, cacao, honey, turkeys, fish, as well as products from industries such as textiles or *mantas* (Fernandez Tejedo 1990, p.35; Landa 1986; Piña Chan 1978). Of these products, textiles constituted the main tribute item during colonial times (Landa 1986, Roys 1972). Pottery, however, is not mentioned by Landa as part of tribute paid. Nevertheless, pottery could have been used to transport tribute items such as honey. During early colonial times, pottery may have been part of the tribute paid to the *encomenderos* as shown by a copy of the declaration of what the town of Motul supplied in 1552 (Fernandez Tejedo 1990, p.52), which included 18 *cántaros* (water carrying/storage ceramic pots) and 26 *ollas* (open-mouth cooking pots).

Another component of tribute was in the form of labour, which could be communal labour. However, not all-compulsory communal labour was part of tribute. This is shown in the small peasant village of Chan Kom (Redfield and Villa Rojas 1962), Yucatán, population 250, and in many towns in Yucatán, through a rotating system in which some of the work needed was performed by boys and adult males involved in communal work. This work was not remunerated, and it was not voluntary either; it was enforced with the punishment of prison. At Chan Kom, both individuals (private houses) and the community (roads and *cuartel*) benefited from communal work.

2.1.3 Markets and other Economic Aspects

According to Roys (1972), when the Spanish arrived, the Maya society was apparently divided into three main classes: nobles, commoners, and slaves. However, Maya society may have been more complex than that (Fox et al. 1996b). For instance, Fox et al. (1996b, p.798) propose that given that the ways in which kinship crosscut through the social organisation are not clear, it may be inappropriate to divide Maya society strictly by classes.

During the Postclassic, of these social classes, the nobility or ruling class was also the main long-distance or inter-regional merchants (Roys 1972, p.33; Piña Chan 1978, p.44; Chapman 1959, p.40). We know that the son of the Cocom was trading in Honduras at the time of the Mayapán massacre (Landa 1986). Chapman (1959, p.10) divided Maya commercial activity into two institutions, long-distance exchange and markets. Chapman considered them to be separate because they differed in many ways (Chapman 1959, p.10), including the type of place where the trading took place, the people involved in the transactions, and the products traded (Chapman 1959).

Yucatecan products exported for long-distance exchange were mainly salt, mantas, slaves, honey, beeswax, (Landa 1986; Roys 1972, p.51-54; Vasquez de Espinoza 1948, cited in Piña Chan 1978, p.42), cochineal, cacao, achiote, and índigo (Vasquez de Espinoza 1948, cited in Piña Chan 1978, p.42). The elite impetus, as interpreted by Landa (1986), for this trading was to obtain currency in the form of cacao beans and stone beads, dyes for the body, or red seashells (Landa 1986) with which to buy luxury display items such as feathers or jewels. These transactions occurred outside Yucatán borders, such as Ulúa and Nito (Honduras), in these predetermined places (Chapman 1959, p.50).

There are no rivers in northern Yucatán, and there were no beasts of burden. Therefore, products to and from inland communities in this area were transported on the backs of porters who, according to Sahagún (1946 cited in Chapman 1959, p.47), were people of the lowest social status, forming long human caravans crossing the land. When possible, large canoes were used to travel around the peninsula, or the rivers found to the south. To give an idea of the distances covered by peddlers on foot, Hammond (1978) documented an overland route from Cobán, highland Guatemala to the lowlands in Belize. This was done at a time in which peddlers set out on foot carrying 45 kg of small products on their backs supported by a band across the forehead (tumpline or *mecapal*). The trip took one week with daily distances travelled of around 32 km, staying overnight at the different towns in between.

Large markets have been described in the northeast part of the peninsula at the towns of Chauaca (Landa 1986, pp.232-233), Cachi (Oviedo 1851-55, cited in Piña Chan 1978), Ecab (Mártyr de Anglería 1892, cited in Piña Chan 1978). However, there are no specific descriptions of inland markets. They are mentioned, nevertheless, even if in passing such as when Landa says that the products of the land are stored in granaries until it is the right time to sell. According to Roys, larger towns likely had markets (Roys 1972, p.52). Given that Tomás López (1552-1553) ordered that every town should have a market and that every transaction should happen at the market, it is doubtful that all small communities had markets (Roys 1972, p.52). It can also be inferred from the ordinance that transactions, or a portion of the transaction, occurred without a marketplace, perhaps performed by peddlers similar to the cobanero traders just described.

Based on tribute paid to the Spanish through the *encomienda* system, Piña Chan (1978) compiled a list of products from different areas of Yucatán. The products include salt, fish, honey, wax, cotton, slaves, cacao, copal, cochineal, flint. The products are regional, produced primarily in certain areas (Piña Chan 1978, Figure 12). The north-central area where most sites in this study are located, including Mayapán, appears without any product. Itinerant merchants may have existed to supply local markets (Roys 1943, p.51).

There are no records of pottery being traded at local markets, but it is presumed that it was (Roys 1972, p.46) given the reference in Landa (1986) to pottery and woodworking as important occupations and objects of trade. Judging by early colonial times, most probably all commoners farmed (Landa 1986; Roys 1972, p.46) attending other industries such as pottery or barter only when land obligations allowed them.

2.2 Current Research

Given the scarcity of regional analyses for the northern Late Postclassic, those conducted for the Late Classic periods in the southern lowlands and Palenque have been a source from which hypotheses can be drawn for this period and others. Excellent summaries of these investigations already exist, such as those from Rice (2009b) and Foias (2004). This research focuses on utilitarian ceramics, and this is the emphasis of the summaries and discussion of research presented in this section.

In the Maya area, direct evidence for pottery production is very scarce. When found, it has been related to ceremonial items such as censers (Masson and Peraza Lope 2014) or prestige items such as the Classic Period polychrome vessels. Polychrome vessels have usually been found in Classic palace contexts without direct evidence of forming or firing (Halperin and Foias 2010; Reents-Budet et al. 2000; see Halperin and Foias 2010 for proposed direct evidence for production).

Only a few structures in the southern and northern lowlands have been ascribed to pottery firing. The most cited example is from the Late Classic site of K'axob in Belize. At this site, connected underground pits (López Varela et al. 2001) have been found together with sherds that look like tools for working clay. However, Prudence Rice (2009b, p.127) is not convinced by this evidence pointing out that signs of firing such as smoke, ash or wasters are not mentioned in the report. Large pits with ash and burn soil are reported At Laguna de On, Belize (Masson 2015) as possibly related to ceramic production. The findings of Smyth and Dore (1992) at the Terminal Classic centre of Sayil (Puuc area, Yucatán), of an apparently non-domestic area (lacking *metates*, or grinding stones) and with a high incidence of overfired sherds (wasters), indicated to the authors the existence of a likely ceramic production area.

On the other hand, Rice describes evidence found by Freter (1996 cited in Rice 2009b, p.138) of the firing of utilitarian items near clay sources in the rural areas around Copán. The common pattern among the presumed firing areas comprised a test pit that yielded abundant ceramics, 5 per cent or so of misfired or warped sherds (wasters), greenstone tools for burnishing, the presence of unfired clay and carbon flecks, and probable areas for firing located near (50-100 m) the residences showing burned and hardened soil areas. Freter concludes that utilitarian production occurred in rural areas in the courtyard of residences built close to the clay source without kilns, investment, or central control.

With the lack of direct evidence, indirect evidence such as anomalous sherd density in the ceramic assemblage or pottery and clay compositional analyses have been used to suggest locations of production (Fry 1980; Howie 2012; Rands and Bishop 1980). In isolation, unusually high concentrations of pottery are considered weak evidence because they may occur for a variety of reasons, such as feasting, storage, dumping, or rubbish disposal. As stated by Costin (1991, p. 33), indirect data 'rarely yield unequivocal evidence for the context, scale, or intensity of production.' On the other hand, indirect data can provide information of the relative standardisation, the number of potters' groups, and probable zones of production (Costin 1991, p. 33).

2.2.1 Late Classic at Tikal and Palenque

Fry (1980) used a combination of methods that included the examination of fall-off pottery distribution curves and paste inclusions to study the production location and distribution of pottery from Tikal, current Guatemala (Figure 1-1). Although monochrome and unslipped types of vessels were produced from different sources around the periphery of Tikal, both serving vessels and wide-mouthed jars were found in considerable amounts in the central area of this site, indicating to Fry that they were taken to central Tikal for market exchange. Based on the distribution of pottery and the comparison of clays and sources, Fry concludes that the organisation was complex (Fry 1980, p.11) and that the patterns observed are explained by the existence of a market system (Fry 1980, p.16). Ceramics were produced in peripheral centres around a core centre and brought to it through a decentralised market system.

He also inferred that the core centre did not produce or distribute ceramics, with the exception of some serving vessels; it mostly consumed ceramics. That a high proportion (around 40 per cent) of the assemblage at minor centres in Tikal peripheral areas comprises serving and polychrome vessels corroborated to Fry that market exchange explains the patterns, arguing that gift giving, for instance, could not explain the significant amount of vessels encountered. Similarly, at Palenque, Campeche (Figure 1-1), Yucatán, Rands and Bishop's (1980) study of ceramic production and distribution, through petrographic and chemical analysis, indicated to them that the amount of pottery brought into Palenque was far greater than the locally manufactured and exported pottery. Similar to Fry (1980), Rands and Bishop conclude that Palenque was a consumer of ceramic items from the agriculturally poor surrounding areas.

Stanton and Gallareta (2001, p.232) challenged these interpretations. The same patterns indicating decentralised market exchange at minor centres can also be interpreted as redistribution by local elites. The presence of all types of peripherally made ceramics at the core centre can be explained as pottery transferred as part of tribute-paying, with a portion going back to the minor centres. They also propose production for feasting to explain the amount of serving vessels found, emphasising that market exchange and redistribution are not mutually exclusive.

Pottery production seemed organised (or specialised) at the community or zonal level (Rands and Bishop 1980, p.42) by form and/or functional classes, based on a compositional division corresponding to figurines and some serving vessels. Specialisation by ware and form is also suggested for the northern Terminal Classic by neutron activation analysis (Smyth et al. 1995) and petrographic analysis (Smith 1971b; Shepard 1958) of Puuc area wares in Yucatán.

2.2.2 The Late Postclassic

Research on ceramic production for the Late Postclassic is sparse and lacks the multi-site and encompassing approaches that exist for the Classic period. Nevertheless, limited but significant research has provided a major contribution.

Direct evidence for production during the northern Late Postclassic is limited to cases of ceremonial items, such as censers or figurines (Hare and Masson 2010). For instance, At Mayapán, Hare and Masson (2010) found a Late Postclassic effigy censer ceramic workshop in a house attached to a palace; it contained fine artisan stucco tools and numerous moulds made with distinctive, unfinished effigy pieces.

Local production during the Late Postclassic, at least in some northern Yucatán areas, is supported by a study (Kepecs 1999) using, mostly, visual hand-specimen examination of Late Postclassic red-slipped pottery from sites in the Chikinchel region, located close to the north-eastern area of the peninsula. The study also included thin-section petrographic analysis of some samples selected by ware. Kepecs found a prevalence of one paste type at each of the three main regional centres, with each paste type being different at each centre, suggesting to her a production that is probably local and distinct from the production centres that supplied the western area (Mayapán). However, alternative interpretations can be drawn. Particularly, the vessels sampled and qualified as locally made could have been imported from another site.

With regard to Mayapán, Masson and Peraza Lope's (2014) in-depth analysis of Mayapán is essential for understanding craft production at this site as a Late Postclassic social and urban entity. Evidence for the production of tools and decorations from lithic and shells is found in many residences, with Brown (1999) also reporting four lithic workshops. Given that the debris indicates that items were produced within the households, Masson and Peraza Lope (2014) are interested in identifying households producing above their own needs, or production for surplus. They identify localities with production for surplus (Masson and Peraza Lope 2014, p.320) by densities one standard deviation above the Mayapán mean. They report surplus production for some items manufactured from materials such as chert/chalcedony, obsidian, or shell within a total of 34 households (Masson and Peraza Lope 2016, p.269). They are all located within the wall enclosing Mayapán. In many cases, the different crafts overlap in a given household.

Extensive research at Mayapán (Masson and Peraza Lope 2014) has not found evidence of direct supervision of production, or that the elite collected, stored, or redistributed products. The exceptions are censers and copper bells. Debris density and comparison with other sites indicate that lithic and shell production were part-time affairs that occurred in a domestic context. Many of these households were multi-crafting households (Masson et al. 2016). Masson et al. (2016) observed that the crafting households were among the wealthiest of the commoner class in Mayapán. This situation has been observed in other Maya and central México areas.

The area around Mayapán and neighbouring sites is a limestone platform (chapter four) in which no chert/chalcedony, obsidian, or shell sources have been found. Chert/chalcedony preforms likely arrived from mines in the Puuc area (as documented by Landa 1986), located 60 km from Mayapán, and most of the obsidian came from Ixtepeque, Guatemala (Masson and Peraza Lope 2016, p.254), located almost 700 km to the south. Marine shells were also imported.

Regarding ceramic production, direct evidence in the form of moulds for production of ceremonial items such as figurines and censers has been found (Masson and Peraza Lope 2016, p.248). However, moulds are portable objects, and the possibility that the location of production is somewhere else cannot be ignored. Masson and Peraza Lope (2016, p.252) found that copper working and effigy censer production are associated exclusively with elite contexts.

Concerning utilitarian ceramics, no direct evidence for production has been found at Mayapán. Two houselots were identified with a concentration of sherds at least one standard deviation above the mean (Masson and Peraza Lope 2014, p.261), structures Q-176 and Q-40a. For instance, surface collection from Q-176 yielded 23 per cent of the total surface collection from the settlement zone (Masson and Peraza Lope 2014, p.261). As mentioned earlier, unusual concentrations of pottery may occur for a variety of reasons including feasting, storage, dumping, or garbage disposal. However, red-slipped Mama vessels (which comprise many forms) and one type of unslipped vessel, a striated open-mouth jar (*olla*) thought to be for cooking, comprise 84 per cent of the assemblage found at this facility. The anomalously high sherd concentrations, which signalled a production facility to Masson and Peraza Lope (2014, p.261), and the limited number of ceramic types (which can signal specialisation) found at this house may be interpreted as a potential ceramic production location. At the lack of other evidence, to obtain a clearer picture, these proportions would have to be compared to proportions at other households to determine whether it is atypical. However, Q-176 appears to be typical, and its assemblage proportions are not distinguished from other households (Masson and Peraza Lope 2014, p. 6). Nevertheless, alternative explanations to the existence of a workshop for the production of utilitarian vessels can be given. For instance, the same pattern can be explained if large amounts of food were prepared (using *ollas*) and consumed at Q-176, such as during feasting or ceremonies, over the period of its occupation.

When kilns are lacking, such as is the case in the Maya area, an alternative indication of ceramic production would be a significant presence of wasters (fragments of overfired vessels). It will be shown in subsequent chapters that it is unlikely that wasters will ever be found for the Late Postclassic utilitarian ceramics of northern Yucatán. Experiments performed (sections 9.4.1–9.4.3) with utilitarian ceramic sherds consisting of re-firing them at a temperature just above the temperature of calcination (transformation into quicklime) demonstrated the destruction of the sherds within a few days.

Concerning long-distance movements of vessels, NAA analysis of censers and red-slipped samples from Mayapán (Cecil 2012) is starting to clarify whether we are dealing with movement of vessels or traditions, at least over a very large area. Petrographic and NAA analysis of samples from Mayapán, Santa Rita Corozal (Belize), and Petén area (Southern Lowland) found that Mayapán-style censers and red-slipped samples from these sites do not share chemical groups, but that they shared forms and style characteristics.

2.3 Explanatory Models of Ceramic Production Organisation

Based on the ethnohistorical and archaeological records, models have been proposed to explain the organisation of ceramic production during the Late Postclassic. These models are outlined and discussed in this section. At the completion of this study, they are compared to the results, aiming to identify the models that better fit the results. Although the list of models discussed in this study is not all-inclusive, when taken in totality, the models deal with a wide spectrum of components of the socio-economic systems. Their discussion in the last chapters of this document has the potential of informing not only these models but also perspectives about the wider socio-economic context of the Late Postclassic.

2.3.1 The Mercantile Production-Distribution Trajectory

This view of Late Postclassic production and distribution is linked to the hypothesised trajectory of ancient societies proposed by Rathje (1975) and Sabloff and Freidel (1975) that takes ancient societies from a stage of monumental construction and elaborate projects to a stage characterised by mass replication (Rathje 1975, p.414-415). If enough local demand exists, local variety accompanies these developments. Rathje (1975) proposed that Maya society during the Late Postclassic in northern Yucatán was in such a developmental stage and that the Late Postclassic pottery production and distribution characteristics included the presence of:

1. Mass production of pottery (Rathje 1975, p.430).
2. Cost-cutting techniques such as specialisation, standardisation (or routinisation), and simplification (Rathje 1975, p.430).
3. Distribution of standardised set of products (Rathje 1975, p.414-415).
4. Overarching production and distribution system (Rathje 1975, p.414-415), instead of local small-scale production units.
5. Land-based regional distribution and large-scale commercial exchange utilising maritime transport (Rathje 1975, p.433).

Under this model, large quantities of standardised vessels were produced (Rathje 1975, p.430). Rathje proposed that standardisation of shapes and sizes would make stacking and transportation more efficient (Rathje 1975, p.430) and that not only the vessels have a simplified appearance but also the manufacturing processes are simplified (Rathje 1975, p.430). This model proposes ceramic production for local, regional, and international markets (Rathje 1975; Sabloff and Freidel 1975). Feinman and Garraty (2010, p.171) define market exchange as economic transactions in which supply and demand forces, as well as price or exchange equivalencies, exist. As is the case in the Maya area, the archaeological evidence for market exchange has been difficult to find.

Hirth (1998, p.453-454) summarised indirect approaches that have been used to identify market exchange. One is based on physical configurations in particular architectural indications of markets. The contextual approach infers the existence of a market based on societal features that are believed to require a market, such as large city, or as a result of the natural growth of society. Another approach is to infer markets from information on craft specialisation or the scale of production. A greater flow of products than with reciprocal obligations or redistribution is expected in a marketplace, and an increased number of specialist producers points to market institutions. Another approach is based on the spatial distribution of commodities at different levels of consumption, from the household to the state level, which is thought to be distributed more widely in a market environment.

To this list, Stark and Garraty (2010) add the comparison of site-level artefact assemblages such as the one performed by Fry (1980) at various locations of the Tikal greater area. Stark and Garraty (2010, p. 40) point out the problem of more than one distribution mechanism (redistribution and market exchange for instance) resulting in similar spatial artefact distribution. They (Stark and Garraty 2010, p. 42-52) summarised current approaches that have tried to alleviate this problem. The approaches are based on extensive and systematic sampling because, as expressed by Hirth (1998), to discriminate market exchange from other types of exchange, it is required to have data at the household level.

Ratjhe (1975) hypothesised that, during the Late Postclassic, the expanded market gave rise to mass production not only of pottery but also other Yucatecan goods. The material remains of that period, therefore, would be characterised by:

1. Wide regional distribution across sites and remote regions of a common paste composition.
2. Uniformity of manufacturing techniques as the result of cost-cutting techniques such as mass production.

2.3.2 A Dichotomous Ceramic Economy

The results and propositions from ceramic production and distribution studies such as Fry (1980) or Rands and Bishop (1980), discussed earlier, represent a dichotomous model of the ceramic economy. The main tenet of this model is, on the one hand, a decentralised market system in which the core centre is a consumer with no role or a small role in redistribution, and a periphery that produced utilitarian (or subsistence) items with a surplus to be exchanged at local and sub-regional markets. On the other hand, the elite controlled the production and distribution of prestige goods by attaching the producers to their residences.

Given the lack of direct evidence for ceramic production, the distributional patterns that Fry and Rands and Bishop encountered can explain more than one mode of exchange. For instance, as mentioned earlier, Stanton and Gallareta (2001, p.232) differ from the interpretation that the patterns described in Fry (1980) for the utilitarian ceramics reflect a market economy. As Renfrew (1975, p.10) has stated, when looking at spatial distribution in isolation, market exchange does not differ from redistribution.

Evidence supporting the dichotomous model would have Mayapán as a pottery consumer centre and dispersed pottery production at the minor centres. Pottery produced at the minor centres should be found at Mayapán. Although other interpretations can be given to these patterns, the criteria outlined represent the necessary requirements to support this model.

2.3.3 The Tributary Mode of Production

According to the informants to Landa (1986), the lords at Mayapán required sheets of cotton (mantas), hens, honey, cacao, and copal as tribute. Elite at Mayapán and the different centres extracted goods, in which pottery may have been involved as tribute, intensifying rural production (Rice 1981). Researchers have observed that some contemporary peasants intensify production and create a surplus only if coerced by the powerful political or ritual elite (Smith 1974, Wolf 1966; cited by Rice 1981).

Barrera Rubio (1984) has proposed that tribute was concentrated at the Maya centres, and the centres monopolised critical raw materials, jobs, and manufacturers. Landa (1986) tells us that the Lords' entourages who live at Mayapán did not pay tribute. Therefore, functionaries and perhaps some of the artisans did not pay tribute. According to some researchers (Barrera Rubio 1984; Fernández Tejedo 1990), the governmental apparatus transferred tribute to trade with the exterior world for products that would be monopolised by the elites.

In order to consider this model based on tribute and given that, according to the accounts, the head of the different houses or lineages resided at Mayapán, at a very minimum, pottery from the minor centres should be found at Mayapán.

2.3.4 Production for Rituals, Ceremonies, Gifting, and Feasting

Other models emphasise ritual performance and a ritual mode of production (Wells 2006, p.283). Ritual is often understood in terms of religious ritual, but in general, it can be characterised by the 'performance of more or less invariant sequences of formal acts' (Rappaport 1999, p.24; cited by Wells 2006, p.284). For instance, Redfield and Villa Rojas (1967, p. 197) describe the adorned, large clay vessel used in the town on Chan Kom, Yucatán during a post-wedding ceremony. Thompson (1958, p.119) describes similar vessels used in north-central Yucatán in similar ceremony. Ritual production can include prestige and utilitarian items for gift-giving, feasting, religious ceremonies, and other rituals and ceremonies.

Ritual economy is a framework that sees some specialty products as socially valued objects (Wells 2006, p.285), and the group as continually negotiating and materialising its values through rituals (West 2006, p.285). Demand for food and items for ceremonies played a role in scheduling or regulating production and consumption, and surplus production was needed to produce items for the variety of rituals. For instance, an alternative explanation to Fry's (1980) market exchange explanation, proposed by Stanton and Gallareta (2001, p.232), is that serving vessels at the core centre were produced to be broken during feasting or ceremonies.

Pottery was and is an integral part of the many fiestas and ceremonies of Yucatecan Mayas, as described for instance in the town of Chan Kom (Redfield and Villa Rojas 1967, p.193). Landa (1986, p.70 - 90) compiled a Maya calendar including fiestas and ceremonies from which it is evident that large amounts of pottery required to be replaced predictably. Much of the ceramic production was likely damaged or broken as a result of these festivities. In the month of *Pax,* all the lords from the minor centres went to the major centre for ceremonies that included eating and drinking for five days. At the return to their towns, they prepared their own fiestas in which drinking and copal burning lasted until the month of *Pop* (3 months). In the month of *Pop*, for the new year celebration, all utilitarian things including dishes and *vasos* (jars) were renewed. They will take the old things and discard them out of town (with very significant implications for archaeologists' frequency counts). Unfortunately, Landa did not describe pottery production and who and how dishes and jars were replaced. In this month also, all the males will get together at the temple for days of drinking and eating. In the month of *Yax*, they will get new censers and clay effigies. Throughout the months, Landa described fiestas and ceremonies with days of eating and drinking, and in some months, discarding and replacing of ceramics.

In a model proposed by Friedel (1981), Maya centres were also pilgrimage shrines. Elites organised periodical festivals, which were then the occasion for the ritual visitation or pilgrimages to the centres. In addition, with the occasion of the festivals and pilgrimages, at the centres periodical markets (fairs) supplied a variety of goods to the centre and to the dispersed Maya population, articulating the local economies with the regionals ones (Friedel 1981, p.378). In this scenario, given the periodical nature of the festivals, ceramics were produced for festivities and rituals. They were also produced for the market, either as trade items or as containers. At the focal centres of peregrinations, ceramic production for ritual and markets was scheduled predictably.

Ethnographic work shows examples of periodical ceramic production for ritual purposes. San José Petén is the town to which, according to interpretations of Native writings, some of the Itza from northern Yucatán migrated during the Terminal Classic to Early Postclassic. San José Petén is, most likely, culturally related to Yucatán (Reina and Hills 1978, p.142). At this town, pottery is fired primarily close to the *milpas* (cropping fields) throughout the countryside. However, there is a dual orientation to pottery production (Reina and Hills 1978). This pottery is for domestic consumption, but once a year pottery is produced in the *pueblo* (town) to be used in a ritual with the skull during the Day of the Dead. For this ritual only, *ollas* are made, and the ritual food is prepared in these *ollas*. When done with the rituals, the *ollas* are then used throughout the year (Reina and Hills 1978). The spatial distribution that rituals such as this create for these *ollas* is similar to distribution, for instance, that market exchange would create, in which *ollas* were produced at one centre but found at many other centres.

However, some characteristics in pottery distribution may help to identify, for instance, gift-giving and feasting for reciprocal exchange (Hirth 1998, p.455). Low volume of commodity movement, small sphere of distribution, and heterogeneity of the sources characterises gift giving (Hirth 1998).

2.3.5 Calendrically Shifting Production Location

The organisation of production could have been more complex than researchers envision. Rice (2009b, p.119) has said that during the Late Classic multiple types of ceramic production organisation coexisted and that the organisation varied dependent on the type of vessels, the intended use, or the intended users. For instance, Rice proposed a ceramic production organisation for elite items based on a calendrically based shifting of production location from centre to centre and another organisation for ordinary items. This organisation parallels the interpretation by Edmonson (1979) of Late Postclassic political organisation in which centres did not permanently hold power but rotated it according to calendrical cycles. The main centres may have rotated according to the *may* or 256 years. The books record another rotation in which each of the thirteen most prominent centres among the provinces became the 'seat of the *katun*'. The seat of the *katun* became a capital for a period of a *katun* (20 years), getting tribute rights and administrative powers (Edmonson 1979, p.3). Ceramic production location, as proposed by Rice, rotated accordingly, at least for ceremonial vessels.

The archaeological record of paid tribute to the seat of the *may* or *katun* that involves pottery, probably as containers (for honey, for instance), would not be that different from the archaeological fingerprint of market exchange. At the centres that are seat of the *katun*, feasting or production for rituals could be recognised if some types of vessels predominate, such as ceremonial, serving, or cooking vessels.

2.4 Summary

The current research and the models presented highlight the difficulties and ambiguities of interpretation when, as in the case in Maya area, direct evidence for production is lacking. The situation is made more difficult because, during the Late Postclassic, colonial and ethnohistorical records mention several coexisting exchange mechanisms, while the quantities involved or the relative importance in the economy was not described. The models for ceramic production, distribution, and exchange presented in this chapter usually emphasise some of these mechanisms more than others. At the completion of this study, the results are compared to these models, aiming to identify the models that better fit the results.

Chapter 3

CERAMICS OF THE LATE POSTCLASSIC

This chapter presents the topic of this research, the Late Postclassic ceramics of northern Yucatán. The chapter starts by outlining the ceramic spheres, region, and period of this investigation. It continues with a brief description of the ceramics included in this study.

3.1 The Northern Yucatán Late Postclassic Ceramics, Period, and Region

3.1.1 Tases and Eastern Tases Ceramic Spheres

During the Late Postclassic, ceramic differences between the north-central and north-eastern zones of the peninsula were relatively minor compared with the extent of differences in art and architecture. This has led most researchers to believe that these areas, though closely related, belong to two different cultural spheres that have been called Tases and eastern Tases. When compared to the Terminal Classic, the Tases ceramic sphere is characterised by new ceramic types (Peraza Lope *et al.* 2006), the almost exclusive presence of calcareous rocks as inclusions, and, usually, soft paste with medium to coarse texture.

Major differences exist related to the distribution of the ceramic groups, with some groups dominating in a given zone. In most inland sites, only one red-slipped (Red Mama or Mama group) and one unslipped ceramic group (Unslipped Navula) dominate the assemblages [this classification was devised by Smith (1971a) for Mayapán, see next section]. Although many vessel forms exist, each basic vessel form within these groups varies little from site to site. The eastern Tases red-slipped vessels, or Payil group (more on this group in the next sections) are thinner, finer, and more compact than Mama. Payil sherds are found throughout the coast, including the Chetumal area and northern Belize. In addition to Payil, the unslipped Navula and Vista Alegre (a local eastern unslipped group) are also common in the coastal area.

Although the ceramics of the period lack finely decorated or elaborated items, Payil was most likely an elite item. Ochoa (personal communication, August 2012) has observed that eastern red-slipped Payil is usually found in ceremonial and elite contexts. Kepecs (1999, p.511) makes a similar observation remarking that Payil sherds were found mainly on the central platforms of the main centres of the Chikinchel region.

Without detailed mineralogical and chemical analyses, it has been difficult to determine whether stylistic similarities of red-slipped pottery in the inland or coastal areas are related to many producers sharing rigid standards of how to make the pots or to the existence a few producers. Similarly, without detailed mineralogical and chemical analyses, it has not been possible to investigate whether stylistic variability within the unslipped ceramics (for instance, vessels with plain versus striated surface) matches a varied composition.

3.1.2 The Period and Region

The sites examined in this study are located in north-central and north-eastern Yucatán. In these areas, during the Late Postclassic, the Tases and eastern Tases ceramic complexes dominated (Peraza Lope *et al.* 2006). Traditionally, the start of the northern Yucatán Late Postclassic is marked by the beginning of construction at Mayapán. This date was recently moved down to around AD 1100 (Masson, Hare, and Peraza Lope 2006, p.189; Masson and Peraza Lope 2014, p.63, 69), coinciding with the appearance in northern Yucatán of Tases most common ceramics (Red Mama and Unslipped Navula ceramic groups). Therefore, the Early Postclassic has been reduced to around AD 1000-1100 and appears to be a transitional period positioned after the end of the Terminal Classic but before the beginning of construction at Mayapán.

3.1.3 Ceramic Classification

Smith (1971a) classified the ceramics from Mayapán using the type-variety system. This classification provides a way to describe and classify the observed variety in the ceramics while enabling the communication in the language used by most ceramists of northern Yucatán material. The type-variety system is composed of a hierarchy of wares, types, and varieties. The whole array of types and varieties defined for each ware are present at Mayapán. This is not the case at other sites, in which only a few types are usually present. Groups of sites that present similar ceramic groups are collected into a single culturally significant ceramic sphere.

3.1.4 Origins of Northern Yucatán Late Postclassic Ceramics

3.1.4.1 Red-slipped Wares

Some researchers (Brown 1999a; Smith 1971a, p.203) do not see much influence of the previous northern Terminal Classic ceramics, the Sotuta (Chichén Itzá) and Cehpech

(Puuc area), in the development of the northern Late Postclassic (Tases) red-slipped ceramics. Smith (1971a, p.254) proposed that migrants, the Itza probably, are at the source of the coarser, limestone-only ceramics from Mayapán. Smith thought that at some point in these peregrinations, the Itza copied the Sotuta (Terminal Classic) vessels but continued using their 'decadent' (limestone-only, coarser paste) manufacturing techniques for the construction of the pottery clay, creating what will be known as Peto Cream Ware.

The implication from Smith's proposition is that the limestone-only slipped Postclassic ceramics of northern Yucatán are descendants of Peto, the first 'decadent' ceramics. Smith (1971a) believed that the red-slipped ceramics from Mayapán did not start at this site, and must have developed somewhere else before the beginnings of Mayapán.

A study that may have a bearing on the topic of the shift to limestone-only temper has been mostly overlooked. A distinctive feature of Terminal Classic Cehpech and Sotuta slipped ceramics is the presence of volcanic ash. Except for Thin Slate Ware from Uxmal, the distinctive slates (a type of slip) of the northern Terminal Classic contain volcanic ash (Smith 1971b, p.169-172).

However, petrographic analysis (Chung et al. 1995, Tabla 2; Varela and Leclaire 1999) of Sotuta and Cehpech slates (Terminal Classic) recovered from 25 northern Yucatán sites showed that the analysed sherds that are not from the main sites of Chichén Itzá or in the Puuc area are composed of carbonates only, with only a few containing volcanic ash. This is to say that Sotuta and Cehpech ceramics (by style and form), at centres distinct from the core centres of Chichén and the Puuc area, predominately have carbonate inclusions. At centres away from the core centres, it appears that ash-tempered Sotuta and Cehpech ceramics coexisted with local calcite-tempered Sotuta and Cehpech ceramics. Locally made calcite tempered Terminal Classic slates goes against the broadly accepted notion of the Late Postclassic ceramic composition representing an abrupt break with the previous period (in which finer and volcanic inclusions were the most common).

Researchers have entertained the idea that Mayapán Red and Payil ceramics may have originated in Belize or the Petén area. For instance, based on findings at Lamanai, Belize, Pendergast proposed (1981, p.48-49) that red-slipped ceramics with affinities to Mayapán developed before the 12th century somewhere else and were later adopted at Mayapán. Radiocarbon analysis yielded a date of A.D. 1140.

At the time of these findings (1981), Mayapán chronology dated the founding of this site to around A.D. 1250. The chronology of Mayapán has now been moved down because the first buildings may have been built as early as A.D. 1100 (Milbrath and Peraza 2003, Peraza Lope et al. 2006). This prompted researchers (Lope Peraza et al. 2006, p.173) to propose that red-slipped ceramics as found at Lamanai are not earlier than Mayapán ceramics (Lope Peraza et al. 2006, p.173). However, new radiocarbon reanalysis (Hanna et al. 2016) of the red-slipped Lamanai ceramics with affinities to Mayapán have moved down the appearance of the Lamanai ceramics to the late 10th century, earlier than the dates from Mayapán.

3.1.4.2 Unslipped Wares

Regarding Postclassic unslipped vessels, some researchers see a continuity of Terminal Classic Sotuta and Cehpech unslipped forms, particular the striated jars, until the Late Postclassic (Brown 1999a, p.295, 314). Furthermore, Brainerd (1958) who classified ceramics of northern Yucatán earlier than Smith (1971a), grouped Mayapán unslipped vessels with previous ones given their similarities.

3.2 Description of the Late Postclassic Northern Yucatán Ceramics

This section presents a brief description of the wares involved in the current research: Mayapán Red, Mayapán Unslipped, Tulum Red, and Peto Cream Wares. Precise detailed descriptions of these wares have been provided by the works of Smith (1971a, 1971b) and Brown (1999).

Of the ceramic types and varieties of these wares (Smith 1971a, p.135-136), the description focuses on the types and varieties sampled for this study. Concerning the forms sampled, sherds mainly from jars but also some deep dishes (*cajetes*) were selected for analysis.

3.2.1 Mayapán Red Ware

Mayapán Red is a red-slipped, smoothed and burnished ware (Figure 3-1, 3-2, see Table 8-18 for details about the sampled sherds) without many defects, such as spalling or popping (Smith 1971a, p.23). Forms such as jars (Figure 3-1, Figure 3-3), deep dishes or *cajetes* (Figure 3-4), bowls, tripod grater, basins, and restricted orifice bowls are common (Smith 1971a, p.203). As observed by Smith (1971a, p.23), variations in colour within the same vessel are common, including red (10R 4/8; 2.5YR 4/8; 10R 4/6, 3/6, 5/6; 2.5YR 4/6; 7.5R 3/8), orange (2.5YR 5/8; 5YR 4/8; 10R 5/8; 5YR 6/8), reddish brown (2.5YR 3/6, 5/6; 5YR 5/6, 5/8), brown (2.5YR 5/4; 5YR 4/4), and grey (fire clouding).

This ware has little in common with the Sotuta and Cehpech slipped wares from northern Yucatán previous periods (Brown 1999a; Smith 1971a, p.202). It shares some forms and overall colour with the Payil group of the eastern area, differing on some of the forms and thickness, compactness, and composition of the paste (Brown 1999a, p.328).

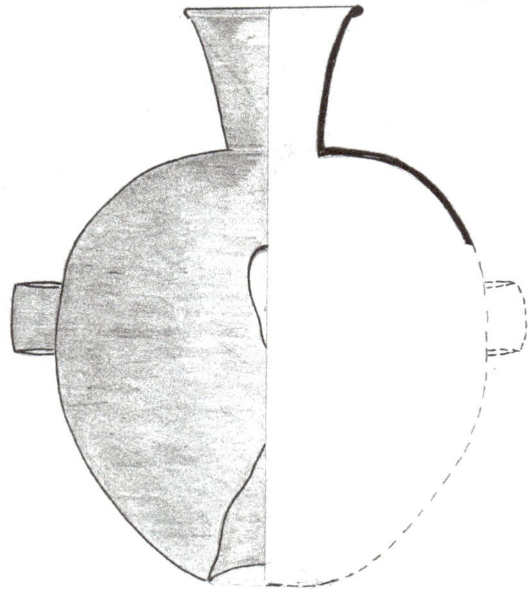

Figure 3-1. Mama jar (drawing by author)

Figure 3-2. Top: Mama variety samples (Mayapán); bottom: Cancún variety (Cobá)

Robert E. Smith (1971), who devised the most commonly used ceramics classification for Mayapán using the type-variety system (1971a, 1971b), defined two ceramic groups for this ware, the Red Mama and the Panabchen groups. The Red Mama group comprises ceramics found in Mayapán early and late stratigraphic lots. The Panabchen group comprises new forms and decoration that appeared in the late stratigraphic lots of Mayapán.

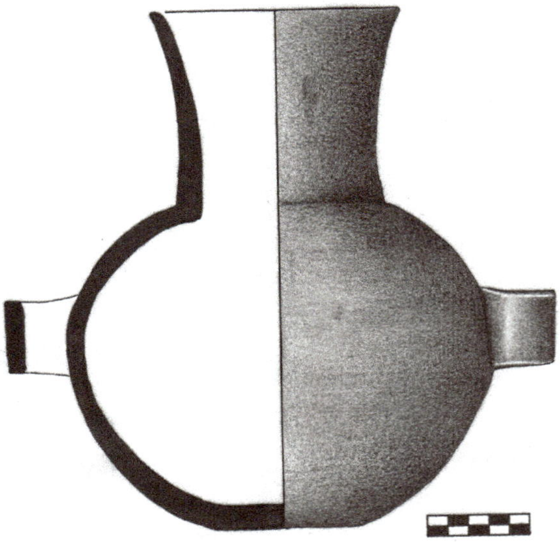

Figure 3-3. Mama jar (drawing by Luis Flores Cobá)

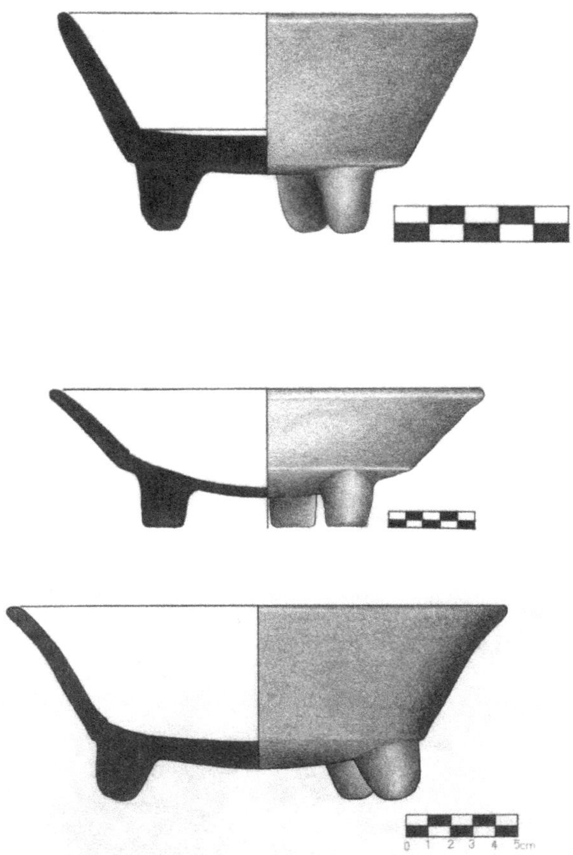

Figure 3-4. Mama tripod dishes or cajetes (drawing by Luis Flores Cobá)

However, given that, usually, stratigraphic divisions have not been found at other sites, researchers have conflated these differences into one group, and the typological distinctions made by Smith have fallen into disuse (Brown 1999, p.297). A simplified terminology in which the groups are named Mama, Navula, or Payil is usually used (Peraza Lope 1993; Cruz Alvarado 2012, p.12).

Of the Mama group, this study focuses on plain, slipped-on-exterior jars and cajetes. Cajetes can be slipped-on-interior. According to Smith (1971a, p.203), the main Mama forms are high and low-neck jars. Smith (1971a, p.79) considered jars one of the most, if not the most, important type of vessels at Mayapán because of their relative abundance (41 per cent of his collection) and the array of functions for which they are suitable.

Although these jars and cajetes are usually classified as Mama (Mama Red type), a Cancún (Mama Red type) variety is sometimes used for Mama-like vessels from the eastern areas of the peninsula. Mama and its eastern variety differ on the pink colour and hardness of the paste, with Cacún being pink and harder.

Red-slipped jars and cajetes showing other variations in decoration (such as Papacal Incised type:Papacal variety, Dzonot Applique type: Dzonot variety, or Mama Red type:Unslipped-exterior variety) were not selected for petrographic or chemical analysis (for reasons summarised in detail later within chapter five).

Regarding their intended use, a considerable amount of red-slipped jars is found by the sinkholes common in the area (Cruz Alvarado 2017, personal communication), or cenotes. Cenotes reach the water table indicating an association with water that can be functional (carrying, storing) or ritual. Ethnographic work in Yucatán (Thompson 1958) at a time in which forms similar to that of the Postclassic were in use shows them used for water carrying and storage.

The second form included in this study is the *cajete* (Figure 3-4), a deep dish that can occur with tripod supports. Although the emphasis of this study is on jars, some cajete sherds were selected for analysis because the functions of jars and cajetes may differ substantially, and they may present differences in production technology.

Figure 3-5. Navula samples

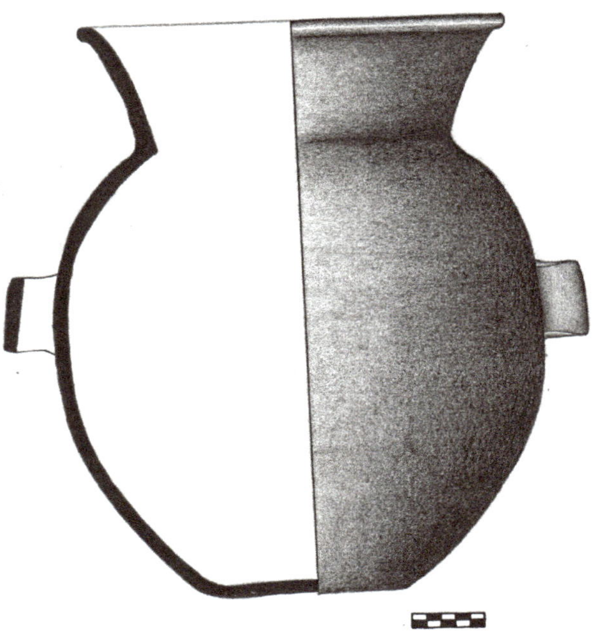

Figure 3-6. Navula jar (drawing by Luis Flores Cobá)

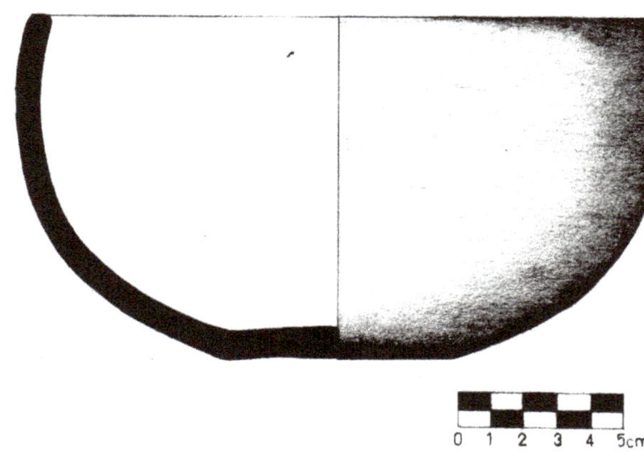

Figure 3-7. Navula cajete (drawing by Luis Flores Cobá)

Of the Navula group, this study focuses on two types. One are jars of the Navula Unslipped Type: Navula variety. These are plain, undecorated utilitarian jars. Some unslipped, undecorated cajetes of this variety were also selected. The second type comprises unslipped, open-mouthed jars showing exterior striation known as Yacman jars or *ollas,* of the Yacman Striated type: Yacman variety (Figures 3-5 to 3-7).

Yacman vessels are a ceramic variety showing light striations (Figures 3.8, 3-9) over most of the body, usually up close to the shoulder. Most are open-mouthed (the mouth is at least one-third of the maximum diameter) with a low-neck and globular body, or *ollas* (Figure 3-9). They most likely were cooking pots, based on descriptions in ethnographic work (Thompson 1958) and observation such as that of Brown (1999a, p.326) of a high frequency of burnt, carbon-coated bottoms in his collection.

3.2.2 Mayapán Unslipped Ware

This is an unslipped, smoothed but never burnished ware (Figures 3-5 to 3-9). It is probably the most abundant Tases ware, and together with Mayapán Red Ware makes up most of the assemblage from north-central sites.

Most researchers have divided these unslipped ceramics into utilitarian and ceremonial vessels (Smith 1971a; Ochoa 2007), based on the probable intended function of the vessels. Some of the forms associated with Mayapán Unslipped Ware are open-mouth jars, deep dishes or *cajetes*, censers, flat vase dishes, and restricted orifice bowls (Smith 1971a).

3.2.3 Tulum Red Ware

Of the eastern Tases ceramic sphere, this study focuses on the Red Payil (Smith 1971a, p.136) or Payil group (Cruz Alvarado 2012, p.6), illustrated in Figure 3-10a and Figure 3-10b, of the Tulum Red Ware.

Figure 3-8. Yacman samples (Teabo)

Figure 3-9. Top: small, reconstructed Yacman jar or *olla* (from Brown 2017); bottom: Yacman jar (drawing by Luis Flores Cobá)

Figure 3-10. Top: Payil (Mayapán); bottom: the incised variety, Palmul (Chac Mool)

The Payil group comprises a fine-grained, red-slipped, smoothed and burnished ceramics rarely found outside the eastern zone, with the exception of Mayapán. Mayapan Red Ware and Tulum Red Ware are both Postclassic ceramic wares sharing many visual characteristics, such as vessel forms and general slip colour.

Tulum Red Ware differs from Mayapan Red Ware in the fineness, hardness, and composition of the paste (Smith 1971a, p.30). Tulum also differs in the uniformity of the red colour (10R 4/6, 4/8, 3/6) of the slip varying little within and between vessels, the care taken during smoothing, the postfire incising (as opposed to prefire), and the style of decoration design (Smith 1971a, p.30).

3.2.4 Peto Cream Ware

This is cream slipped ware. At Mayapán, this ware is more common in the early deposits (Smith 1971a). Smith (1971b, p.254) considered that this ware overlapped during the end of the Terminal Classic and the Late Postclassic and served as a bridge between these periods.

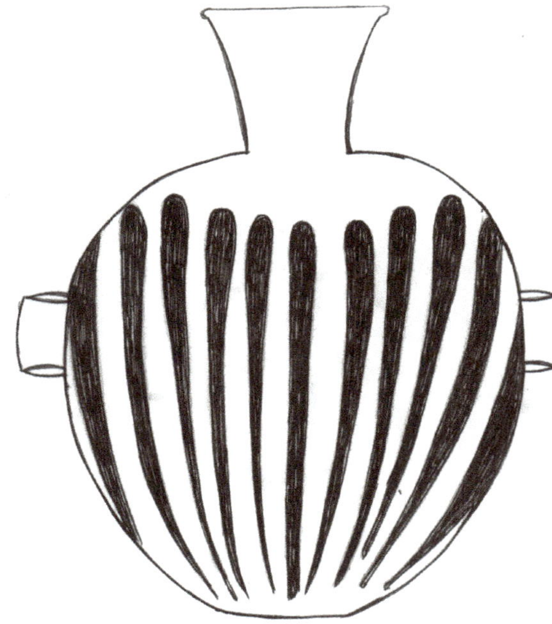

Figure 3-11. Xcanchakan, Peto Cream Ware (drawing by author)

Figure 3-12. Xcanchakan sample

Smith offered two reasons to the notion that Peto started during the Terminal Classic. One is the ceramic associations, which indicated to him an earlier beginning for Peto than for Mayapan Red. The second is that Peto shares forms and decoration with Sotuta (Chichén) ceramics and presents a method of manufacture with only coarse carbonate rock temper (as opposed to volcanic ash during the Terminal Classic) that presaged the composition to come during the Late Postclassic. A very early and sealed deposit at Mayapán contains Peto Crean, Sotuta, and Cehpech ceramics (Milbrath and Peraza Lope 2003, p.6), which indicates that Peto is earlier than Mayapán Red. That Peto, Sotuta, and Cehpech ceramics may have overlapped is suggested by other evidence such as radiocarbon dates of AD 900–1000 from Isla Cerritos (Robles 1987, p.105–106 cited in Milbrath and Peraza Lope 2003, p.5). This is a location on the east coast, and Peto may have been earlier on the coast than in the inland areas of Yucatán (Ochoa 1999).

Chapter 4

THE GEOLOGICAL SETTING

This chapter presents the geological environment and raw materials that surrounded the potters who made the vessels in this study. It is crucial to understand the geological environment because the physical, petrographic (in hand specimen and thin section), and chemical characteristics of these vessels are compared to the raw materials collected. The technological choices made by potters are, by necessity, related to the availability and properties of the raw materials chosen (Howie 2012; Sillar 2000, p.5) in the construction of the pottery clay. Nevertheless, the availability and suitability of raw materials are not the only factors driving the selection made by potters (Sillar 2000, p.5).

This chapter starts by introducing the general geological setting of northern Yucatán, including the Chicxulub meteor that hit Yucatán and its effects on the morphology and sedimentation of the region. It continues with the main rocks, unique features, and clays within the region.

4.1 The Geological Setting of the Study Region

The Yucatán Peninsula is a marine platform that has been covered with water multiple times throughout its geological history (López Ramos 1975). The study region is located in the northern part of the peninsula. This is an area without surface rivers due to the fractures and depressions in the carbonate rocks (Instituto Nacional de Estadística, Geografía, e Informática [INEGI] 1983). Sinkholes reaching the water level, or cenotes in Maya language, are common.

Complex processes have interacted in Yucatán including geological changes, soil formation, redeposition of soil material toward karstic depressions (Cabadas et al. 2010), weathering, and the strike of a large meteor (Hildebrand et al.1991) at the end of the Cretaceous near the current town of Chicxulub (Figure 4-1), Yucatán. These processes have been instrumental in the creation of the current-day soils of Yucatán (Pope et al. 1996) and surface features.

The morphology of Yucatán was affected by the meteor (Pope et al. 1996). The impact created a crater and a semi-circular depression, or 'moat,' 3-5 m deep located at a radius of 83 km (Pope et al. 1996). Figure 4-1 shows an aerial image of the Yucatán peninsula showing a faint but unmistakable contour of the Chicxulub crater. At the edge of the crater, there is a semi-circular area with high permeability coinciding with the semi-circular depression (Pope et al. 1996) that contains a ring of sinkholes (Figure 4-1) that separates the basin of the crater from the area outside the ring. Sinkholes reaching the water table are known as cenotes (from Maya language). The area outside the crater has many cenotes, so many that it is called the 'pockmarked' terrain (Perry et al. 2002).

The sedimentation patterns in northern Yucatán are closely linked to the impact of the meteorite. Sedimentation and rock formation inside the basin are different from the outside. During the Late Tertiary, while the sedimentation took place within the crater's basin, the area outside the ring may have been exposed and underwent weathering and karstification made evident by the cenotes (Perry et al. 2002; Pope 1997).

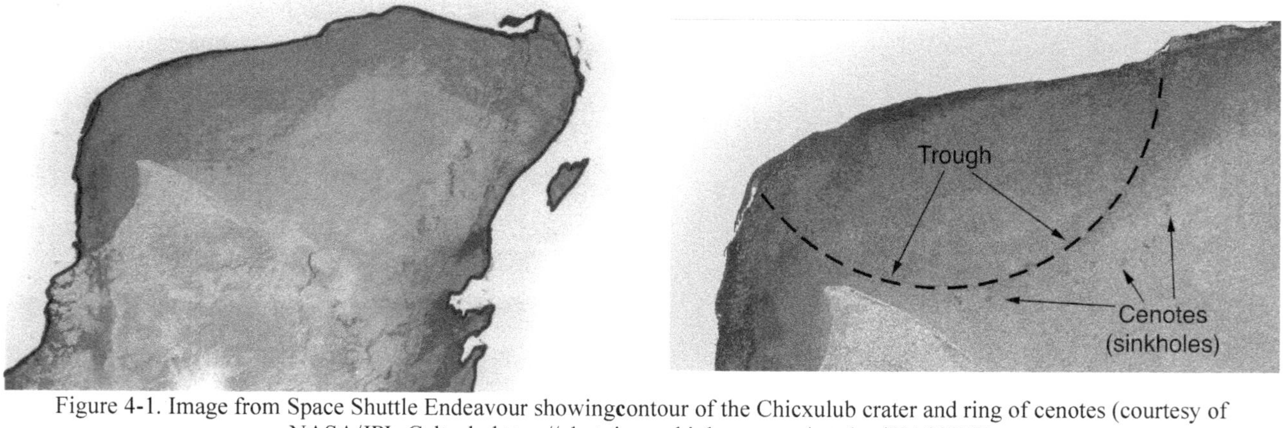

Figure 4-1. Image from Space Shuttle Endeavour showingcontour of the Chicxulub crater and ring of cenotes (courtesy of NASA/JPL-Caltech, https://photojournal.jpl.nasa.gov/catalog/PIA03379)

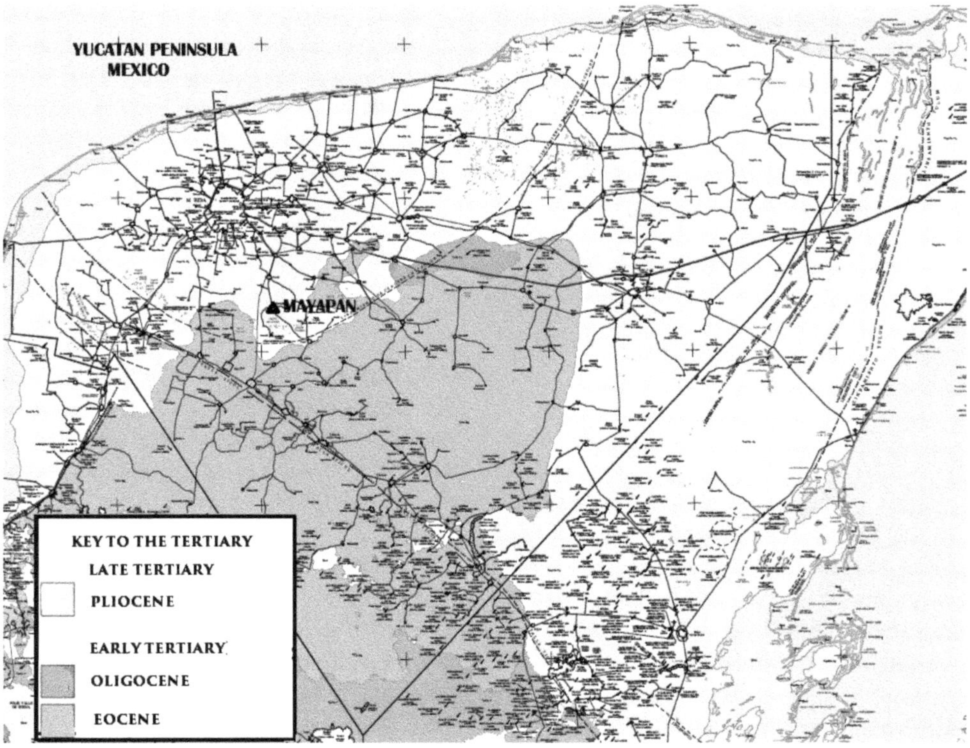

Figure 4-2. Geological map of Yucatán (from Servicio Geologico Mexicano, available at https://mapserver.sgm.gob.mx/Cartas_Online/geologia/peninsula_3.pdf)

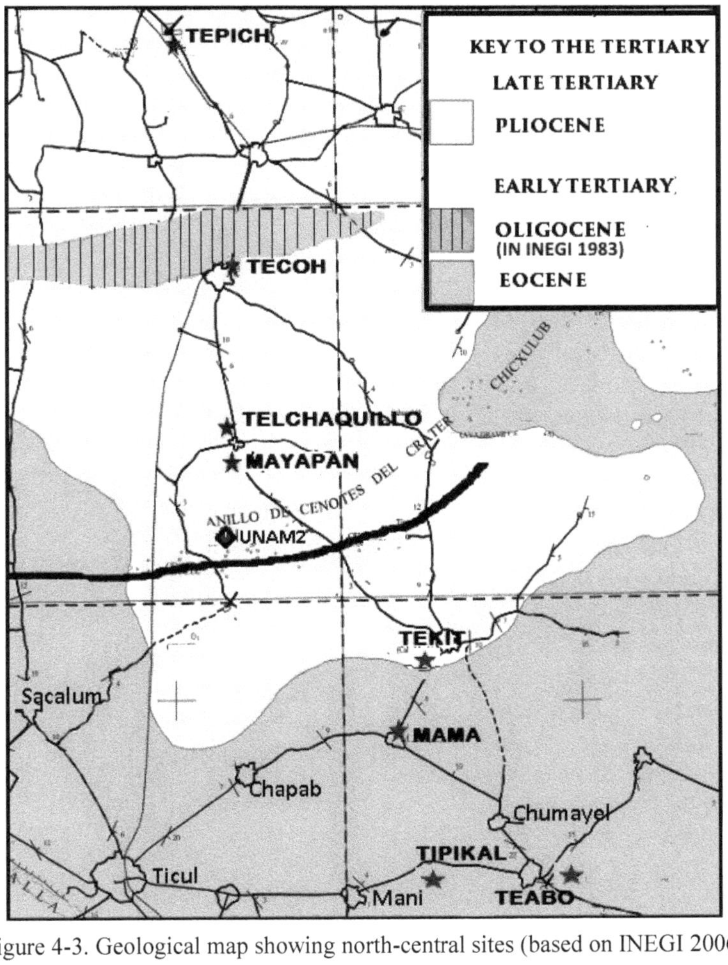

Figure 4-3. Geological map showing north-central sites (based on INEGI 2006)

Regarding the soils, age differences are associated with differences in soil characteristics (Pope et al. 1996). Pope found that what most explains differences in Yucatecan soils is age.

The area inside the basin is mapped by Pope and INEGI as Late Tertiary, while the area south and outside the ring is mapped as Early Tertiary. Different dates inside and outside the cenote ring were corroborated by the dating of surface carbonate rocks by Gilli et al. (2009), using $_{87}Sr/_{86}Sr$ isotope techniques, concurring with the dates proposed by Pope.

As shown in Figure 4-2, the Late Tertiary (off-white) covers the area inside the crater and arches around, comprising almost all of the north part of the state of Yucatán, most of the eastern half of the same state, and most of the state of Quintana Roo (Perry et al. 2009). The oldest exposed rocks in this area comprise Early Tertiary (Eocene) carbonate rocks (INEGI 2006) located south and southeast of the cenote ring (Figure 4-2).

Most sites in this research are located in the north-central area, north and south of Mayapán, forming close to a straight line. Their locations in relation to Figure 4-2 are shown in detail in Figure 4-3. In relation to these sites, the Early Tertiary area includes the sites of Tekit, Mama, Tipikal, and Teabo. The Early Tertiary may also comprise a small-restricted Oligocene area in the Tecoh area (INEGI 1983). The Late Tertiary (Pliocene-Miocene) area includes Mayapán and sites located north of it: Telchaquillo, Tecoh, and Tepich. In addition, three sites, Culubá, Cobá, and Chac Mool are located in the eastern part of northern Yucatán.

4.2 The Rocks within the Region

The Yucatán peninsula is mainly composed of carbonate rocks over a much older basement of terrigenous origin (López Ramos 1975). Cores taken from wells drilled in northern Yucatán, mainly by PEMEX (Petróleos Mexicanos) and by the National University of México (UNAM) Chicxulub drilling programme (Urrutia-Fucugauchi et al. 2004), have provided samples for the identification of surface mineral phases. Based on these programmes, limestone, dolomitic limestone, and dolostone are the rocks that commonly crop out in that area (Leftticariu 2003). A carbonate rock is classified as a limestone when calcite dominates, as opposed to dolostone when dolomite dominates. Mostly two minerals with very similar crystal structure form the carbonate rocks of Yucatán: calcite ($CaCO3$) and dolomite ($CaMg(CO3)2$). In carbonate sediments, dolomite is usually associated with calcite (Nesse 1986, p.144), but it can be found in isolation (Adams and MacKenzie 2001, p.132).

The lithological columns for the sampled wells in Leftticariu (2003, p.3, Figure 2) located in north-central Yucatán (Leftticariu 2003, Figure 1), show in detail the composition of the rocks at or near the surface. One well in particular, UNAM2, is located close to Mayapán (Figure 4-3). The samples taken were subjected to a battery of tests. The results indicate that the crater was a distinct sedimentary basin throughout the Tertiary; basin roughly delimited by the cenote ring (Leftticariu 2003, p.2). In this basin, pelagic limestones predominate. Dolomitised limestone and limestone with dolomite may be more common outside the crater. In addition, layers of marls in these older rocks are described in INEGI (1983, 2006).

Although dolomite layers may be more common outside the basin of the crater than inside (Leftticariu 2003, p.2), dolostone and dolomitic limestone were found close to the surface at -20 m, in samples taken from well UNAM 2 (Leftticariu 2003, Figure 2).

4.3 Other Surface Features

The relief of the northern Yucatán Peninsula developed from the limestones and processes including a coastal plain and a karstic plain (Bautista and Zinck 2010, p.2). The coastal plain extends along the coast, and it is less than 10 m above sea level. The karstic plain has a topography that varies from flat to undulating at 10 – 60 m above sea level (Bautista and Zinck 2010, p.2) that in the north-central area is less contrasting.

4.3.1 Cenotes

In the karstic plains of northern Yucatán, cenotes are common features. Cenotes are sinkholes reaching the water table, described earlier and illustrated in Figure 4-4.

Figure 4-4. Entrance to cenote (sinkholes reaching the water table)

4.3.2 Calcrete (Caliche) Layer

In addition to cenotes (Figure 4-4), a characteristic surface feature in northern Yucatán is a calcrete, or caliche, layer (Perry et al. 2002). Calcrete is created by the dissolution and re-deposition of the calcareous materials (INEGI 1983) forming a hard carbonate layer. This hard layer (Figure 4-5) is more developed within the crater's ring, the Chicxulub Sedimentary Basin, in which it can reach up to 3 m thick (Perry et al. 2002). Carbonate rocks (limestones, dolomitic limestones, or dolostones) underlie the caliche layer.

Figure 4-5. Exhausted sascabera under the hard caliche layer

4.3.3 Sascab

Underlain the caliche layer or between rock or clay layers, there are relatively thin lenses of unconsolidated carbonate powder, pebbles, and boulders known as *sascab* (Perry et al. 2002). It is very common in the peninsula (Figure 4-6). The composition of *sascab* varies according to the associated rock layers (Littmann 1958). Littmann showed that *sascab* is an unconsolidated version (as observed in Figure 4-7) of the nearby rocks and that *sascab* located near limestone contains limestone, while *sascab* found in pockets associated with nearby dolostone is composed of dolomite. In the same manner, a clay layer underlying the *sascab* layer affects the composition of *sascab*.

4.3.4 Altillos and Planicies

Two forms occur throughout the karstic area, namely the mounds or *altillos* and the flat areas or *planicies*. The soils in the mounds have a large amount of stones and organic material (Bautista and Zinck 2010). The *planicies* are the flat areas from which the water filters to the underground, and the soils are washed away containing the least amount of carbonate minerals. Black soils occur on mounds, and deep red soils on the *planicies* (Bautista and Zinck 2010, p.2). In general, the carbonate rocks are found at the surface and under the shallow soils within the area.

Figure 4-6. *Altillo* exposed by road construction, with dark soil on top and pockets of unconsolidated rock, or *sascab*

Figure 4-7. Detail of sascab texture.

4.4 Clays within the Region

The Yucatán Peninsula has witnessed geological, extra-terrestrial, sedimentation, and soil formation processes and events. They have resulted in a variety of pedogenetic, detrital, and direct crystallisation clays (Isphording 1984, p.60).

Given that north of the Champoton River (area including the whole state of Yucatán), there are no surface rivers and all drainage is subterranean (Isphording 1984, p.61), the clays do not contain significant detrital material. The exceptions are some clays found at fault basins of the eastern area (smectite) or some flat closed basins in the interior (Isphording 1984, p.61; see Isphording 1984 on how the clays may have been formed).

As described in the next paragraphs, kaolinite, kaolinite-rich clayey thick red soils, montmorillonite, kaolinite-montmorillonite, palygorskite, and sepiolite have been documented in the northwestern area. In the eastern area, the clays are dominated by smectite, while kaolinite is present but not common (Isphording 1984, p.65), and palygorskite has not been found.

4.4.1 Kaolinite

The soils formed on limestone in Yucatán are rich in well-crystallised kaolinite (Cabadas et al. 2010), indicating to Cabadas et al. a high degree of weathering. Kaolinite is associated with the pedogenetic soils of the northeast coastal plains and the northwestern plains (Quiñones 1975; Isphording 1984; Cabadas et al. 2010). The soils of these areas are similar in that they are composed mainly of weathered limestone, organic material, clays, and oxides. The mineralogy of the clays includes poorly crystallised kaolinite, boehmite, and traces of talc and chlorite (Isphording 1984, p.61). De Pablo-Galan (1996, p.100) examined 90 per cent pure disordered kaolinite around Sierra del Ticul representing mature pedogenetic soils (De Pablo-Galan 1996, p.100). Associated with this kaolinite are minor quartz and calcite (De Pablo-Galan 1996).

4.4.2 Red Soils and Clays

Red and reddish brown materials are found in some of the low or *planicies* areas. A study by Cabadas et al. (2010, p.1441) of thick red soils in three different areas of Yucatán found that these soils are dominated by clayey fine material (usually over 50 per cent with one location, Kantunil Kin, having 88-99 per cent). These clays owe their red or reddish brown colour to red iron oxides and brown ferruginous nodules. They are mostly friable or slightly plastic, usually without sand inclusions other than ferric concretions (Isphording 1984, p.61, 65). The clay minerals in these red soils were, in order of abundance, kaolinite, vermiculite, mica, and goethite and gibbsite. Illite and palygorskite are absent.

The amount of organic matter in unprocessed red clays from the north-central regions may be high. In a study (Shang and Tiessen 2003) that measured the organic material in black and red soils of the Hocaba community, it was found that they contained 15 per cent organic material.

4.4.3 Palygorskite

Palygorskite is common in northwestern Yucatán. It has been found in several places close to the surface (Sánchez del Rio et al. 2009). Isphording and Wilson (1974, p.484) consider palygorskite as the dominant clay not only in the Ticul area but also in other areas of the peninsula.

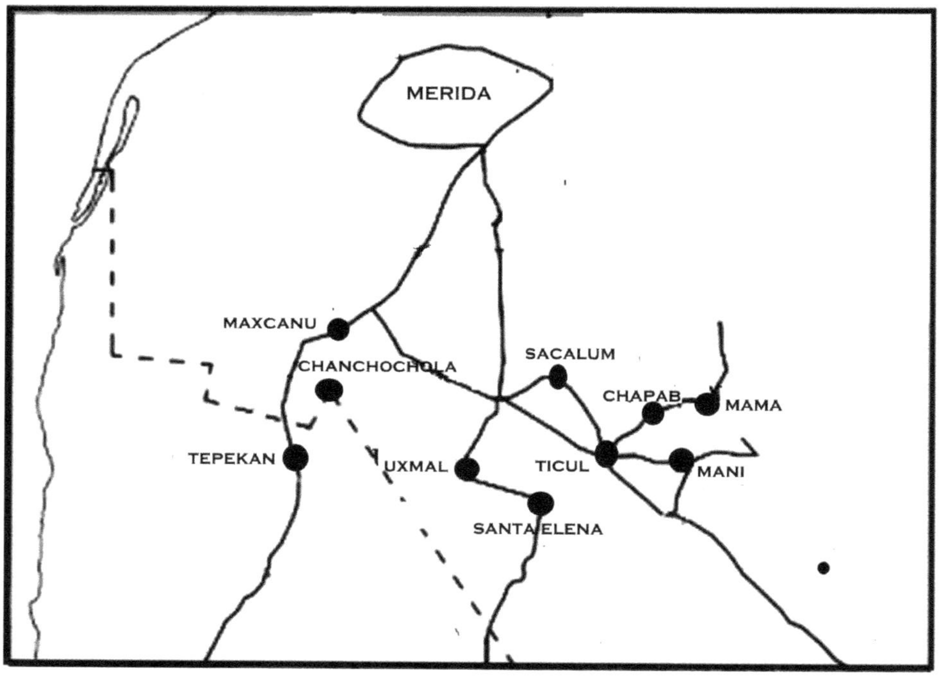

Figure 4-8. Locations of palygorskite clays (data from Arnold 1971; de Pablo 1996, p.94; Isphording 1984, p.68, Fig. 4; Morales Valderrama 2005; map and data from Sánchez del Río et al. 2009, Fig. 1)

Figure 4-8 summarises the locations at which palygorskite has been found. Sánchez del Río et al. (2009) analysed samples from these locations finding that these deposits do not contain smectites, carbonate minerals, or significant quartz amount. (Isphording 1984) explains palygorskite formation by direct precipitation. He also compares Yucatan tertiary basins to Tertiary basins of West Africa in which a seaward succession of montmorillonite, palygorskite, and sepiolite was formed by direct crystallization, and that the type of clay depends on factors such as the relative concentration of alumina.

Sánchez del Río et al. (2009) also collected two samples of sascab, one from a sascabera south of Maxcanú, Chanchocholá, and another from a pottery shop at Ticul. The analysis showed that they are palygorskite marls, the first containing dolomite and the other calcite and dolomite. Figure 4-10 illustrates a palygorskite marl sascabera.

Currently, there are no known palygorskite sources within the crater's basin (Figure 4-8). Most palygorskite clays from Yucatán are hard as a rock when dry but plastic and soft when wet (Isphording 1984, p.67, 69), while sepiolite clays are non-plastic (Schultz et al., Isphording 1984, p.67-69).

Figure 4-9. Entrance to palygorskite-sascab mine in the vicinity of Chapab (Figure 4-8), Yucatán

4.4.4 Smectite

Smectite has been documented in samples from Maxcanú, Figure 4-8 (Sánchez del Rio et. al 2009). These clays are plastic and of grey to black, or dark brown, colour (Isphording 1984, p.65), containing abundant silt to sand size quartz, magnetite, and ilmenite (black opaque).

4.4.5 Kaolinite-montmorillonite

The presence of kaolinite-montmorillonite has been documented (Schultz et al. 1971) in locations of northern Yucatán, such as Yo' K'at (Arnold 2005a, 2008) close to Ticul, and at Tepekan, and Becal (Campeche).

4.4.6 Montmorillonite

Arnold (1971, Table 1) has documented this clay in the vicinity of Ticul, found in samples taken from sascaberas in that area. The samples are composed of clay and calcite. In one instance, the clay is 100 per cent montmorillonite. In the other, 85 per cent is composed of montmorillonite, and for the rest the results were inconclusive.

4.4.7 Montmorillonite-kaolinite

The clay analysis of a marl (plastic sascab) collected from a sascabera near Mayapán (#11 in Table 5-4) found that the clay portion is composed of montmorillonite, with kaolinite as a minor component. The analysis of this sample was performed by PMET Laboratories (Shannon 2017, report included in Appendix A for ease reference).

It is important to mention that quartz is found as a minor component naturally present in soils and clays from the north-central area, and the north-western area in general (Shang and Tiessen 2003, p.192; Sánchez de los Ríos 2009). For instance, patches of red soils located in the community of Hocaba, located in north-central Yucatán 56 km south of Mérida, were examined finding that quartz is concentrated in the silt fraction (Shang and Tiessen 2003). In the clay fraction of these red soils, kaolinite was the most abundant mineral. The analysis of multiple palygorskite clay sources by Sánchez del Rio (2009) throughout central Yucatán showed that traces of quartz are commonly found in these clays.

4.5 Summary

Carbonate rocks are the most common rocks in Yucatán. They are composed mainly of calcite and/or dolomite. Common in northern Yucatán is an unconsolidated version of the carbonate rocks. It is locally called *sascab*. It is readily found almost at surface level in lenses and pockets. These pockets are locally known as *sascaberas*.

The impact of a meteorite affected the morphology and sedimentation patterns of Yucatán. The impact formed a variety of features such as the basin of the crater and a ring of sinkholes (cenotes) around it (Figure 4-1). During the Late Tertiary, sedimentation took place within the basin of the crater, while the area outside the ring, dated as Early Tertiary, may have been exposed to the corresponding weathering (Perry et al. 2002; Pope 1997). These differences in dates have been corroborated with isotope techniques. Research has shown that these age differences are associated with differences in soil characteristics, with Pope et al. (1996) finding that what most explains differences in Yucatecan soils is age. A variety of clay deposits including kaolinite, kaolinite-montmorillonite, smectite, and palygorskite are reported in the literature. However, references in the literature of clay deposits located within the basin are sparse.

Chapter 5
RESEARCH METHODS

The ultimate goal of this study is to contribute to a better understanding of the Late Postclassic social and economic environment. In chapter one, hypotheses and research questions were proposed to investigate regional patterns in ceramic production technology that might reflect production organisation, social groups or divisions, and interactions including exchange and traditions. These are key components of the Late Postclassic Maya socio-economy. The expectation of learning about these many aspects of Maya society through the study of ceramic technology rests on a theoretical framework that considers technology as a cultural production. Under this framework, technology is conditioned to not only the constraints of the environment but also to the wider symbolic, economic, and political context embedding the artisans (Dobres 2010; Lemonnier 1993; Jordan 2015). This is the central tenet of the research methodology of this study.

This chapter starts by presenting the theoretical foundation for the research methodology including three concepts that proved to be useful in this study: technological choices, techniques, and the *chaine opératoire*. Next, three hypotheses proposed for guiding this research in addressing the research questions are discussed. Within the conceptual framework, a combination of analytical methods, concepts, and constructs was used to test the hypotheses and address the research questions including the use of the *chaine opératoire*, test cases, ceramic experiments, petrographic, chemical (NAA), statistical, and surface features analyses.

5.1 Technology as Cultural Production

The notion of technology as cultural production is closely related to the process of learning how to implement technological solutions. These solutions are physical objects and part of the material culture (Jordan 2015, p.1). Usually, people do not invent technological solutions by themselves just by trial and error: most of the time, people learn from other people (Jordan 2015, p 1). A new potter is made through a long learning process. It involves person-to-person contact, transfer of traditions, sharing or demonstrations of material culture, in which not only the mode of execution is conveyed to the potter but also the appropriate behaviours, movements, mental models of the finished work, traditions, symbolism, and cultural meanings (Wendrich 2012, p.2-4; Jordan 2015).

That ancient technologies are socially embedded (Whitbread 2001, p.450) is demonstrated by ethnographic work showing the relation between the way things are done, or technologies, and cultural and symbolic perceptions of the world (Arnold 1971, Gosselain 2000, Thompson 1971, Morales Valderrama (2005), Mahias 1993). Van de Leeuw (1993), for instance, showed that the way potters conceive a pot is a factor in the choices of technical processes.

Technological choices are related to the concept of technological style. The basic stages in the manufacture of ancient ceramics are the same worldwide because of the nature clay (Rye 1981, p.3). They include the selection of raw materials, the preparation of the pottery clay, forming, drying, and firing (Rye 1981, p.3). In addition to or within these necessary steps, potters can choose from a variety of valid alternatives with regard to clays, tempers, shapes, thickness, and techniques of construction, and firing (Sackett 1990:33) that would produce a successful vessel. Sackett (1990) defines technological style as the selection, within the alternatives present to potters, of functionally equivalent alternatives. Therefore, style is a given manner of doing something, which is specific to a time and place.

Two concepts used in the approach in this study need to be defined: techniques and *chaine opératoire*. A technique is a series of actions to achieve a particular goal such as collecting and processing the clay or slipping (Sillar and Tite 2000, p.4). The sequence in which techniques are executed to transform the raw materials into the final product is known as the *chaine opératoire* (Sillar and Tite 2000, p.4, van der Leeuw 1993, p.240).

How does the study of techniques and *chaine opératoire* lead to a better understanding of ceramic production and organisation as well of the socio-economic organisation? One reason for the usefulness of the *chaine opératoire* is that it provides a method to examine the choices at every stage of the production process (Wendrich 2012, p.257) and facilitates the identification of variations. Techniques are usually grouped (Rye 1981, p.4) based on their places in the sequence of production, such as those associated with material processing.

Reconstructing this sequence allows identifying invariants or steps that cannot be modified (van der Leeuw 1993, p.240) without compromising the final product, and variants that allow more freedom. This sequence provides a road map from which techniques used among groups of pottery can be identified and compared.

Another reason for using a *chaine opératoire* is that it may uncover social divisions. Different behaviours and differences in the environment may result in technological variations in the pottery. Therefore, patterned similarities in ceramic technology express 'the tradition to which the producer group belongs and therefore discriminate between main technical traditions' (Roux and Courty 2005, p.202). The choices made by potters with regard to materials, the techniques applied, and the sequence of steps followed for the construction of the end product are conditioned by the wider cultural, symbolic, economic, and political environment surrounding the artisans (Dobres 2010; Lemonnier 1993, p.3; Gosselain 2000). Material analysis, therefore, provides a route to access past societies (Whitbread 2001, p.450).

The reasons for that are related to the processes of knowledge transfer. The learning of a craft and knowledge transfer is dictated by social factors such as family, traditions (Wendrich 2012, p.11; Jason 2015), or gender (Wendrich 2012, p.11). Therefore, the technological style of one member of the group resembles that of other members (Jeffra 2015, p.142). For instance, Gosselain (2000, p.189) observed that networks of interaction and social boundaries tended to be reflected in the spatial distribution of the different styles and ways of making pottery, the *chaine opératoire*.

The artefacts (or material culture), then, are more than just the result of technology for they embody traditions and social context. Therefore, the study of how ceramics were made, the choices made from what may have been available to potters, and variations in technology have, as a result, archaeological value for they may point to divisions in the potters' environment, such as technological traditions, or production groups and zones (Roux and Courty 2005, p.202). These groups may have implications and are useful in this study for the understanding of ceramic manufacture, organisation, exchange, and other interactions.

Another area in which a *chaine opératoire* is useful for this research is in the design and evaluation of experiments on local raw materials. Jeffra (2015) has advocated the use of the *chaine opératoire* in experimental archaeology as a form to bridge the results of experimental data and the social environment. This allows making inferences about the social context.

However, there are many limitations as to the use of the *chaine opératoire* in archaeology. One such limitation is that the work of the archaeologist differs from the ethnographer in that the actions made by potters are not visible in the archaeological record (Rye 1981, p.4). The archaeologist has to be content with observing the physical attributes of the pottery and making inferences about the actions and techniques that may have produced them (Rye 1981, p.4). Another major limitation when examining pottery is that the observed attributes could have been the solution to multiple technological problems. Given the premise that artefacts reflect the broader cultural context, this ambiguity implies that attributes could have been the physical representation of many types of social meanings (Lemmonier 1993, p.16). In addition, some actions do not leave traces in the archaeological record and others are erased. Lastly, while the *chaine opératoire* refers to a particular sequence of manufacture for whole pots, the archaeological material many times is made of small pieces of different pots.

5.2 Outline of the Methodology

To test the hypotheses and address the research questions a regional- and technology-based approach for the reconstruction of pottery manufacture is needed. Given that the *chaine opératoire* is used for whole vessels, a variation of the *chaine opératoire* method used by Roux and Courty (2005) in South Levant is useful for this study. Their approach starts with the raw materials and ends with the finished fired sherds, instead of the finished fired pots. Roux and Courty (2005, p.202) first grouped the sherds according to surface forming or finishing characteristics such as presence and type of slip. These groups were further broken down into petrographic groups based on petrographic analysis that discriminated the samples by the materials selected and clay processing techniques based on the properties of the fine portion of the pottery clay and properties of the coarser inclusions such as their type, morphology, and abundance.

In the current study, the reconstruction of pottery manufacture involves the hierarchical or cumulative study of raw materials and pottery attributes, the inference of techniques, *chaines opératoires*, and traditions using the implementation used by Roux and Courty (2005). The current research is structured around two test cases comprising one set of sites in the north-central area of Yucatán and another in the eastern area (Figure 5-1). The steps in the analytical method are outlined below and discussed in the next sections:

1. inspection of the available collections;
2. sampling of pottery vessels from existing collections;
3. collection of raw materials from the north-central area;
4. preliminary survey of pottery samples and formulation of hypotheses;
5. petrographic analysis of hand specimens into broad groups;
6. petrographic analysis of thin sections and characterisation into fabric classes;

7. chemical analysis of pottery samples and clays and characterisation into chemical groups; and lastly
8. examination of patterns and interpretation of ceramic production patterns.

5.3 Sampling Strategy

Testing of the hypotheses involves the identification and comparison of patterns in ceramic technology using a regional approach. The sampling strategy was designed to support the identification of these patterns.

5.3.1 Sites Selected

The research is structured around two test cases of selected sites from the north-central and eastern areas of northern Yucatán (Figure 5-1), designed to test the hypotheses (outlined in section 1.4 and described in section 5.5) at short- and long-distance levels. The test cases will be referred as Case A and B (Table 5-1). Figure 5-1 illustrates the sites and test cases included in this research.

5.3.1.1 Case Study A

Case study A comprises the sites located within the north-central area of the peninsula, as shown in Figure 5-1. This area is expanded Figure 5-2. In a north to south direction, the sites are Tepich, Tecoh, Telchaquillo, Mayapán, Tekit, Mama, Teabo, and Tipikal. A brief introduction to these sites is presented next.

Tepich presents monumental architecture (Garrido 2003, p.20). It is located 15 km southeast of Mérida (Garrido 2003, p.49). There is no mentioning of Tepich as a town in Roy's (1957) compilation.

Tecoh is located 29 km southeast of Mérida (Garrido 2003, p.51) inside the town of the same name. The town of Tecoh was an important town that first appeared in the early records when its cacique Don Juan Cocom signed the Treaty of Maní (Roys 1957, p.39), but it is Roy's contention that the name Cocom is common in Yucatán and that the relation with the powerful branch of the Cocom is not clear (1957, p.39). Tecoh most likely was under Mayapán rule, but it is not clear if the Cocom was the head lineage at this town.

Telchaquillo is located 1.5 km north of Mayapán inside the town of the same name. According to Roy (1957, p.67), Telchaquillo was an unimportant town bordering an archaeological site named Hoal with architecture similar to that of Mayapán (this project includes Hoal ceramics as part of Telchaquillo ceramics).

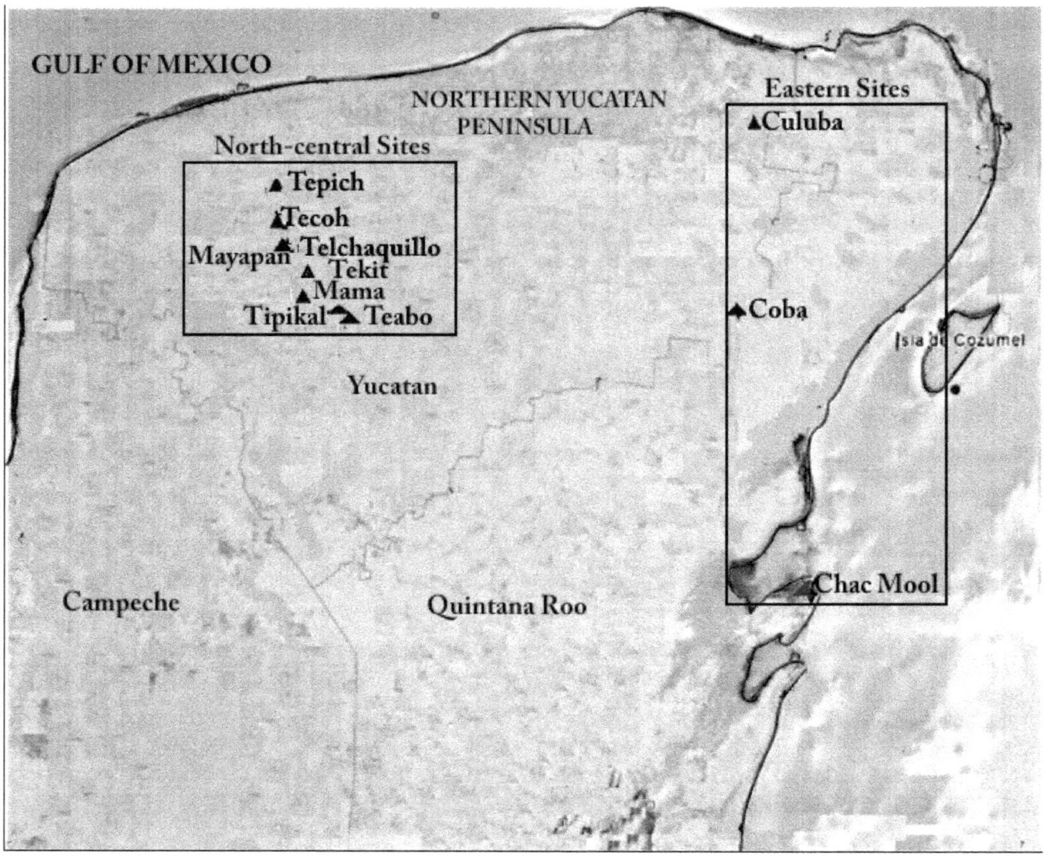

Figure 5-1. Sites and test cases in this study

Table 5-1. Case Study in Each Zone and List of Sites within each Case

Case Study	Zone	Sites (Figure 5-1)
A	Mayapán and neighbouring sites located in north-central area of Yucatán Peninsula	Tepich, Tecoh, Telchaquillo, Mayapán, Tekit, Mama, Teabo, and Tipikal
B	Mayapán and distant sites including eastern sites	Mayapán and Chac Mool, Cobá, and Culubá

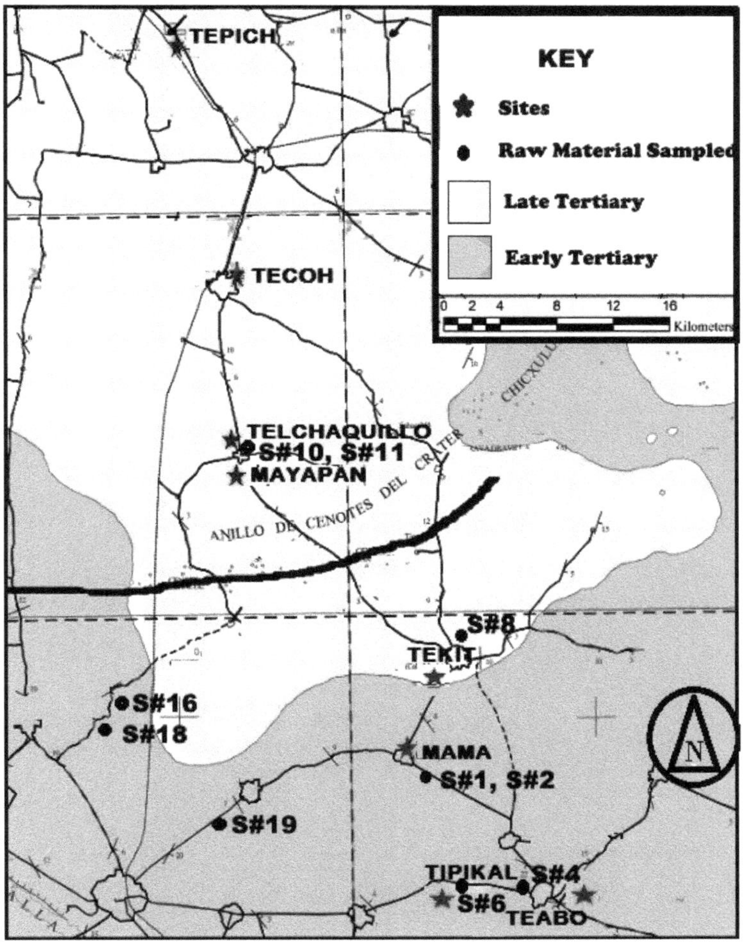

Figure 5-2. Locations of north-central sites and area of raw material collection (base map from INEGI 2007)

Tekit is located 55 km southeast of Mérida. Tekit was under Mayapán's jurisdiction (Roys 1957). According to Roys, the Cocom family was the head lineage at this town during Mayapán times.

Mama appeared in the tax lists and the Treaty of Maní (Roys 1957, p.68). It was under Mayapán's jurisdiction and, most likely, the Cocom was the head lineage at this town during Mayapán times.

Teabo was most likely also under Mayapán's jurisdiction, and the Cocom family was the head lineage in this town—for the natives told an old story of a captain called Cacomcat that constructed important buildings (Roys 1957, p.68). The town was populated with very noble people (Roys 1957, p.69).

Tipikal was under Mayapán's jurisdiction (Roys 1957). There are no indications of who may have been the head lineage at this town.

Illustrating the trajectory of the sites regarding ceramics, Figure 5-3 shows the frequencies of ceramics found at each of the north-central sites discriminated by period (data from Cruz 2012). This figure shows that since early times, communities had existed around the sites in this study, in particular at Tepich and Tipikal. Based on the relative ceramic presence, it is possible that, during the Late and Terminal Classic, Tepich and Tipikal were sizable towns. Teabo, Mama, and Tekit also had a significant presence of ceramics, and some people were already living in the area of Mayapán, Telchaquillo, and Tecoh. The ceramic materials found in all these sites during this period belong to the Cehpech horizon (Uxmal, Puuc area is the type site for Cehpech ceramics, while Chichén is for Sotuta ceramics) almost exclusively (Cruz 2012, p.38 and the Sotuta ceramics at these sites is negligible (Cruz 2012, p.38).

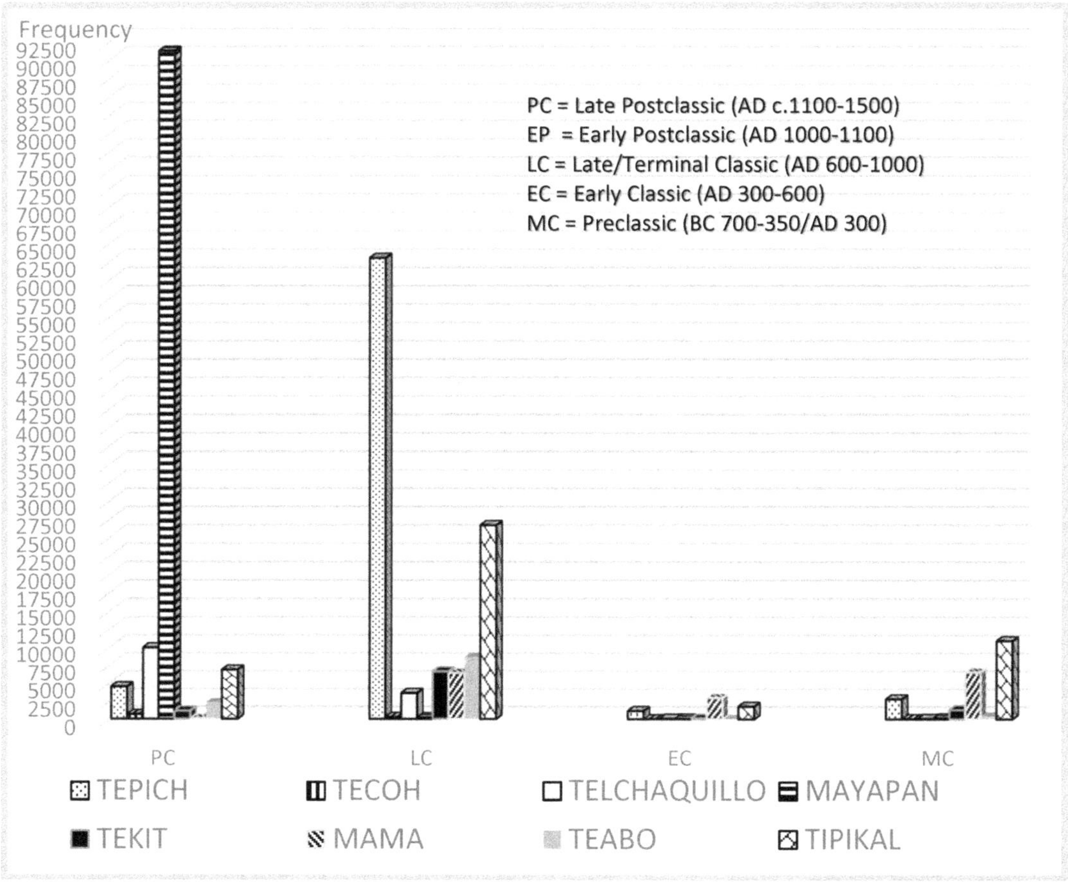

Figure 5-3. Ceramics frequencies (data from Cruz Alvarado 2012, Tabla 1, p.33; dates from Cruz Alvarado 2012; Masson and Peraza Lope 2014, p.63, 69; and Masson, Hare, and Peraza Lope 2006, p.189)

Table 5-2. Aerial Distances from Mayapán to North-central Sites

Mayapán	Tepich	26 km
Mayapán	Tecoh	11 km
Mayapán	Telchaquillo	1.5 km
Mayapán	Tekit	18 km
Mayapán	Mama	22 km
Mayapán	Tipikal	30 km
Mayapán	Teabo	33 km

During the Postclassic, the ceramic frequencies of most of the sites are much diminished particularly at Tipikal and Tepich in comparison to the previous period. The exceptions are Telchaquillo and, of course, Mayapán.

The north-central sites, with the exception of Mayapán, are considered small sites (Garza and Kurjack 1980), with only Tepich and Tipikal presenting monumental architecture (Garrido 2003). The north-central sites are relatively close to each other. Table 5-2 shows the distance between Mayapán and each of the centres. Maya porters could have covered these distances in one or two days: modern peddlers covered from 20 to 40 km in one day (Hammond 1978).

5.3.1.2 Case Study B

Case study B comprises Mayapán and sites located to the east: Chac Mool, Cobá, and Culubá. A brief introduction to these sites is presented next.

Culubá is located within the Tases (western) ceramic sphere (Figure 5-1) relatively near the northern coast. The ceramics from this site show a strong presence of Cehpech ceramics during the Late Classic (Lope Peraza and Barrera Rubio 2006). Cehpech, as found in eastern sites, differs somewhat from the Cehpech from western sites. It is then called eastern Cehpech, which is also found at Cobá (Lope Peraza and Barrera Rubio 2006). The Terminal Classic and its Sotuta ceramics, however, presents the largest concentration of ceramics, and a close relation with Chichén Itzá is suspected (Lope Peraza and Barrera Rubio 2006). Tases ceramics are also present but to a lesser degree (Lope Peraza and Barrera Rubio 2006).

Chac Mool (Figure 5-1) is located on a long strip of land between Bahía de Ascensión and Bahía del Espíritu Santo that is more like an island delimited by the Caribbean Sea on one side and semi-lakes, islands, and mangroves at the back (Terrones 2006). Bahía de Ascensión was the place from which the principals on Chichén embarked in their travels to Honduras seeking items such as cacao and feathers (Fray Alonso Ponce, cited by Terrones 2006, p.16). Chac Mool may very well have been a port in the trade network between the Caribbean and the Maya area (Terrones 2006, p.17).

Considerable amounts of Xcanchakan and the trade ware Silho Naranja were found at Chac Mool. Also present were ceramics that are common in Belize such as Paxcaman and ceramics common in Guatemala and El Salvador such as Tohil plomizo (Terrones 2006). During the Postclassic, the architecture changes into a style found at many other coastal towns called the eastern coast style. The ceramics for that period comprises the eastern Tases Payil (and the incised variety Palmul), and the Tases Mama, Navula, and censers (Terrones 2006, p.19).

Cobá is located 50 km inland from the eastern Yucatán coast in the area that later will be called the province of Ecab. It is, however, located within the Tases cultural sphere. Cehpech ceramics dominated at Cobá for the duration of the Terminal Classic (AD 900-1050), and the presence of Sotuta ceramics (Chichén) is minimal (Robles 1990, p.39). Although it is located only 50 km from the coast, the ceramic patterns observed at Cobá for the Late Postclassic differ greatly from eastern sites.

5.3.2 Samples Selected

This section presents the criteria followed for the selection of the ceramic samples included in this study.

5.3.2.1 Ceramics Selected

Sample selection aimed to provide a combination of ceramic groups and vessel forms from each site. The red-slipped groups included are Mama and Payil. As mentioned in chapter three, Mama sherds selected for analysis fall mainly in the Mama variety of the Mama group. This variety does not include ceramics with decorations such as incising, appliques or with unslipped exterior. Of the many varieties in the Navula group, the plain Navula variety (Navula Unslipped type) and the striated Yacman variety (Yacman striated type) were selected for analysis.

These varieties are specifically targeted because they are some of the most common ceramic varieties in northern Yucatán in numerical terms, and they occur at several sites throughout the region, enabling the examination of ceramic technology in similar types of vessels at different sites. In addition, they correspond to a mixture of unslipped and slipped, utilitarian, and elite classes, expected to increase the likelihood of uncovering differences in ceramic technology and composition. A few samples of other varieties less commonly found outside Mayapán were included for comparison purposes. The Table 5-3 shows the number of samples taken discriminated by ceramic variety and site.

Using the type-variety classification as a starting point has many advantages. One advantage is that it groups the ceramics by stylistic and textural categories such as surface finish, modifications to the surface such as striations, general grain size texture, and compactness, and in some cases by vessel form (e.g., Yacman open-mouth jars). Furthermore, the type-variety system corresponds with conventional Mesoamerican assemblage descriptions for it is customary to express the results in term of this system (Brown 1999a; Cruz Alvarado 2012; Kepecs 1999; Smith 1971a), making research accessible and meaningful to other researchers in the field.

Three forms were selected for study based on ethnographic work on the function of Yucatecan vessels (Rendón 1947, 1948a, 1948b; Thompson 1958). The selected forms are: jars, thought to have been used for storage and cooking (*ollas*); slipped jars (*tinajas* or *cántaros*) likely used for storage and water carrying; and some deep dishes known as *cajetes*, which based on the form may have been for serving.

5.3.2.2 Sampling Strategy

The samples were located in two Mexican states, Yucatán and Quintana Roo. This brought about two approaches to sampling.

North-central sites and Culubá, Yucatán State. Figure 5-1 shows the location of these sites within the Yucatán peninsula, and Figure 5-2 shows the location of the north-central sites in more detail. Due to storage limitations, most archaeological ceramics in northern Yucatán, after classification and quantification, are returned to their find spot and buried, keeping in storage a small selection of the original collection. The relative proportions of different varieties and vessel forms found in the excavations are not maintained in storage.

Most samples were provided by the excavations of Arqlgo. Carlos Peraza Lope (Cruz Alvarado, W. A., 2012, Garrido Euán, M.A., 2003), Arqlgo. Alfredo Barrera Rubio, and Dr. Clifford T. Brown (Brown, C.T. 1999a, 1999b). These samples were judgementally selected. This was based on criteria distinct from composition and

Table 5-3. List of Samples for Hand-specimen Analysis within Sites

Site	SLIPPED				UNSLIPPED			
	Mama (jars, cajetes)	Cancun (jars)	Xcan_ chakan (jars)	Payil (cajetes , bowls)	Navula (plain: jars, cajetes)	Yacman (striated: jars)	Other	Total
NORTH-CENTRAL								
	9				6	6		21
Tecoh	8				8	5		21
Telchaquillo	11		6		10	5		32
Mayapán	54		8	10	26	22	(*) 20	140
Tekit	5				3	5		13
Mama	11				5	4		20
Tipikal	9				11	10		30
Teabo	8				5	6		19
EASTERN								
Coba		9			6			15
Culuba	10	9		2	6	7	(**) 8	42
Chac Mool	1		4	6	4	3		18
Total	126	18	18	18	90	73	28	371

(*) Mayapán: Unslipped Exterior = 3, Polbox = 5, Tecoh = 4, Sulche = 1, Kukula = 4, Chen Mul = 2, Moyos = 1
(**) Culubá: Vista Alegre

focused on sampling sherds from the same types of vessels (same ceramic varieties) and forms (jars or cajetes) across all the sites.

This strategy ensures that technological attributes and techniques are compared in equal terms. In addition, this selection met restrictions on the size of the specimen (large enough to provide material for chemical and petrographic analysis) and INAH, Yucatán prohibition of the use in destructive analyses of parts of vessels considered diagnostic of the form.

Chac Mool and Cobá, Quintana Roo State. Samples from these two sites were selected by the excavation archaeologists, Arqlga. María José Con (Con Uribe, M.J., 2011) and Arqlgo. Enrique Terrones Gonzáles (Terrones González, E., 2006) and handed to this author.

At eastern coastal sites, Tases ceramics (mainly Mama and Navula groups) are rare, while the Payil group dominate the slipped ceramics, and a local type called Vista Alegre dominates the unslipped ceramics. In contrast, at Cobá, during the Late Postclassic, Payil presence is minimal. At Cobá, the Mama group dominates the slipped ceramics, while the unslipped jars are dominated by Yacman jars.

The restrictions and strategy to the sampling result in the fact that the proportions or percentages in this study relate to the sample collection taken and not to the site. The samples, however, had the potential of providing much-needed information about the variability of compositions and techniques within and between ceramic varieties and across the sites through the region.

5.4 Collection of Raw Materials from the North-central Area

Raw materials were collected from locations around the north-central sites. Figure 5-2 illustrates these sites, and Figure 5-4 shows in detail the locations of raw materials collection in relation to the sites. The eastern sites, Chac Mool, Cobá, and Culubá, were not included in this selection.

The objectives with the collection of raw material are many, including to

1. identify materials available to the ancient potters;
2. study their suitability for the construction of the vessels sampled;

3. examine the relation of raw materials to pottery to draw inferences about choices made by potters; and
4. investigate provenance.

Similar to pottery samples, the raw materials were examined in hand specimens, chemically, and petrographically. The results are presented in the corresponding chapters. This section presents the information about each sample taken, including the coordinates of the find spot and description of the material as observed in hand specimens.

The study area can be described (section 4.3) as having flat areas (or *planicies*) framed by very low-slope hills (or *altillos*). Cuts into these slopes for the construction of the roads have exposed the soil layers and carbonate rocks. The documented pottery clay mines are located further to the south or west of the north-central sites in this study. (more on these location in section 9.11). However, clay inside a cenote within Mayapán has been collected before (Don Fernando Flores, personal communication Feb 2016). Unfortunately, the entrance to the cenote was plugged with natural beehives.

Potters rarely would walk more than seven km for raw materials (Arnold 2005b). A prospection for clay deposits within 10 km from Mayapán and other north-central sites was undertaken, but it was infructuous. The search for a documented clay deposit located five km east of the town of Mama (Arnold et al. 2007; Morales Valderrama 2005, p.131) was unsuccessful, probably because it is submerged most of the year (Morales Valderrama 2005, p.131). Clay samples were taken from a documented palygorskite deposit (Sánchez del Río et al. 2009) located 20 km south-east from Mayapán, and 13 km east of Mama, in the Ticul-Chapab road (sample # 19, Table 5-4).

In contrast to clay deposits, sascab (unconsolidated carbonate rocks) is ubiquitous around the north-central sites. A simple test of plasticity (described below) showed that sascab is of two types (although both types look similar to the naked eye):

1. carbonate-rock sascab, a non-plastic and granular material, forming pebbles and granules; and
2. marl sascab, a more homogeneous material when in place and plastic when wet.

In this project, the proportion of clay in the samples was not measured, and the term marl is used for carbonate material that, based on its plasticity, contains a proportion of clay. Table 5-4 contains the list of all the raw material samples collected, coordinates, and a summary of the samples' physical characteristics. Plasticity and shrinkage are meaningful for plastic samples only. Plasticity was determined by wetting the clay and testing whether a coil can be rolled up twice without breaking. Shrinkage is given as a percentage based on the difference in the length of a straight wet slab when and its length when dry.

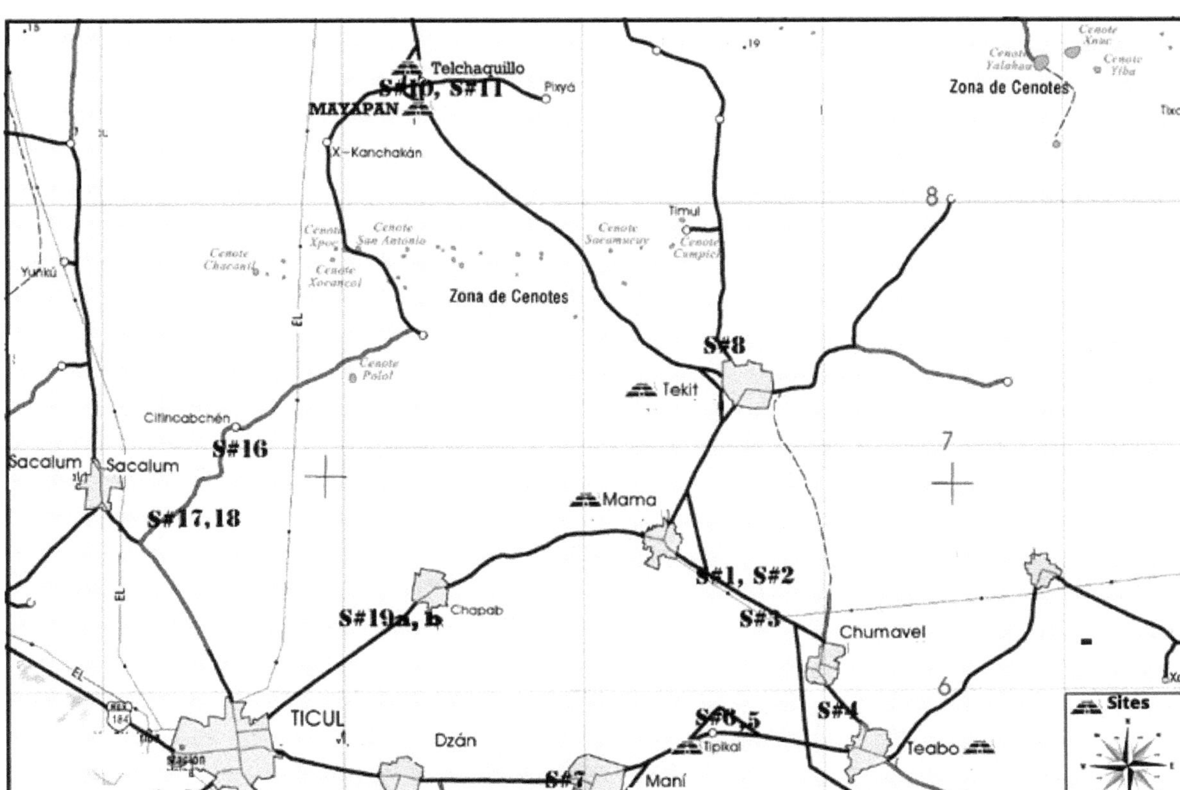

Figure 5-4. Expansion of Figure 5-2 to show more detail of the locations of collection of raw materials (base map from INEGI 2007)

Table 5-4. Raw Materials Collected and their Characteristics

Raw material sample# (MURRID)	Coordinates	Location	Type of material	Physical Characteristics
#1 (CGS187)	N 20° 27' 40" W 89° 20' 15"	Mama-Chumayel, Mama A (right side looking at Chumayel, Mama-Chumayel road, *altillo* about 1.85 km south of the exit to Mama)	Marl sascab, plastic	a) Raw: Colour = 2.5 YR 8/1 White; Shrinkage = five per cent; Plasticity = Yes
#2 (CGS188)	N 20' 27' 40" W 89° 20' 16"	Mama-Chumayel, Mama B (left side Mama-Chumayel road, altillo on front of #1)	Marl sascab, plastic	a) Raw: Colour = 2.5 YR 8/1 White; Shrinkage = three per cent ; Plasticity = yes
#3	N 20° 27' 38" W 89° 20' 13"	Mama-Chumayel road (right side looking at Chumayel, around 1 km south of sample #1, red soil from planicie)	Red clayey soil, plastic	a) Raw: Colour= ; shrinkage = seven per cent; plasticity = yes
#4 (CGS189)	N 20° 24' 27" W 89° 18' 25"	Teabo A (altillo about 200 mts north Tipikal/Teabo junction, right side), sascabera.	Marl sascab, plastic	a)Raw: Colour ; Shrinkage = zero per cent; Plasticity = low
#5	N 20° 24' 35" W 89° 20' 58"	Tipikal structure (soil, around 10 mt from structure)	Red-clayey soils, no plastic	a) Raw: Colour= strong brown
#6 (CGS190)	N 20° 24' 36" W 89° 20' 58"	Tipikal structure (soil, around 30 mt north of estructure)	Red-clayey soil, plastic	a) Raw: Colour = reddish brown; Shrinkage = seven per cent ; Plasticity = yes
#7	N 20° 22' 15" W 89° 24' 02"	Mani-Oxkutzcab road (soil from naranjal)	Red-clayey soil, plastic	a) Raw: Colour = red; Shrinkage = seven per cent ; Plasticity = yes
#8 (CGS191)	N 20° 32' 38.69" W 89° 21' 11.85'	Old Road Tekit-Tecoh, Tekit (red soil from small altillo, exit to Rancho S. Antonio)	Red clayey soil, plastic	a) Raw: Colour = 2.5 Y 5/8 red; Shrinkage = 13 per cent ; Plasticity = yes
#10 (CGS192)	N 20° 38' 24.81" W 89° 28' 01.45"	Sascabera Mayapán A (road Tecoh to Mayapán, btw north and south entrances to Telchaquillo, 0.7 km from south entrance)	Carbonate-rock sascab, no plastic	Raw: Colour = 2.5 YR 8/1 White; Plasticity = n/a; particle size 2-4 mm
#11 (CGS193)	N 20°38'24.47" W 89°28'01.13"	Sascabera Mayapán B (road Tecoh to Mayapán, btw north and south entrances to Telchaquillo, 0.7 km from south entrance)	Marl sascab, plastic	a) Raw: Colour = light reddish brown 5YR 6/4; Shrinkage = 12 per cent; Plasticity = yes/high
#16 (CGS194)	N 20°29'47.68" W 89°33'03.02"	Road Citincabchen (altillo in old road Telchaquillo to Sacalum, around 2.6 km south from Citincabchen)	Marl sascab, plastic	a) Raw: Colour = 2.5 YR 8/1 White ; Shrinkage = n/a ; Plasticity = yes
#17	N 20° 28'49.58" W 89° 34'08.73"	Road Citincabchen (altillo in old road Telchaquillo to Sacalum, around 3.4 km south from Citincabchen)	Carbonate-rock sascab ('cuut'), no plastic.	Raw: colour = 2.5 YR 8/1 White; Particle size approximately 1-2 mm
#18 (CGS195)	N 20° 28' 29.57" W 89° 34' 28.56"	Road Huanabchen-Sacalum (intersection roads to Sacalum/Xcanchakan/Ticul)	Marl sascab, plastic	a)Raw: Colour = 2.5 YR 8/1 White; Plasticity = yes
#19a (CGS196)	N 20°26'29" W 89°28'45"	Mine Ticul-Chapab, Chapab a (road Ticul-Chapab, right side looking at Chapab, 7 km from Calle 24 with 13)	Clay, plastic	a) Raw: Colour = light grey, btw 10YR 8/1 and 7/1#; Plasticity = Yes
#19b (CGS197)	N 20°26'29" W 89°28'45"	Mine Ticul-Chapab, Chapab b (road Ticul-Chapab, right side looking at Chapab, 7 km from Calle 24 with 13)	Clay, plastic	a) Raw: colour 2.5Y 8.5/1 white; Plasticity =very plastic

The rest of this section describes the collected raw materials.

Red colour soils and clays. Red soils are found in the *planicies* (section 4.3). Although there are no north-central pottery samples in this study that match the deep red or brown red colour of these soils, the samples were collected to investigate the availability of potential raw materials that can be used for pottery making and to determine the range of choices that potters may have had. The red raw material samples were found in the flat areas or *planicies* (Figure 4-8), with coordinates found in Table 5-4.

The samples taken include some without coarse materials (#3 and #7, Table 5-4) and others containing very coarse angular carbonate fragments (#6, #8, Table 5-4). The red soils collected are plastic when abundant water is added. After manually removing pebbles (no other treatment was performed), briquettes and small vessels were formed (such as the one in Figure 5-5) for firing tests and for thin sections.

Figure 5-5. Small container made with dark red soil

Off-white-to-yellow marl sascab. Samples #1, #2, #4, #16, and #18 (Table 5-4) are sascab samples that showed to be plastic. Figure 5-6 shows marl sascab pockets in an *altillo* under the dark soil layer, exposed by a road cut.

Marls were collected to study their suitability as raw materials for pottery clay following processing by the potter. These marls are composed of a mud or clay portion with embedded very coarse (>1 mm) carbonate rock fragments.

After manually removing pebbles, briquettes for firing tests and for thin sections were made from these samples without any further processing. Figure 5-7 illustrates a small vessel made from one of these raw materials samples (#1).

Figure 5-6. Two sascab marls. (a: top): sample #2, (b: bottom): sample #1, card = 15 cm

Off-white-to-yellow rock sascab. Samples #10 and #17 are non-plastic carbonate rock sascab (Table 5-4). They were collected to investigate their potential as a source of the observed white carbonate inclusions in the pottery samples. The two samples contain fine to very coarse carbonate rock fragments. They were collected from the sides of road cuts.

Light greenish grey clay. One sample (#19a) of a light greenish grey clay was collected 20 km south from Mayapán in the vicinity of Chapab, on the road between Chapab to Ticul, at seven km from the intersection of Calle 13 and 24 (Table 5-4 and Figure 5-4). This mine is documented by Sánchez del Río et al. (2009) as a 100 per cent palygorskite deposit. Arnold (2005a) also documents that sascab is collected from this mine by Ticul potters to obtain *sak lu'um, or* sacalum. At this town, it is considered an indispensable ingredient in the preparation of temper (sacalum mixed with calcite or dolomite) to improve the strength of the pottery clay.

Figure 5-7. Small jar prepared with sample #1

Light reddish brown marl sascab. A sample (# 11, Table 5-4) of a light-brown marl sascab (Figure 5-8) was collected in the vicinity of Mayapán, at 0.7 km north from the south Telchaquillo exit, on the left side looking to the south (Table 5-4). This is an old sascabera (Don Fernando Flores, personal communication, Feb 2016). In a subsequent visit to this place, it had been destroyed and levelled by a new expansion to the road. A simple test of plasticity indicated that this sample is a marl (marl sascab), and briquettes and a small vessel were made from it.

5.5 Preliminary Survey of Pottery Samples

A preliminary survey of some of the selected pottery samples using a hand lens was conducted to get a sense of the variation of the pottery samples. This survey indicated that patterned differences in the compositions of slipped and unslipped samples might exist in the pottery samples.

The patterns observed in the preliminary survey are:

1. In the north-central area, the macroscopic composition of red-slipped samples appears more uniform throughout the sites than for the unslipped ones; and
2. Samples classified as Payil, the eastern variety of red-slipped ceramics commonly found at eastern sites, and also found at Mayapán, have similarities to a few other samples from the north-central sites and could have been locally made.

Figure 5-8. Marl sascab sample (#11, Table 5-4) near Mayapán (white card = 15 cm)

Based on these observations, three hypotheses were proposed with the purpose of guiding this research in addressing the research questions. The hypotheses were outlined in chapter one, and they will be discussed in this section. The hypotheses are specific cases of the research questions designed such that in the process of collecting the information to test them, data are gathered to address the research questions.

5.5.1 Hypothesis A: Unslipped Jars

Hypothesis A: Unslipped jars (plain and striated, or Navula and Yacman) found at the different north-central sites were locally produced at minor centres and at Mayapán.

This hypothesis is based on the observation made during the preliminary survey of a variety of inclusions within the unslipped samples. As a result of the lack of direct evidence for ceramic production, to test Hypothesis A it is necessary to establish local production: the chemical compositions of the vessels or the technology (petrography) applied in their construction is different at each site. This is to say that a division of composition by site exists. This is a strong argument, but it is not sufficient to support Hypothesis A because other explanations can be given for the same pattern. Different fabrics at different sites may be due to sites importing pottery from different sites located somewhere else. The argument for local production is more convincing if pottery and local clays match, whilst acknowledging that the pottery may have been made of clay from a different location with similar chemical composition.

If the technology is different at each site, chemical compositions of the vessels could have followed a different pattern. The chemical groups can still give us information about the sources of the raw materials. If the chemical groups are also different at each north-central site, then it is possible that raw materials were sourced at each location. If instead there are only a few chemical groups, then, local production (as suggested by the divisions by site of the petrographic fabrics) may have occurred in combination with other scenarios. Such scenarios include a region with very little chemical variability, or potters sharing raw material sources and building their ceramics according to technological norms that differ from site to site. Hypothesis A is rejected if the same composition is found at more than one site.

5.5.2 *Hypothesis B: Red-slipped Mama Jars*

Hypothesis B: In the north-central area, red-slipped pots (Mama) were produced at one production locality that supplied the rest of the north-central centres.

This hypothesis is based on observations during the preliminary hand-specimen survey of two macroscopic fabric groups that include most of the red-slipped sherds of the north-central area. To test this hypothesis, it is first necessary to establish whether there is chemical or petrographic variation between sites and between ceramic classes. Evidence that will support this hypothesis is composed of a combination of conditions. One necessary criterion that needs to be met is that the compositions of Mama vessels, which are found at several sites, fall within a single chemical group and one petrographic group. However, because this pattern may be created by a widespread technological tradition (one petrographic group) in a geologically homogeneous area (one chemical group), more evidence for production of Mama at one location is needed. To establish that Mama was produced at the specific site, further evidence such as local clays that match the pottery, or Mama fabrics that match those of vessels found at the site but not elsewhere is needed. Hypothesis B is not supported if there are many chemical or petrographic groups within Mama.

Hypothesis B puts forward the notion of centralisation of red-slipped ceramic production to explain red-slipped vessels' homogeneity, at least in the north-central area. It is important to mention that if the conditions are met, and the results are consistent with the hypothesis, there is room for other interpretations to the findings. The same pattern of one or a few petrographic and chemical groups over a wide area is given by massive import of Mama from somewhere else. The idea of importation would be lessened, and Hypothesis B would be strengthened if there were a match between the chemical group(s) containing Mama sherds and the clay from the region. An alternative explanation for one chemical and petrographic group is one of local production under chemical homogeneity across the region and a widespread technological tradition. If Hypothesis B holds true, it can be interpreted as uniformity in style and composition that could have been driven by elite control, or by consumer demand for exchange, payment of tribute, aesthetic uniformity, or other perceived needs for social uniformity.

5.5.3 *Hypothesis C: Eastern Red-slipped Payil Items*

Hypothesis C: Payil, the fine-grained red-slipped ceramic type commonly found at eastern coastal sites, when found at north-central sites, was locally made.

This proposition is based on paste characteristics of samples categorised as Payil but found at Mayapán and have fabrics similar to a few unslipped samples considered probably to be local to the north-central area. Payil is commonly found at eastern coastal sites and has many similarities to Mama. When macroscopically examined, they differ in that Payil presents a fine-grained, crystalline, compacted, usually reddish-grey fabric, and red refired colour. Payil samples found at Mayapán refired to a red colour, and red refired clays are found in the north-central area. Payil samples have single translucent to opaque crystals, and some unslipped samples found at the north-central sites contain discrete translucent crystals. Hypothesis C is supported if local production is established or the compositions of samples classified as Payil match the composition of locally produced samples.

5.6 Petrographic Analysis

Petrography was chosen in this study because it can discriminate pottery clay fabrics based on the composition of rock fragments and other materials. What is more, it can discriminate based on many other variables including grain size, crystal texture, or structural features of the clay portion (clay matrix or micromass) porosity and voids' shapes. These characteristics of a petrographic analysis are relevant and useful to this study: despite the mineralogical uniformity within the study area, limited but significant research of Mayapán ceramics (Sánchez 2009, 2013) has shown that textural characteristics can discriminate late Postclassic pastes. This ability makes petrography a valuable tool to study ceramic technologies and uncover technological traditions. For instance, the structure of the clay matrix or the characteristics of the inclusions may result from potters' actions such as the selection of certain materials, discarding of others, crushing or sieving during the construction of the vessels. Decisions may reflect the group and technological tradition to which the potters belonged (Roux and Courty 2005, p.202) and their wider social context (Dobres 2010; Lemonnier 1993, p.3; Gosselain 2000).

5.6.1 *Petrographic Analysis of Hand Specimens*

The objective of the petrographic analysis of hand specimens (as mentioned earlier, hand-specimen refers to a sample as found: it has not been subjected to further processing) is to characterise (Whitbread 1995, p.372) the samples into broad groups that can be duplicated by other researchers. It is low cost, requiring little training in a technology using a combination of naked eye, hand lens, and stereomicroscope. The term 'characterisation' as adopted in this study includes identification of inclusions and classification (Whitbread 1995, p.372) of samples into fabric classes. The term 'fabric' is the term preferred by ceramic petrographers for the 'clay body' or paste. The groups defined by this analysis are the basis for the sample selection for the thin-section and chemical analyses. The results of hand-specimen examinations are broad macroscopic fabric groups, which are presented in chapter six.

5.6.2 *Petrographic Analysis of Thin Sections*

The polarising microscope used in ceramic petrography makes possible the characterisation of samples based on the mineralogical composition and texture of inclusions within the ceramic fabric. Petrographic research in the general geographic area has shown that despite the mineralogical homogeneity in the area, fabrics can be discriminated by textural characteristics. For instance, a petrographic analysis of ceramics from the Puuc area performed by Shepard (Smith 1971b, p.171) characterised the fabrics based, in part, on carbonate micromorphology, including clear calcite, grey limestone, sparite, calcite, or cryptocrystalline calcite. Kepecs' (1999) analysis in the Chikenchel area characterised the samples into five fabrics. A petrographic analysis of samples from Mayapán (Sánchez Fortoul 2013) characterised fabrics based on micromorphology (micrite, sparite, grey micrite, crystalline calcite). These analyses are useful in that they have shown that variability exists, in particular, related to the micromorphology of calcite and dolomite. The studies are, however, restricted to one site or a limited geographical area and cannot address questions about ceramic production organisation, regional exchange, traditions or other interactions properly. The method followed for the analysis of each thin section and the description of the fabric class can be found in Whitbread (1995, Appendix 3).

5.7 Chemical Analysis

The chemical analysis provides data on elemental concentrations (Whitbread 2001, p.451) and enables incorporation of the clay portion in the characterisation (Whitbread 2001, p.187) of the samples. However, the results of this analysis do not identify the mineralogy of inclusions or clay. Petrographic and chemical analyses complement each other because petrography makes it possible to identify rocks, inclusions in general, and textures.

In addition to the characterisation of pottery and raw materials samples, a chemical analysis can provide significant information about provenance. Early in the research, it became apparent that, due to the geological homogeneity in the region and the widespread distribution of pottery with similar inclusions, petrography, although an ideal tool to define technological classes, is less likely to discriminate samples in terms of materials procurement. For these reasons, this study includes chemical analysis (NAA) owing to its ability to detect trace elements that can be applied in characterising the samples and in the study of procurement zones, even in areas of apparent geological homogeneity.

Chemical variability of pottery and clays can then point to differences in resource zones. Current research shows that chemical analysis is a viable approach for this project having the potential of discriminating distinct zones in north-central Yucatán. The analysis of modern pottery using neutron activation from different Yucatecan communities (Arnold et al. 1999, 2000) showed that different communities (Tepekan, Ticul, Mama and Akil) produce pottery with different chemical patterns. In another study, Arnold et al. (2007) used INAA and distinguished provenance groups when analysing samples of the same clay type, palygorskite, based on relative concentrations of Rb versus V,Ni, or Mn. It was expected then that chemical analysis is capable of discriminating northern Yucatán pottery into chemical zones

Chapter 6

RESULTS OF HAND-SPECIMEN ANALYSIS

The hand-specimen analysis is the foundation of the technological analysis performed in this study. The attributes of the pottery samples were examined aiming to describe and classify, or characterise, the samples into broad petrographic groups with similar characteristics. As is presented in subsequent chapters, samples taken from these groups underwent further chemical and petrographic thin-section analyses, ensuring compositional variability in the selection process for these analyses. Furthermore, hand-specimen characterisation can serve as a guide to and be replicated by other researchers, given the relatively small financial and technical expertise that it requires.

A total of 376 samples was selected (sampling strategy in section 5.3) from eight north-central sites, or case study A, including Mayapán and neighbouring centres, and three eastern sites, Chac Mool, Cobá, and Culubá. The hand-specimen analysis does not require complex preparation of the samples. The main equipment consisted of a hand lens, portable stereomicroscope, portable test kiln, Mohs scale, and pliers. The physical attributes of the sherds (Orton et al. 2008, p.231-242; Peacock 1977, p.29; Whitbread 1995) were examined, and the samples characterised into broad macroscopic fabric groups. Regardless the structure of the selection of the samples, the resulting groups are technological in nature created unrelatedly of the contextual information such as their find spot, ceramic variety, or form of the samples.

This chapter is divided into two sections. In the first section, the attributes examined and the scales and measurements used are described. The second section contains the results of this analysis including a description of the types of inclusions and fabrics found.

6.1 Attributes, Scales, and Measurements

The attributes examined should be observed on a fresh break of the sherd. The attributes that pertain to the inclusions observed are the following:

Type of Inclusion. This refers to the type of mineral or material (for instance, shells or organic matter) identified. When identification is not possible, the particle is described by salient characteristics such as colour or crystal structure.

Percent Abundance. It is estimated for each type of inclusion as a range of variation of the percentage of the whole area occupied by the inclusion in the fresh break. This is a visual task aided by comparison charts for the visual estimation of percentages of mineral content (Matthew et al. 1991). Table 6-1 contains the scale and labels used for the description of percent abundance of the inclusions (Kemp 1985, p.17, cited in Whitbread 1995, p.379)

Table 6-1. Frequency Label used for Range of Variation of the Inclusions (Kemp 1985, p.17, cited in Whitbread 1995, p. 379)

Frequency Labels	**Range of Variation (percentage of total area)**
Predominant	> 70 per cent
Dominant	50 – 70 per cent
Frequent	30 – 50 per cent
Common	15 – 30 per cent
Few	5 – 15 per cent
Very few	2 – 5 per cent
Rare	0.5 - 2 per cent
Very rare	<0.5

Particle Size. When needed, the size of individual grains was estimated visually on the fresh break. The microscope reticle was often used to corroborate measurements. A modified Wentworth (1922, Table I) scale for the classification of particle sizes in which sizes less than 0.25 mm are conflated into a fine size is used to describe the particle size:

Fine sand = < 0.25mm
Medium sand = 0.25 – 0.50mm
Coarse sand = 0.50 – 1.0mm
Very coarse sand = > 1.0mm

Rounding. Rounding of a particle refers to the prevalent rounding as defined in Orton (2008, Figure A.5). The scale used is in Orton (2008, Figure A.5), including angular, sub-angular, rounded, sub-rounded, irregular, flat.

The attributes that pertain to the whole sherd are fired colour, re-fired colour, hardness, fracture, thickness, texture (or grain-size), and sorting.

Fired colour. This is the original colour of the clay of the pottery under natural light. It is usually described using Munsell charts (Munsell 2009).

Re-fired colour. This attribute refers to the colour of a pottery sample after being fired again in the laboratory (Whitbread 2007 et al, p.180). The reason for re-firing ceramics is that differences in the original firing conditions, such as temperature or atmosphere (Whitbread 2007 et al., p180), may mask underlying colours of the clay. Refiring all the samples under the same conditions at a temperature higher than the original eliminates these differences (Whitbread 2007 et al., p180). The samples were refired at 900°C. The Figure 6-1 shows on its right side a sherd with its original fired colour, and, on the left, after refiring Destruction of the samples by calcination (more about this process in section 9.6) showed that 900°C is higher than the original firing temperature. Differences in the re-fired colours correspond to differences in the relative concentrations of iron (Whitbread 2007 et al., p187-188) and can point to broad clay differences.

The re-fired colours obtained were red (R), white (W), and light reddish brown or pink (B). The range of variation for these codes is found in Table 6-2.

Table 6-2. Colour Codes of Re-fired Sherds

Re-fired Colour	Code	Munsell colour
White	W	5yr 8/1 to 8/2; 2.5yr 8/1 to 8/2; 7.5yr 8/1, 8/2, 8/3; 10yr 8/1, 8/2, 8/3;
Light-reddish brown, pink	B	5yr 8/3 to 8/4; 2.5yr 8/3 to 8/4; 5yr 7/3 to 7/6; 5yr 6/3 to 6/6; 2.5yr 7/2 to 7/6; 2.5yr 6/2 to 6/6
Red	R	2.5yr 7/8,6/8,5/8,4/8; 2.5yr 5/6, 4/6, 3/6;

Hardness. At first, this attribute was rated in comparison to the hardness of a 'typical' Mayapán sherd. A characteristic of most sherds from Mayapán is that they are very soft, friable, and could be chipped with a nail. The hardness test was initially performed using Sánchez's nails. These nails did not last, and to be consistent throughout the test, a Mohs test was used for all the sherds. Sherds that could not be scratched with Mohs # 2 are considered hard (hardness = h), and the rest, soft (hardness = s). Some sherds were found to be atypically hard.

Fracture. The fracture can be described as conchoidal, smooth, hackly, or laminated (Orton et al 1993, p.235). Colour, fracture, and hardness are attributes that could be a function of firing temperature or other firing conditions (Orton et al 1993, p.68).

Texture or grain size. The overall texture of the sherd or the fresh break section was estimated using the charts in Matthew et al. (1991) for the visual estimation of percentage. A code for fine, medium, and coarse was assigned to each sherd.

In this study, a fabric has a fine texture as long as the coarsely grained (> 0.5 mm) inclusions are constrained to an abundance of very few (Table 6-1) or less of the overall abundance of the inclusions, and the medium grained (0.25-0.5 mm) particles have an abundance of few or less of the overall abundance of inclusions.

A fabric has a medium texture if it has common or more overall abundance of medium-grained particles, and, at the same time, the coarse inclusions are constrained to very few or less of the overall abundance.

Lastly, a fabric has a coarse texture if it has few or more overall abundance of coarse inclusions. Table 6-3 summarises the key to determine the textures and the codes assigned to the samples.

Table 6-3. Key to Estimate the Overall Texture

Prevalent texture	Code	Key
Fine	f	very few or less coarse-size particles and few or less medium-size particles
Medium	m	very few or less coarse-size particles and common or more medium-size particles
Coarse	c	Few or more coarse-size particles

Sorting. Sorting or size homogeneity was estimated using inclusion sorting charts in Figure A.6 of Orton et al. (1993).

6.2 Types of Inclusions Observed

The inclusions identified through hand-specimen analysis include white and grey micritic limestone, fragments of a sparry carbonate rock (calcite or dolomite), dark particles, single glassy crystals. These inclusions and the code assigned to them are listed in Table 6-4.

Table 6-4. Types of Main Inclusions in Hand specimens

Inclusion Code	Short Description
WM	white, dull and chalky calcareous rock
SP	bundles of microscopic spar crystals
GM	dark grey or dark particles
XT	single crystal

6.2.1 Micrite

The term micrite is used for clusters of calcite crystals in which each crystal is less than 0.005 mm (Adams and MacKenzie 2001, p.6). In hand-specimen samples, micrite is white or very light grey, opaque, dull, and chalky.

Under the hand lens or binocular microscope, from low magnification, such as ten times natural size or 10X, to high magnification or 30-40X, micrite looks uniform, chalky, without internal structure (Figures 6-1 and 6-2) as a result of its microcrystalline nature in which individual crystals cannot be distinguished. Micritic inclusions that look white are called micrite in this study, to differentiate them from dark micrite, because dark particles may have looked different to ancient potters.

Two instances of micritic samples are illustrated in Figures 6-1 and 6-2. The identification of the samples is given in the figures. This number can be used in Table 8-18, containing the list of pottery samples, to obtain more information about them.

Figure 6-1. Micrite inclusions in hand specimen (#111 in Table 8-18, Mama from Mayapán, left: fired, right: same sherd when refired changed to white colour)

Figure 6-2. Micrite inclusions in hand specimen (#161, Mama from Mayapán)

As found in nature, these microcrystalline carbonates can be present in large masses forming calcareous rocks and in discrete calcareous aggregates of different origin, including calcification of organic and inorganic grains and marine skeletons.

6.2.2 Sparry Grains

Particles composed of clusters of sparry crystals were observed in some of the samples. The term sparite, spar, or sparry calcite is used for crystals over 0.01 mm (Adams and MacKenzie 2001, p.6; Folk 1980, p.156; Greensmith 1989, p.106). There is no set limit for a largest sparite size, although Folk (1980, p.156) states that it could range up to one mm and more.In fact, the crystals could be calcite or dolomite crystals, but in hand specimens, they cannot be identified with certainty. The small individual crystal size prevents sparry crystals from being distinguished with the naked eye. The binocular microscope was used to this end, using magnification at least of 30-40 times the natural size. Even at this magnification, only crystals equal or larger than 0.040 mm were resolved with the microscope used in this study. This implies that, in hand specimens, a portion of the sparry samples was misclassified as micrite. Figure 6-3 shows grains (bottom centre) at this magnification and silty discrete crystals in the matrix. These grains are expanded in Figure 6-4 to distinguish the small (around 0.040-0.060 mm) crystals better.

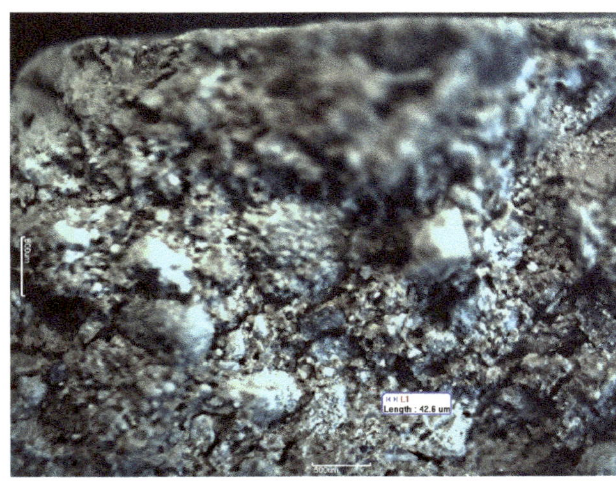

Figure 6-3. Particles composed of sparry crystals (#272, Yacman from Tipikal)

Figure 6-4. Expansion of Figure 6-3 to show individual crystals within particles (#272, Yacman from Tipikal)

Figure 6-5. Sparry fabric showing granular porous grains and powdery crystallites (#273, Yacman from Tipikal)

Figure 6-6. Dark particles (#155, Mama from Telchaquillo)

Figure 6-7. Dark areas in matrix (Mama from the site of Mama)

Sparite particles in which the individual small crystals have a granular appearance are sometimes referred as saccharoidal or sucrose calcite (López Ramos 1975, p.266).

Figure 6-8. Sample showing translucent, angular inclusions (#281)

Figure 6-9. Translucent and angular, smooth particles (#60, Navula from Mayapán, from Sánchez Fortoul 2009, Fig. 6.4)

Although the binocular microscope was needed to resolve the individual crystals, in some (but not all) of the samples, two characteristics observed with 10-20X hand lens are good indications that the inclusions are clusters of sparry crystals.

One characteristic is that the particles exhibit a granular or porous texture. The second is that the paste contains abundant semi-translucent to whitish crystallites, apparently from the obliteration of the sparry particles.

These characteristics can be observed clearly in the sample in Figure 6-5 (also Figures 6-7 and 6-11). Similar to micrite, in nature in the Yucatán peninsula, sparite is found in masses and discrete particles, and both are components of the carbonate rocks within the region (chapter four).

6.2.3 *Dark Particles*

Dark, usually grey, soft or indurated particles were observed. Figure 6-6 illustrates this type of inclusions. Their nature cannot be determined with certainty in hand specimens. Grey/brown limestone, oxides, peloids (as fossilised faecal matter), and organic matter are some of the types of particles that look dark in hand specimens.

In the sample in Figure 6-6, some of the dark particles were plucked out from the matrix and treated with hydrochloric acid, which reacts with calcite. The particles dissolved leaving a dark soft residue, indicating that these particles are fragments of carbonate rocks. It is then probable that the dark colour, at least in the treated particle, results from the presence of organic material in the carbonate rock. The individual crystals of the dark particles could not be resolved with the binocular microscope.

The sample in Figure 6-7 presents long irregularly shaped diffused areas of a dark colour matrix. This is an example of what should not be classified as a dark particle fabric. Although there are defined areas of dark matrix, they do not correspond to individual particles. Instead, they appear related to organic material. The sample was classified as sparry: the white carbonate grains appear granular or porous and there are powdery particles sprinkled on the fresh break indicating that the white larger inclusions are, most likely, clusters of small sparry crystals.

6.2.4 *Single Crystals*

Translucent to semi-translucent, mostly angular, particles were observed in the samples, in particular from Culubá and Chac Mool. The term single crystal is used in this study to refer to discrete individual crystals to differentiate them from grains composed of an aggregation of crystals, such as sparite and micrite. Figures 6-8 and 6-9 illustrate the appearance of single crystals in the samples.

6.3 Hand-specimen Fabrics

The samples were characterized into macroscopic fabric classes following a series of steps described in section 6.9. Table 6-5 summarises the hand-specimen fabrics found. They are briefly described in the next sections. The detailed description of the petrographic characteristics of each one of these fabrics can be found in Appendix D.

Table 6-5 includes the fabric name, a fabric code, and a short description. The code consists of the main inclusion code (Table 6-4), sample texture (Table 6-3), and refired colour (Table 6-2). The scales and measurements used in the description of the fabrics can be found earlier in this section (section 6.1).

6.4 White Micrite Fabrics

Included in this group are samples for which micrite is the dominant inclusion. The samples were grouped into two fabrics differing mainly on the hardness of the paste.

6.4.1 *Micrite-Medium-W (WM-m-W) Fabric*

Samples presenting this fabric are characterised by the presence of micrite as dominant inclusion, soft paste, medium-grain size (mode), and white refired colour. Figure 6-10 shows a sample with this fabric. This is a fabric with a wide range of variation in many of the attributes (more detail in Appendix D).

Associations. This fabric was found at the north-central sites almost exclusively associated with red-slipped Mama and the plain Navula. This fabric was not found at the eastern sites.

Table 6-5. List of Hand-specimen Fabrics with Fabric Code and Description

Fabric Name	Fabric Code	Fabric Description
Micrite-Medium-W	WM-m-W	White micrite particles, medium grained, refired to white
Micrite-Medium-W-Hard	WM-m-W-H	Sparry/Micritic, medium grained, refired to white, hard paste
Spar-Coarse-W	SP-c-W	Sparry particles, coarsely grained, refired to white
Spar-Medium-W	SP-m-W	Sparry particles, medium grained, refired to white
Spar-Coarse-B	SP-c-B	Sparry particles, coarsely grained, refired to lrb
Dark-Coarse/Medium-W	GM-cm-W	Dark particles, medium to coarse, refired to white
Dark-Coarse-R/B	GM-c-R/B	Dark particles, coarse, refired to red and lrb
Crystal-Fine-R/B	XT-f-R/B	Discrete crystals, fine grained, refired to red or lrb
Crystal-Coarse-R/B	XT-c-R/B	Discrete crystals, coarsely grained, refired to red or lrb
Crystal-Coarse-W	XT-c-W	Discrete crystals, coarsely grained, refired to white

KEY: lrb = light-reddish brown

Figure 6-10. Fabric Micrite-Medium-W (#316, Mama from Telchaquillo)

Figure 6-11. Fabric Spar-Coarse-W, with granular inclusions and silty matrix (#256, Navula from Mayapán)

Figure 6-12. Fabric GM-cm-W (#300, scale = 1cm)

Figure 6-13. Fabric Crystal-Fine-R/B (#136)

Figure 6-14. Fabric XT-f-R/B (Sample #137 Payil from Mayapán)

Figure 6-15. Clear single crystals (#281)

6.4.2 Micrite-Medium-W-hard (WM-m-W-h) Fabric

This fabric is a variant of fabric Micrite-Medium-W (WM-m-W). Contrasting to most north-central fabrics, samples with this fabric cannot be chipped with a nail or scratched with Mohs # 2. The fabric is characterised by the presence of micrite particles (although sparry particles were observed in a few), white refired colour, and medium texture (mode), differentiated from WM-m-W by its unusual hardness (more in Appendix D). This fabric also includes a few hard paste samples with sparry particles.

Associations. This fabric is associated with the site of Mama and to Mayapán to a lesser degree. It was also observed that this fabric was found, mostly, in the red-slipped Mama samples.

6.5 Sparry Fabrics

Samples in this fabric class are characterised by the dominant presence of clusters of sparry crystals of calcite or dolomite. Many times, the particles present a micritic alteration that makes their identification difficult. Samples in which spar dominates are characterised into three hand-specimen fabrics. They are described below.

6.5.1 Spar-Coarse-W (SP-c-W) Fabric

Sherds grouped within this fabric class are characterised by the presence of coarse particles composed of clusters of sparry (SP) carbonate (calcite or dolomite) crystals, soft paste, high porosity, and a white refired colour. Figure 6-11 illustrates this fabric.

Associations. The samples presenting this fabric were found at the north-central sites associated with the unslipped jars included in this study: plain Navula jars and the striated, unslipped Yacman jars. This fabric also occurred in a few Mama (red-slipped) jars.

6.5.2 Spar-Medium-W (SP-m-W) Fabric

Sherds with this fabric are characterised by sparry particles as main inclusion, soft medium-grained paste, and a white refired colour. This fabric differs from the coarsely grained Spar-Coarse-W (SP-c-W) described earlier not only in the predominant grain size but also in the lower frequency of inclusions and in a moderately compacted paste (less porous than SP-c-W).

Associations. This fabric was found at the north-central sites primarily associated with red-slipped Mama and plain unslipped Navula samples, while striated open-mouth jars (Yacman) were largely absent.

6.5.3 Sparry-Coarse-B (SP-c-B) Fabric

The main characteristics of this fabric are the presence of coarsely grained sparry inclusions, probably calcite or dolomite, soft, friable, highly porous paste, and light reddish brown refired colour. The main differences from the equally sparry and coarse Spar-Coarse-W are the light brown or light red fired colour, the light reddish brown refired colour, and slightly higher frequency of inclusions of this fabric.

Associations. This fabric is found at the north-central sites and Cobá associated with unslipped jars: Navula (plain) and Yacman (striated open-mouth). The Yacman samples exhibiting this paste are relatively thin and very uniform in their thickness (6 to 7.3 mm) and in their high porosity.

6.6 Dark Particles Fabrics

This is a group of samples characterised by the presence of dark particles as main inclusions. The samples were characterised into two fabrics, described below.

6.6.1 Dark-Coarse/Medium-W (GM-cm-W) Fabric

The dominant inclusions in vessels with this fabric are dark particles with medium to coarser grain size, and white refired colour. An example of a sherd with this fabric is shown in Figure 6-12.

Associations. This fabric was found throughout the north-central sites, in particular at Tepich. This fabric occurred in all ceramic classes: Mama, Navula and Yacman.

6.6.2 Dark-Coarse-R/B (GM-cm-R/B) Fabric

This fabric is a variant of the Dark-Coarse/Medium-W fabric just described. It differs only in its light-reddish brown or red refired colour. Sherds distinguished by this fabric are characterised by the presence of darker limestone or dark particles, medium to coarser texture, and red and light reddish brown colour.

Associations. In the north-central area, samples with this fabric refired to light-reddish brown and were found at Tecoh, Tekit, Teabo, and Cobá associated with Yacman and Xcanchakan. Samples with this fabric were found at Culubá and refired to a red colour.

6.7 Single-crystal Fabrics

Sherds with this type of fabric are characterised by the presence of usually angular, glassy, semi-translucent to whitish, fine to very coarse single crystals. The crystals are probably composed of calcite, dolomite, or quartz. The samples are characterised into three hand-specimen fabrics, described in the next sections.

6.7.1 Crystal-Fine-R/B (XT-f-R/B) Fabric

Sherds with this fabric are characterised by their fine-grained particle size, shown in Figures 6-13 and 6-14, semi-translucent to whitish angular inclusions, compacted to moderately compacted paste, and red or light-reddish brown refired colour. This fabric includes samples having a hardness from soft to hard. Grey micritic particles may be present. The scales and measurements used in this table can be found in section 6.1.

Associations. In the samples taken this fabric is associated with Payil sherds found at Mayapán and Chac Mool. The exception is a sherd classified as Payil from Culubá that also belongs to this fabric.

6.7.2 Crystal-Coarse-R/B (XT-c-R/B) Fabric

A soft paste, coarsely grained macro-crystalline inclusions, red or light reddish brown (or strong pink) refired colours, and varied harness characterise the sherds grouped within this fabric. Figures 6-15 and 6-16 illustrate two samples classified under this fabric.

Associations. This coarse fabric was commonly found in samples from two of eastern sites: Culubá and Chac Mool. At these eastern sites, this fabric was found in all the utilitarian varieties included in this study: Mama, Navula, Yacman, and Xcanchakan. This fabric occurred in four Xcanchakan samples from Chac Mool and one from Mayapán.

Samples from the eastern coast with this fabric present varied hardness including soft, moderately hard to hard. Samples from Culubá with this coarsely crystalline fabric refire to a deep red colour. They include Mama, Cancún, Navula, and Yacman samples. Samples from Chac Mool refire to light-reddish brown.

In addition, this fabric was found in a few samples of Navula and Yacman found at Mayapán, Telchaquillo, Tekit, and Tipikal. Samples from these four north-central centres refired to light-reddish brown mostly.

Figure 6-16. Angular, smooth whitish inclusions (#279)

6.8 Associations between Hand-specimen Fabrics, Sites, and Ceramic Typology

Table 6-6 summarises the associations between the fabrics, sites, and ceramic varieties. It shows that differences exist between the types of fabrics present in red-slipped Mama samples when compared to the unslipped samples. It also shows differences between the fabrics found in most north-central when compared to the eastern sites (Cobá, Chac Mool, and Culubá).

6.8.1 North-central Sites

Most red-slipped samples of the Mama group from the north-central area were characterised into Micrite-medium-W (although sparry and grey micrite samples were also found), acknowledging that (due to the microscope's resolution) many samples with spar clusters might have been classified as micrite.

Within the samples of plain, unslipped Navulá jars and cajetes, a variety of fabrics were found including sparitic, micritic, and dark particles. No particular type or types of inclusions appear to dominate. In addition, neither medium-grained nor coarse particles dominate the Navula samples.

In contrast, Yacman, the striated and unslipped open-mouth jars, is characterised by its coarse paste. Coarse sparry fabrics or coarse dark-particle fabrics dominate in the Yacman samples. In addition, differing markedly from Mama, fabrics with translucent crystals were found in Yacman samples.

Payil samples from Mayapán contain finely grained crystals. Their type of inclusions, texture, and refired color differ from the fabrics in Mama, Navula, and Yacman.

Table 6-6. Associations between Hand specimens and Ceramic Varieties

		SLIPPED				UNSLIPPED	
Fabric Class Code	**Sites**	**Mama (red)**	**Cancún (red)**	**Xcan chaka (cream)**	**Payil (red)**	**Navula (plain)**	**Yacman (striated)**
Micrite-Medium-W WM-m-W	North-central	X		X		X	x
Micrite-Medium-W-Hard WM-m-W-h	Mama, Mayapan	x					
Spar-Coarse-W SP-c-W	North-central	x				X	X
Spar-Medium-W SP-m-W	North-central, Coba	x				X	
Spar-Coarse-B SP-c-B	North-central, Coba					X	X
Dark-Coarse/Medium-W GM-cm-W	North-central	x				X	X
Dark-Coarse-R/B GM-c-R/B	Teabo, Tecoh, Tekit,	x		x			x
	Culubá	X					X
	Coba		X				
Crystal-Fine-R/B XT-f-R/B	Mayapan, Chac Mool,				X		
Crystal-Coarse-R/B XT-c-R/B	Culubá,	X	X			X	X
	Mayapan, Tekit, Tipikal					X	X
Crystal-Coarse-W XT-c-W	Mayapan,					x	x
	Coba		X				

(*) North-central = most north-central sites including Mayapán.

6.8.2 Eastern Sites

Within the samples from Cobá, Chac Mool, and Culubá, fabrics with relatively large, single crystals are the norm. Contrastingly, the most common fabrics in the north-central samples contain sparry calcite (or dolomite) clusters or/and micrite.

At Culubá and Chac Mool, coarse crystalline fabrics are found almost exclusively in the utilitarian slipped, Mama and Xcanchakan, and the utilitarian unslipped samples, Yacman and Navula. In contrast, finely grained crystals were commonly found in the elite Payil ceramic variety.

6.9 Key for the Assignment of Fabrics: Procedure for the Characterisation into Fabrics

The hand-specimen analysis made apparent that among all the attributes examined, four accounted for most of the variation observed. The fabric groups are based mainly on these four attributes: hardness, type of inclusions, overall texture (grain size mode), and refired colour.

The procedure followed for the characterisation into hand-specimen fabrics is summarised in Figure 6-17. Starting with a sherd, the sherd is tested for hardness with Mohs scale or nails. Next, using pliers, a fresh break is made, removing around one centimetre of the sherd. The sherd area exposed by the fresh break is examined to determine the main types of inclusions and texture. The small piece cut for the fresh break is then refired to determine refired colour.

Table 6-7 functions as a key to follow for the assignment of the broad hand-specimen fabrics. For instance, if the hardness is soft, the dominant inclusions are sparry particles (SP), the overall texture is coarse (texture = c), and the refired colour is white (W), and the fabric is SP-c-W.

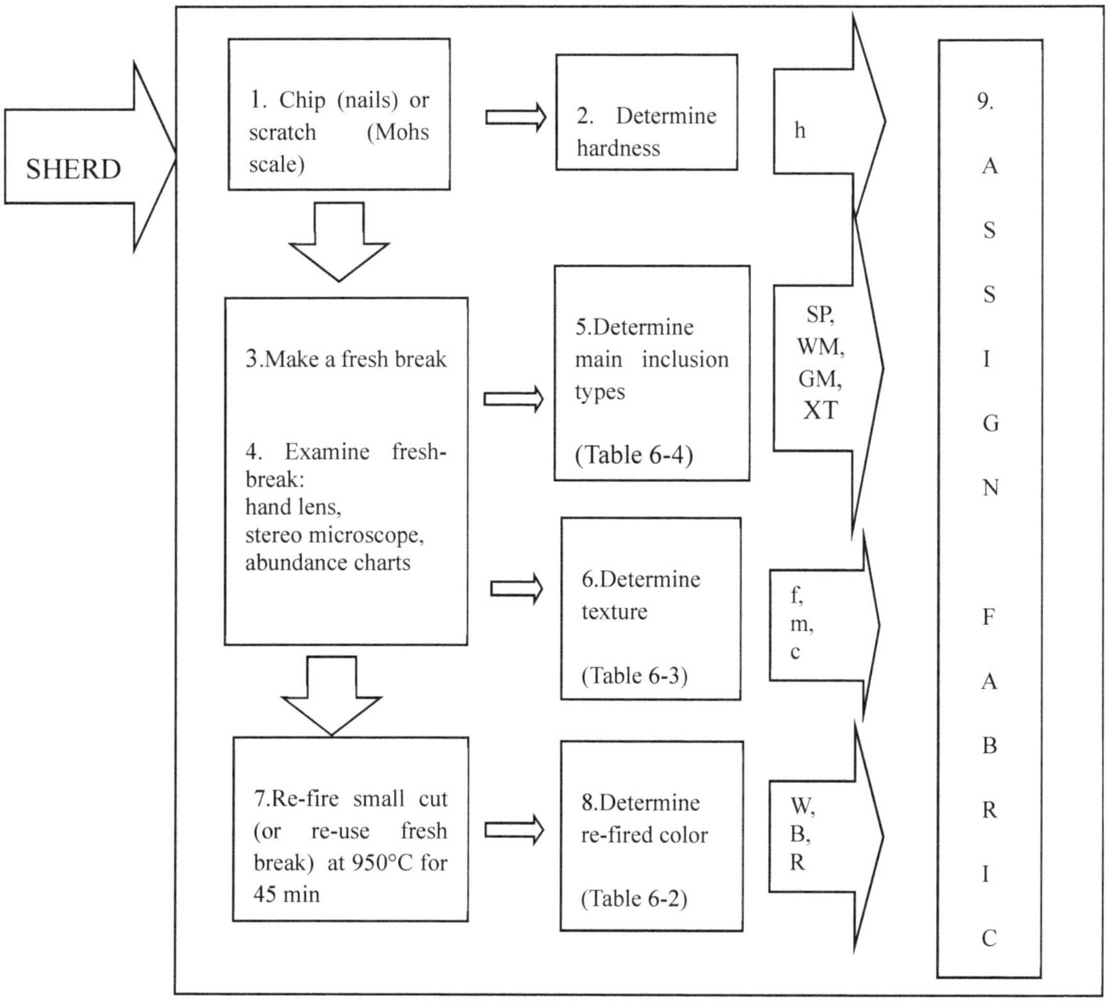

Figure 6-17. Summary of procedure followed for hand-specimen fabric characterisation

Table 6-7. Key for the Assignment and Summary of Fabric Classes

< ------------**Main Inclusions**------------>

Hard	**Main Inclusion**	**Texture**	**Re-fired**	**SP**	**WM**	**GM**	**XT**	**Shell**	**Fabric Code**
Soft	**Spar (SP)**	c	W	X		x			SP-c-W
		m	W	X		x			SP-m-W
		c	B	X			x	x	SP-c-B
	Micrite (WM)	m	W	x	X	x		x	WM-m-W
	Dark Particle (GM)	c,m	W		x	X			GM-cm-W
		c	R, B	x		X	x		GM-c-R/B
Hard	**Micrite (WM)**	m	W		X	x			WM-m-W-h
Varied	**Single Xtal (XT)**	f	R, B		x	x	X		XT-f-R/B
		c	R, B				X		XT-c-R/B
		c	W				X		XT-c-W

Refired colour: W=white, B = light reddish brown; R = red
SP= spar; WM=white limestone; GM = dark particle; XT= discrete crystals; Shell = skeletal
Inclusions: X = dominant or diagnostic (not necessarily dominant) inclusion; x= may be present

6.10 Hand-specimen Analysis of Local Sascab Materials

Hand-specimen analysis, using hand lens and binocular microscope, was performed on two rock (non-plastic) sascab samples (#10, #17, Table 5-4) with the purpose of comparing carbonate fragments found in sascab to the carbonate inclusions found within the pottery samples from north-central sites.

The particles in the pottery samples are composed of sparry crystals. The binocular microscope showed that the particles in raw material sample #10 are composed of rhombohedral sparry crystals. Each crystal has a size around 0.040 to 0.05mm. These crystals are probably dolomite, judging by the characteristic rhombohedral shape observed.

The classification of spar as opposed to micrite was possible when the individual crystallites were at least larger than 0.04 - 0.05 mm. The cluster of crystals appears to be held together by a very fine mud. It was not possible to determine with certainty whether the observed grains in the collected sascab are composed of calcite as opposed to dolomite. A test with chlorohydric acid differentiates calcite from dolomite in hand specimens, but it is not reliable when used with fine-grained dolomite. Sample #17 contains some clusters of euhedral sparry crystals but most of the grains could not be resolved.

Micritic and sparry particles plucked out from pottery samples were also examined using the binocular microscope. The physical attributes of sascab, such as crystal size and the rhombic form of the sparry particles, matched the grains of some of the sparry pottery samples.

6.11 Summary

Hand-specimen analysis showed that variations exist in the composition of the samples. It was observed that the main characteristics that divided the samples for the formation of the hand-specimen fabrics are the dominant inclusions, the texture (or grain size), the refired colour, and the hardness of the paste. It was found that there are patterns to this variability that allowed characterising the samples into ten fabrics.

The results made evident that the fabrics are correlated to the types of vessels and to the geographical location. In the north-central area, most red-slipped samples have medium-grained microcrystalline fabrics. The samples of the unslipped, striated open-mouth Yacman jars always have coarse fabrics in which the main inclusions are sparry clusters, dark particles, or single crystals. These differences between red-slipped and open-mouthed striated jars, a cooking pot most likely, may be related to their intended function. Navula samples, plain and unslipped, on the other hand, do not have a recognisable pattern and do not show a dominant grain size or type of inclusion.

This analysis also showed that the fabrics within the utilitarian vessels sampled from the north-central area differ from the fabrics found at eastern sites (coarse single crystals) for the same type of vessels. On the other hand, Payil, found at Mayapán and Chac Mool, shares the same fine-grained single-crystal fabric.

However, there are many limitations when performing a hand-specimen analysis. The most challenging is that most inclusions cannot be identified, and it was not possible to determine whether the white carbonate inclusions in the pottery samples are calcite or dolomite. There is also the issue of the maximum magnification of the microscope used in this work, 40 times the natural size. Crystals less than 0.040 mm cannot be distinguished with clarity. An analysis using thin sections and the polarising microscope presented in subsequent chapters addresses these issues.

In sum, while hand-specimen analysis was successful in characterising the samples into fabrics and showing that patterns exist, these fabrics, however, are very broad and insufficient to test the hypotheses guiding this research or address the research questions. In the next chapter, the samples are further characterised into groups by their chemical attributes given by differences in raw materials sources or technologies applied.

Chapter 7

RESULTS OF THE CHEMICAL ANALYSIS

This chapter is dedicated to the description of the methods and results of the chemical analysis. The main objective of this analysis is to characterise the pottery fabrics in terms of their chemical attributes. To this end, a combination of petrographic and chemical analyses was used. This approach addresses the challenges to studying the Late Postclassic ceramic economy. These challenges are the result, in a good part, from stylistic similarities within the ceramics, combined with a geologically largely homogeneous northern Yucatán that has hindered research on ceramic production. It was anticipated from the start of this study that there would be little variability among types of rocks in northern Yucatán. It was expected that chemical analysis overcomes these challenges. While petrographic analysis is based on the rocks and mineral inclusions, chemical analysis incorporates the clay component in the characterisation of the fabrics (Whitbread 2007, p.187).

A total of 186 pottery samples and 12 raw materials samples were analysed using Neutron Activation Analysis (NAA). The samples were taken from the different fabrics identified with hand-specimen analysis, focusing on the selection of each fabric from the various sites and ceramic classes. Chemical analysis was performed by Pierce and Glascock (2015) at the University of Missouri's Research Reactor (MURR). Their report is included in Appendix B for ease of reference. The results of this analysis will be re-examined in subsequent chapters (chapter nine and 10) in light of the results of the petrographic analysis (chapter eight).

In addition to the characterisation of the pottery samples, this research makes use of chemical analysis for provenance analysis. Provenance of pottery samples cannot be known from examining chemical groups in isolation. The chemical groups need to be compared to known sources, including raw materials taken from known or probable sources. It is preferable, however, to use pottery, including wasters, found in known or presumed production centres. The reason for comparing pottery to pottery, as opposed to raw materials, is that the raw materials used to manufacture the pottery could have been processed in such a way that changed their chemical fingerprint. The concentrations of elements may be changed by adding temper, mixing clays, or through levigation. A pottery-to-pottery comparison would incorporate changes within the chemical composition in response to techniques of clay preparation. However, a search of the literature showed that there are no known production centres for utilitarian ceramics in the study region for any period (Dahlin and Ardens 2002).

Therefore, the investigation of the chemical compositions of the samples took a different approach that includes the investigation of the structure that the elemental chemical data of pottery samples and raw materials from the region might have and that seeks patterns in the data leading to the characterisation of the samples into chemical groups. Given that the location of production is not known, if raw materials or raw materials with distinctive chemical information are lacking, the characterisation of pottery samples by itself might provide significant information for this research by forming potters' groups of similar chemical characteristics. On the other hand, if raw materials are available that provide distinctive chemical information, the groups formed, in addition to characterising the pottery samples can potentially determine provenance. Because these groups have different chemical compositions, they may represent geological or social divisions such as different geological sources of raw materials or ways of making pottery (traditions).

7.1 The Structure of the Data

The procedures followed at MURR for the processing of the samples and collection of the data are described by Pierce and Glascock (2015) and (Glascock 1992, 2009). Briefly, the procedure includes a series of steps for the removal of small fragments from each sample and the abrasion, washing, drying, weighing, and irradiation of the fragments. After irradiation, the samples are allowed to decay, and the concentration of the chemical elements counted and recorded in parts per million. The 33 chemical elements analysed are: aluminium (Al), barium (Ba), calcium (Ca), dysprosium (Dy), potassium (K), manganese (Mn), sodium (Na), titanium (Ti), vanadium (V), arsenic (As), lanthanum (La), lutetium (Lu), neodymium (Nd), samarium (Sm), uranium (U), ytterbium (Yb), cerium (Ce), cobalt (Co), chromium (Cr), caesium (Cs), europium (Eu), iron (Fe), hafnium (Hf), nickel (Ni), rubidium (Rb), antimony (Sb), scandium (Sc), strontium (Sr), tantalum (Ta), terbium (Tb), thorium (Th), zinc (Zn), and zirconium (Zr).

A combination of analytical methods was used to investigate the structure that the elemental chemical data of pottery samples and raw materials might have and to seek patterns in the data aiding in the characterisation of the samples into chemical groups. These methods (Pierce and Glascock 2015) and the structure in the data that they suggest are the focus of this section.

7.1.1 *Cluster Analysis*

Cluster analysis was used to identify initial groups of samples presenting similar compositions (nickel was not considered due to the presence of many missing values (Pierce and Glascock 2015, p.7). Cluster analysis is not used for final group formation because many disadvantages to its use with highly correlated data have been documented, and compositional data are usually correlated (Baxter 1994, p.167-168).

7.1.2 *Principal Components Analysis*

Principal components analysis, on the other hand, works well in this situation and, in addition, it can show the relationships between the samples (Glascock 1994, p.98), mainly through the use of plots. Principal components analysis reduces the number of variables by calculating 'principal components' (PC), which are linear combinations of the original variables (Formula 7.1). The coefficients (a1, a2, …,n32 in Formula 7.1), or elemental loadings, are based on calculations on the correlations matrix of the data such as new principal component variables are uncorrelated (Baxter 1994, p.49). The goal is that the first few principal components, hopefully, the first two, capture as much variation as possible. The principal components for the 32 elemental concentrations take the following form (based on Baxter 1994, p.49):

$$P1 = a1\ As + a2\ La + \ldots + a32\ V \qquad \text{(Formula 7.1)}$$

$$P2 = b1\ As + b2\ La + \ldots + b32\ V$$

....

$$Pn = n1\ As + n2\ La + \ldots + n32\ V$$

This analysis structures the data enabling the determination of which elements are driving the most variation between the samples. This variation is usually examined in two (first and second principal components) or three (first, second and third principal components) dimensions (Shennan 1998:270-276) because, although multiple principal components may be calculated, any significant variation is found in the first and second, or occasionally third, components.

The calculated P1 and P2 for the pottery samples and raw clay samples are plotted in principal component #1 axis against principal component #2 axis, as illustrated in Figure 7-1. In addition, the loadings (or coefficients) of all the variables (the elements Al, As …Zr) for PC1 And PC2 can be added to the plot representing the relationship between the variables (Glascock et al. 2004, p.100). The length of each variable's vector indicates the contribution of the element. When straight angles separate the vectors of two elements, the two elements are not correlated. An angle of around 180 degrees means high inverse correlation, while a small angle means high correlation (Glascock et al. 2004, p.100). Because two types of information have been plotted, a chart such as Figure 7-1 is called a biplot.

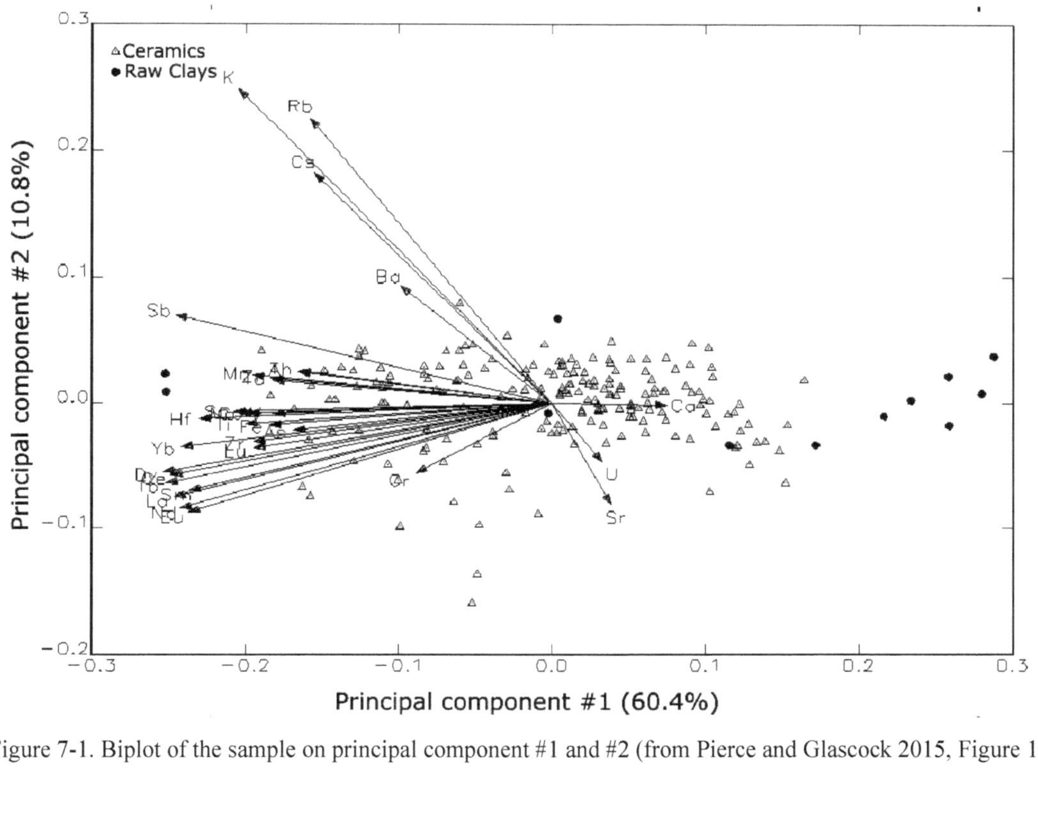

Figure 7-1. Biplot of the sample on principal component #1 and #2 (from Pierce and Glascock 2015, Figure 1)

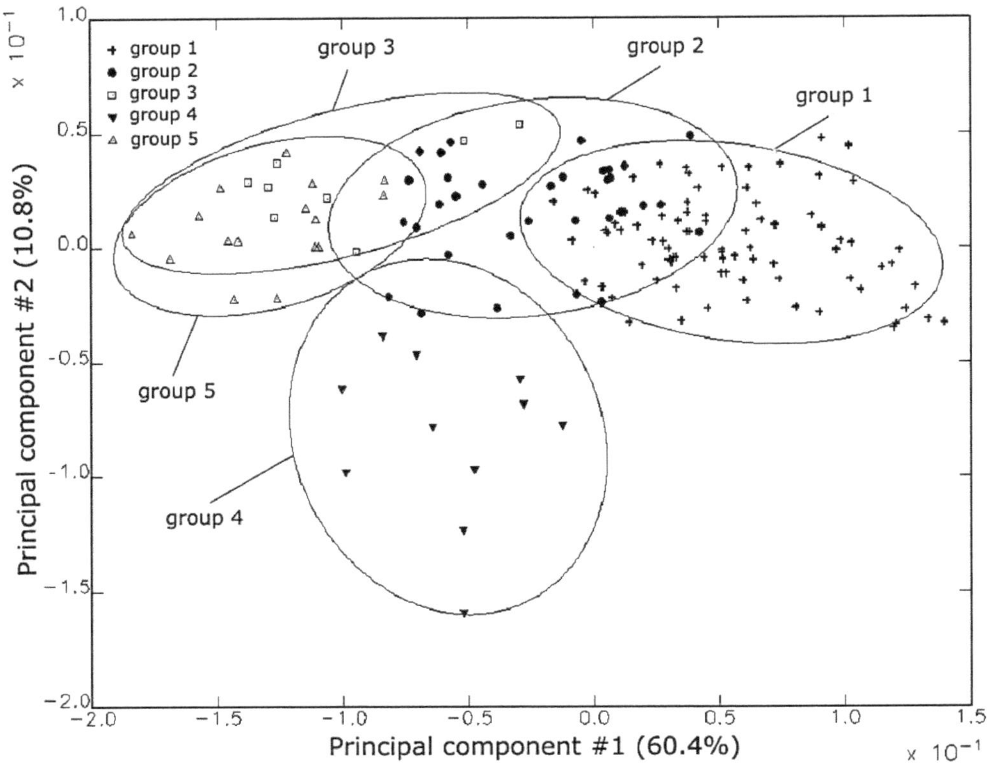

Figure 7-2. Plot of samples in principal components #1 and #2 axes (Pierce and Glascock 2015, Figure 2), or P1 and P2 (Formula 7-1), depicting the samples within each of the five groups resulting from the initial Cluster Analysis.

In Figure 7-1, a relationship among variables can be visualised. Many highly correlated elements (small angle) have an inverse correlation (180 degrees angle) in the principal component # 1 axis with calcium in particular, and strontium, and uranium to a lesser degree. This means that the higher the concentration of calcium, the lower the concentration of these many elements, and conversely. Calcite and strontium are common in calcareous rocks. Most pottery sampled contain carbonate inclusions (calcium or dolomite), and most of the raw material samples collected (section 5-4, Table 5-4) are marls (combination of clay and carbonate mud).

In Figure 7-2, principal components # 1 and # 2 are plotted in the x and y-axis. One group can be distinguished in principal component # 2 (Group 4), but the rest largely overlap in principal components # 1 and # 2.

7.1.3 Bivariate Plots

As a result of the overlap that exists when plotting on principal components, other techniques were used to perform the final group assignments, starting with the use of bivariate plots. They allow exploring the relationships between two variables (elemental concentrations) and the groups. For instance, plotting (in Pierce and Glascock 2015, Figures 3-6) the concentration of antimony (Sb) against chromium (Cr) shows that Groups 3 and 5, which overlapped in Figure 7-2, are distinguishable based on this element. Similarly, Groups 2 and 4, which have some overlap in principal component #1 and #2 (Figure 7-2), are distinguishable when plotted on chromium versus caesium.

7.1.4 Mahalanobis Distance

Mahalanobis distance (MD) was used to confirm final group membership given that overlaps between Groups 1 and 2 and between 2 and 5 persisted, necessitating the application of other means to confirm groups (Pierce and Glascock 2015, p.10). MD is particularly useful when data are highly correlated (Glascock 2004, p.100). In this case, the scatter plot of a set of points forms an ellipsoid (in n-dimensions), and the Euclidean distance used in many multivariate methods is not suitable because it assumes that the scatter is circular (Baxter 1994, p.80). The Mahalanobis distance (MD) between a sample and the group centroid is defined as the squared Euclidean distance between them, divided by the group variance in the direction of the sample (Glascock 2004, p.100). This calculation gives less weight (than the Euclidean Distance) to variables with larger variances and groups of correlated variables (Baxter 1994, p.80).

Information about the correlations between pairs of elements (based on the variance–covariance matrix) is included in MD, allowing the calculation of the probability that a particular specimen belongs to a group. One factor in this probability is the distance of the specimen to the group centroid. The calculation of these probabilities resulted in 81 percent of the samples being assigned to a total of five chemical groups, with the rest remaining unassigned (the samples with the chemical group assignments can be found in Table 8-18). The MD calculations discriminate the groups but do not inform about their characteristics.

7.1.5 Discriminant Analysis

After finding the groups in the chemical data, discriminant analysis (DA) was used to further analyse the data by visualising in a plot the separation of the groups and pointing to the variables that most differentiate a group from the others (Pierce and Glascock 2015, p.14). This can be done because the calculations involved in DA maximise the variations between groups (Pierce and Glascock 2015, p.14). By making use of the variances within groups, the output of DA are new functions in which the calculations have made the within-groups variance small, while making the between-group variance, large (Baxter 1994, p.189).

The first and second DA function values were calculated and plotted in the axes of the discriminant functions #1 and #2, separating the troublesome overlap of Group 1 and Group 2 (Pierce and Glascock 2015, Figure 7) of previous methods. Similarly, discriminant functions #3 and #2, separated Groups 2 and 5 in discriminant function #2 (Pierce and Glascock 2015, Figure 8).

7.1.6 Mahalanobis Distance Test

The groups found were tested using MD calculations to determine the likelihood of each specimen being a member of each chemical group from which that specimen had been removed (Pierce and Glascock 2015, p.16). These tests were successful and confirmed cohesive group assignment (Pierce and Glascock 2015, p.16).

7.2 Location of the Sources of the Chemical Groups

With the objective of suggesting the geographical source of the pottery samples, the compositions of the collected raw materials are compared with the five groups just defined. The collected clays/marls (Table 5-4) are plotted in the principal component #1 and #2 plot of Figure 7-3. From this figure, one marl (plastic sascab) sample, CGS193 (or raw material # 11, Table 5-4), fits within the chemical group 1. The sample was collected near Mayapán.

Aiming to resolve the observed ambiguity in Figure 7-3 (in which it is not clear whether CGS193 belongs to Group 1 or 2), a Mahalanobis distance (MD) was calculated to confirm that this sample belongs to Group 1 rather than Group 2. As can be recalled from chapter five (Table 5-4), clay deposits were not found near (within 10 km) Mayapán or the other north-central sites (a documented deposit five km from Mama was not located). Excluding clayey red soils and one clay sample from Chapab-Ticul road (13 km west of Mama), the samples collected are marls (plastic sascab).

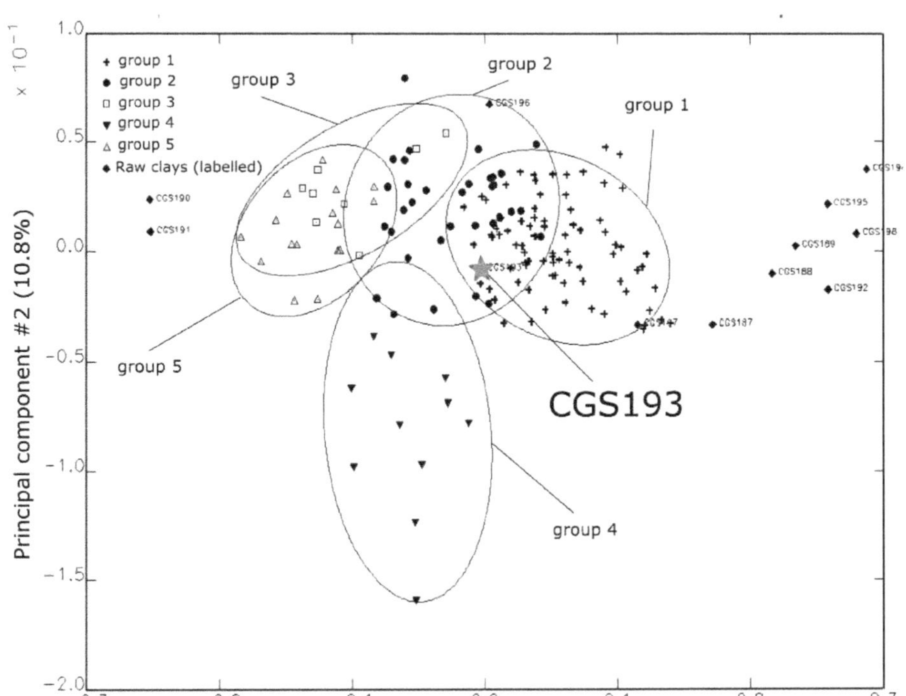

Figure 7-3. Clay samples and the chemical groups (Pierce and Glascock 2015, Figure 9)

The results of the MD comparison show that a marl that matches the composition of the pottery found at Mayapán and other north-central sites within Group1 is located near that site. The results also showed that the rest of the clay/marl samples do not fit well any of the chemical groups (Pierce and Glascock 2015, p.18).

To investigate the potential provenance of the pottery samples, the compositions of pottery samples were compared (Pierce and Glascock 2015, p.19) against all the samples stored within MURR database from the general Yucatán region: 2071 samples, of which only 51 are from Mayapán. Nevertheless, 80 of the samples in this study have compositions closest to the 51 MURR samples from Mayapán than to rest of samples within MURR. Table 7-1 summarises the results. The most similar compositions for samples within Group 4 were found at Laguna de On, Belize, and there are only 57 (of 2071) samples within MURR from that site. A number of pottery samples within Group 5 also have compositions that match MURR samples found in the Belize area, at Santa Rita Corozal. Group 2 and 3 are too varied and no pattern is observed.

Pierce remarked that there is the potential for error (personal communication 11/2015). The calculation of the Euclidian distance uses the place of collection, the find spot, of the samples. It may be different from the place where raw materials were sourced. The results of this calculation can be found in the Table 8-18 for each sample.

Figure 7-4 shows the sites in this research programme and sites of closest Euclidian match. The region of Santa Rita Corozal and Laguna the On in Belize is encircled.

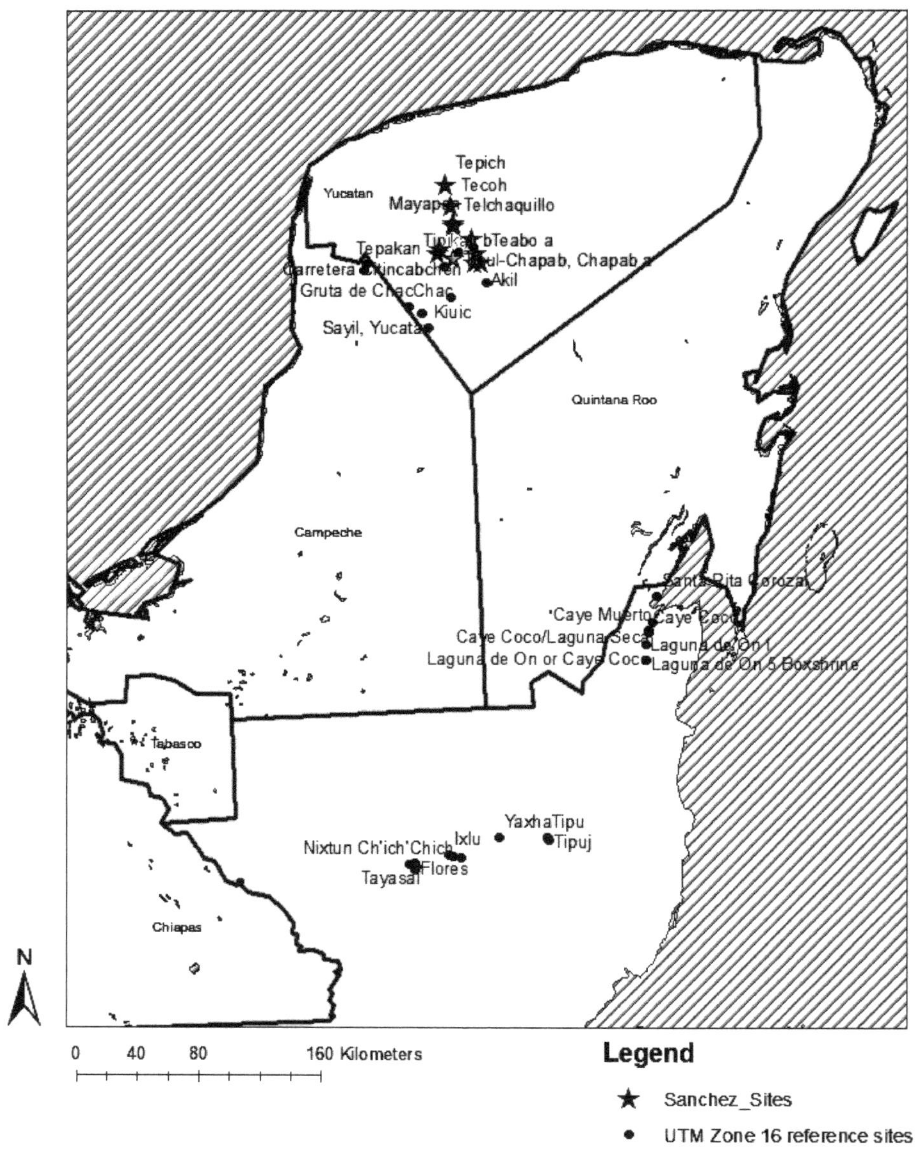

Figure 7-4. Map showing north-central sites and Euclidean matches (from Figure 10, Pierce and Glascock 2015)

Table 7-1. Sites of Closest Compositional Euclidian Match

	Total #	Mayapán	Laguna de On	Santa Rita Corozal	Mama	Kiuic	Caye Coco	Ticul	others
Group 1	82	80				2			
Group 2	34	7	2		3	3		5	13
Group 3	8		2				1	4	1
Group 4	11		11						
Group 5	16			10				3	3

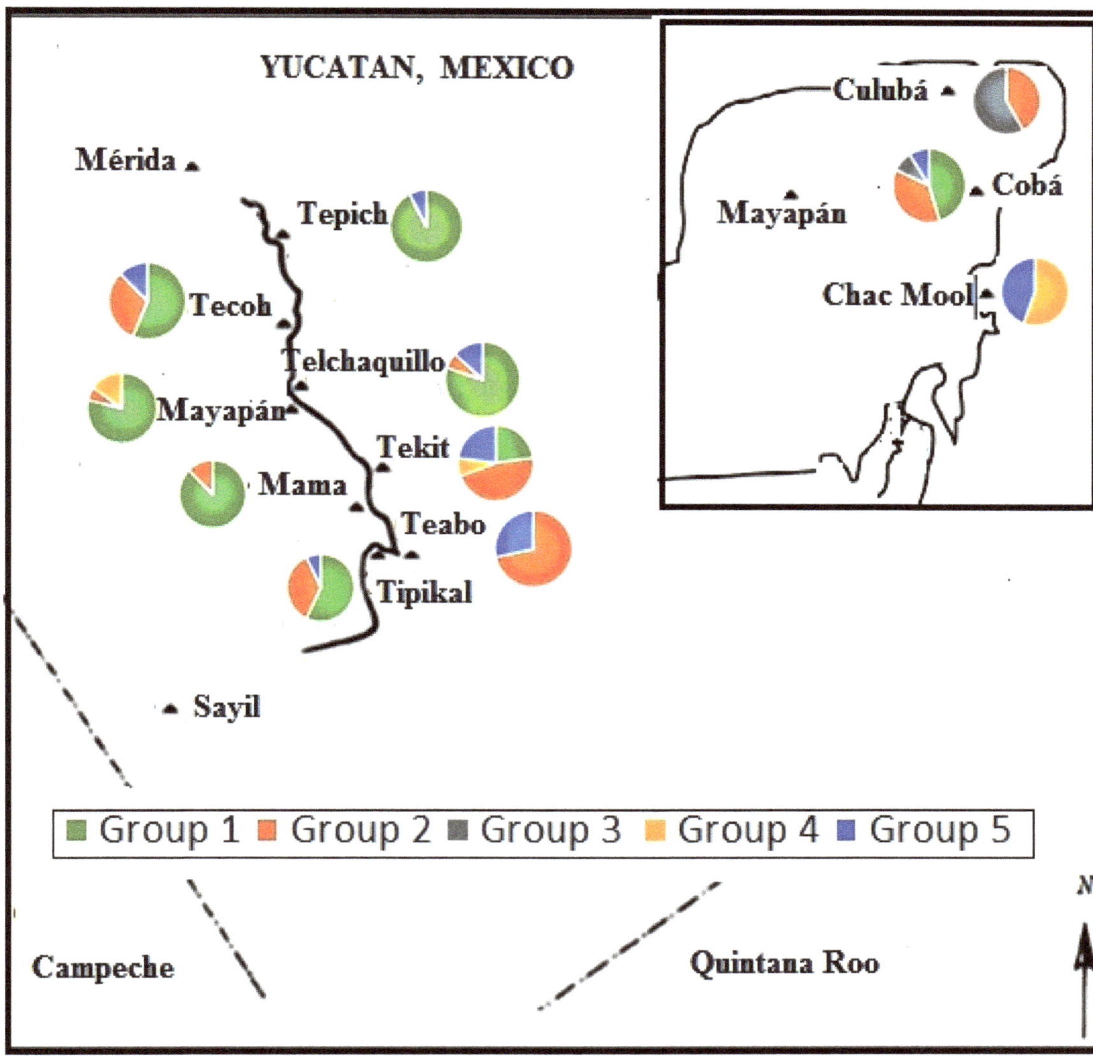

Figure 7-5. Percentage that each chemical group represents of the total chemically analysed samples from each site

7.3 The Chemical Groups

Eighty-one percent of the samples was assigned to a total of five chemical groups, with the rest remaining unassigned. Figure 7-5 illustrates for each site the chemical groups found and the percentage that each chemical group represents of the total number of chemically assigned (some samples do not fit in any group and remained unassigned) samples for that site.

The objectives of the next sections are to discuss and characterise the chemical groups in terms of the elements that distinguish them and, then, draw inferences about the possible zones of raw materials procurement.

Table 7-2 displays the means and standard deviations calculated from the elemental concentrations in each chemical group. Figure 7-5 and Table 7-2 are referenced several times during the presentation of the chemical groups.

Table 7-2. Means and Standard Deviation of Elemental Concentrations

Chemical Group	As	La	Lu	Nd	Sm	U	Yb	Ce	Co	Cr	Cs
Group 1, mean	5.99	8.13	0.14	7.61	1.60	1.98	0.79	19.30	2.00	56.07	1.41
std dev.	3.18	2.68	0.03	2.70	0.42	0.73	0.21	5.66	0.64	22.98	0.56
Group 2, mean	5.41	15.61	0.21	13.87	2.97	1.23	1.59	36.23	2.35	42.15	2.23
std dev.	2.79	5.95	0.09	5.08	1.04	0.41	0.68	9.90	1.08	31.93	1.10
Group 3, mean	14.27	20.76	0.33	19.14	4.32	1.12	2.40	59.69	2.85	28.30	2.77
std dev.	3.41	8.64	0.08	8.24	1.65	0.40	0.69	15.15	0.31	3.28	0.71
Group 4, mean	7.11	29.54	0.28	26.47	5.39	1.69	2.14	70.35	2.94	77.32	0.59
std dev.	1.71	13.17	0.05	11.66	2.50	0.58	0.42	32.21	0.63	26.70	0.20
Group 5, mean	13.40	27.89	0.37	25.25	5.31	1.61	2.80	63.11	6.57	94.29	3.53
std dev.	5.21	7.03	0.08	6.72	1.38	0.78	0.71	13.55	1.98	34.80	0.83

Table 7-2 (continued)

Chemical Group	Eu	Fe	Hf	Ni	Rb	Sb	Sc	Sr	Ta	Tb	Th
Group1, mean	0.29	10319.61	1.72	10.58	14.32	0.39	3.53	358.73	0.44	0.21	5.61
std dev.	0.09	3243.30	0.33	23.29	5.58	0.12	0.92	137.68	0.09	0.06	1.10
Group 2, mean	0.48	13028.78	3.40	7.66	26.51	0.86	5.27	324.75	0.72	0.42	9.04
std dev.	0.18	4021.53	0.76	14.40	11.99	0.30	1.54	214.45	0.18	0.18	2.27
Group 3, mean	0.59	18487.74	6.35	6.52	32.27	3.28	8.78	150.45	1.11	0.66	13.51
std dev.	0.23	1838.64	0.74	11.41	8.86	0.45	1.12	149.04	0.07	0.26	1.01
Group 4, mean	1.04	19870.68	5.66	3.00	4.60	0.52	7.21	591.39	0.76	0.70	8.06
std dev.	0.48	3647.79	1.62	6.45	3.66	0.10	0.98	577.77	0.15	0.27	1.35
Group 5, mean	0.99	28460.63	4.96	35.12	34.16	1.61	9.86	419.05	1.08	0.75	13.63
std dev.	0.25	4689.09	1.17	16.93	6.36	1.20	1.87	303.51	0.21	0.20	2.32

Table 7-2 (continued)

Chemical Group	Zn	Zr	Al	Ba	Ca	Dy	K	Mn	Na	Ti	V
Group 1, mean	19.66	45.54	33716.02	114.03	269632.78	1.26	2652.96	69.08	578.93	1154.08	27.28
std dev.	6.15	9.05	6956.45	42.42	25357.49	0.36	1406.33	21.88	330.47	322.58	9.45
Group 2, mean	34.85	73.93	50654.91	165.61	262836.37	2.57	6183.54	120.18	751.55	1610.34	34.57
std dev.	10.98	17.05	10248.85	48.25	35775.93	1.17	4659.16	87.67	417.14	515.22	14.01
Group 3, mean	57.48	139.05	80806.71	157.48	199986.70	3.85	7470.65	66.91	1440.68	2402.06	70.76
std dev.	7.90	21.87	6648.62	55.94	21084.28	1.37	2760.17	12.25	1026.94	344.10	14.47
Group 4, mean	38.56	134.56	51402.56	88.91	216083.13	3.92	1466.82	83.23	1668.00	2568.01	44.92
std dev.	8.22	31.44	8171.46	45.76	39740.98	1.29	760.57	33.85	783.97	489.34	6.19
Group 5, mean	45.73	114.42	81376.45	210.75	181713.88	4.60	8606.88	305.17	1746.55	2834.67	75.00
std dev.	16.46	26.21	14146.11	92.22	43958.14	1.13	3847.19	103.16	1197.10	626.25	19.92

7.4 Chemical Group 1 (n=82)

This is the largest chemical group, containing 44 percent of the pottery samples. It is fairly homogeneous and clustered in principal components (Pierce and Glascock 2015, p.19). The elemental loadings of principal component #1 distinguish this group from the rest, with the exception of overlaps with Group 2 (Figure 7-2). The variables most driving the variation in principal component #1 in a positive direction are calcium (Ca), strontium (Sr), and uranium (U). Pulling in the negative direction are a series of elements including cerium (Ce), dysprosium (Dy), terbium (Tb), antimony (Sb), europium (Eu), and hafnium (Hf). In Figure 7-1, they appear with a small angle among them, indicating high correlation. Discriminant function # 2 separates Group 1 from Group 2.

Inspection of the list of elements involved in discriminant function #2 (Pierce and Glascock 2015) and the mean concentrations (Table 7-2) (taking into consideration the spread of the data) can point to the most significant elements in separating the groups. Some of these elements are plotted in Figures 7-6 to 7-9. Figure 7-6 shows that Group 1 is low in samarium (Sm) and europium (Eu) when compared to the rest of the groups, with some Group 2 overlap.

As shown in Figure 7-7, Group 1 is low in aluminium compared to the rest of the groups, and low in iron when compared to most Group 2 samples and Groups 3, 4, and 5. Similarly, Group 1 is low in zirconium and relatively low in zinc (Figure 7-8) when compared to the other groups.

Likely, hafnium (Hf) is the element that best distinguishes Group 1 from the rest. Pierce and Glascock (2015, Figure 5) and Figure 7-9 show very little overlap of Group 1 with the rest of the groups when plotted on this element. Group 1 has the lowest, or relatively low, concentrations in elements important in driving the distinctions in principal component # 2 including europium (Eu), samarium (Sm), cerium (Ce), iron (Fe), aluminium (Al), hafnium (Hf), lanthanum (La), zirconium (Zr), and zinc (Zn).

7.4.1 Associations between Group 1, Ceramic Typology, and Sites

In this section, associations between samples within Group 1 and ceramic varieties and sites are examined. Table 7-3 summarises the samples presenting compositions within Group 1, within site and ceramic variety. It is important to recall from chapter five that percentages or proportions of the pottery samples represent the collection of pottery samples taken and do not represent percentages of the sites.

Table 7-3 shows that:

1. Samples from Culubá or Chac Mool do not have compositions within Group 1.
2. Sherds with compositions within Group 1 were found throughout the north-central sites and Cobá (Figure 7-5), with the exception of Teabo.
3. Most of the pottery samples found at Mayapán, Tepich, and Telchaquillo are associated with Group 1.
4. The utilitarian varieties Mama (medium-grained red-slipped), Yacman (unslipped, striated), or Navula (unslipped, plain) are equally represented within Group 1.
5. Payil, the red-slipped variety common at eastern sites characterised by its compact and fine paste, was not found in this group, even when found at Mayapán.
6. Navula samples within Group 1 were found mostly at Mayapán and Telchaquillo, while none was found south of Mayapán. Mama and Yacman samples within Group 1 were found at most north-central sites.

7.4.2 Sources of Group 1 Vessels

Given the lack of direct evidence for production, indirect measures of provenance such as the Criterion of Abundance (Bishop et al. 1982, p.301) were considered. This criterion postulates that a strongly represented ceramic unit at a given site is presumed to be of local manufacture. Based on these criteria, production of Group 1 pottery is related to three sites: Tepich, Telchaquillo and/or Mayapán. For instance, 66 per cent of the samples from Mayapán are found within Group 1. The percentage is higher if only chemically assigned samples are considered (79 per cent) and even more if Payil samples are taken out of consideration (92 per cent) because they are, as discussed in the next sections, foreign. The abundance of Group 1 at Tepich, Telchaquillo and Mayapán is indicating that the geographical scope of raw material sources within chemical Group 1 may be wide and most probably encompassing at least these three sites.

In the case of Mayapán, in addition to the criterion of abundance, two more arguments support the notion of local production in its vicinity. One is that when the Euclidian distance is calculated between samples in this research programme and the rest of MURR database, the most common match for the samples within Group 1 is Mayapán in 80 out of 82 cases, or 97.6 per cent (Pierce and Glascock 2015, p.19). The second argument is the match in the composition of a marl (#11) from Mayapán and pottery within Group 1, and Group 1 closest Euclidian match is Mayapán.

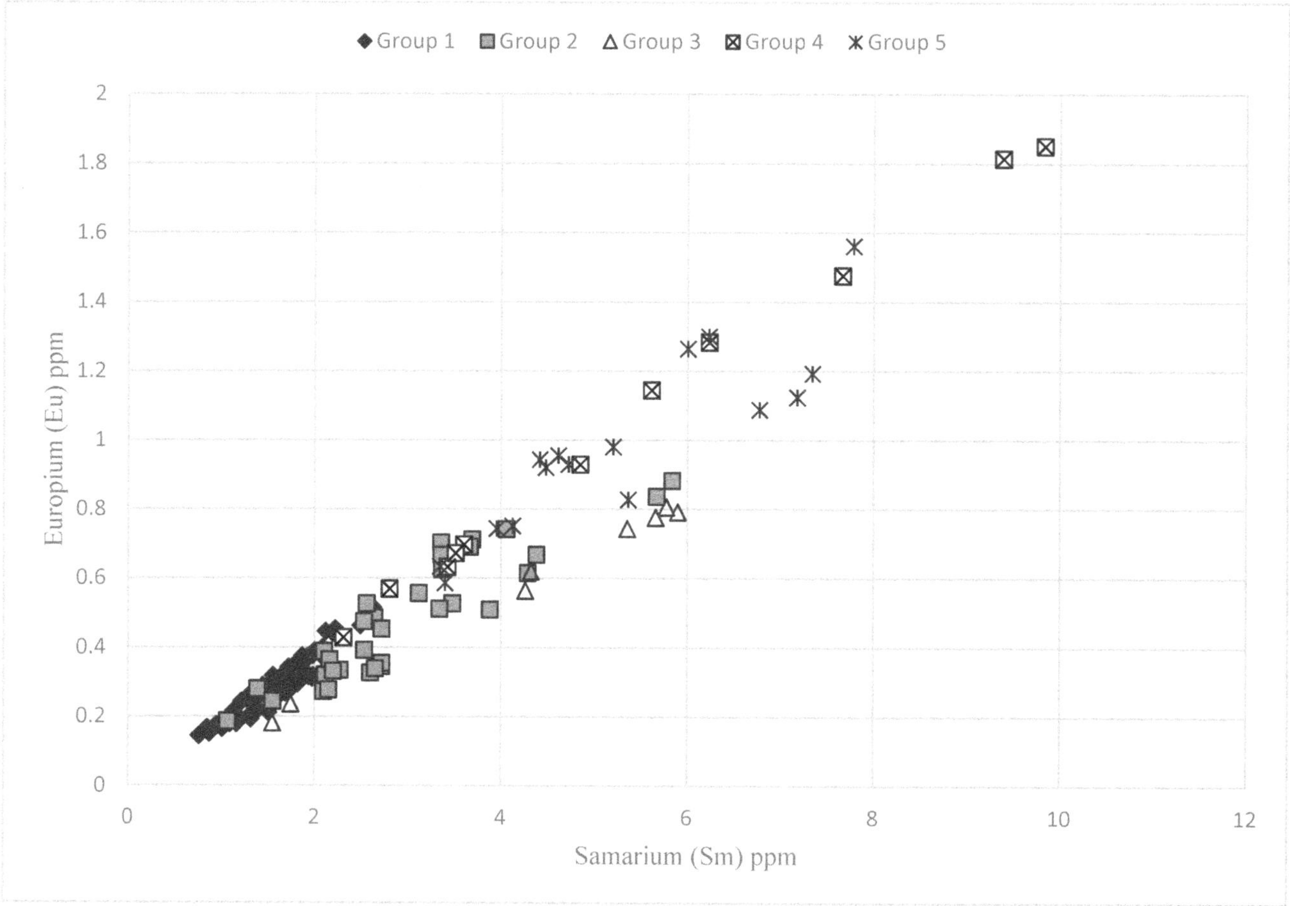

Figure 7-6. Europium (Eu) against samarium (Sm)

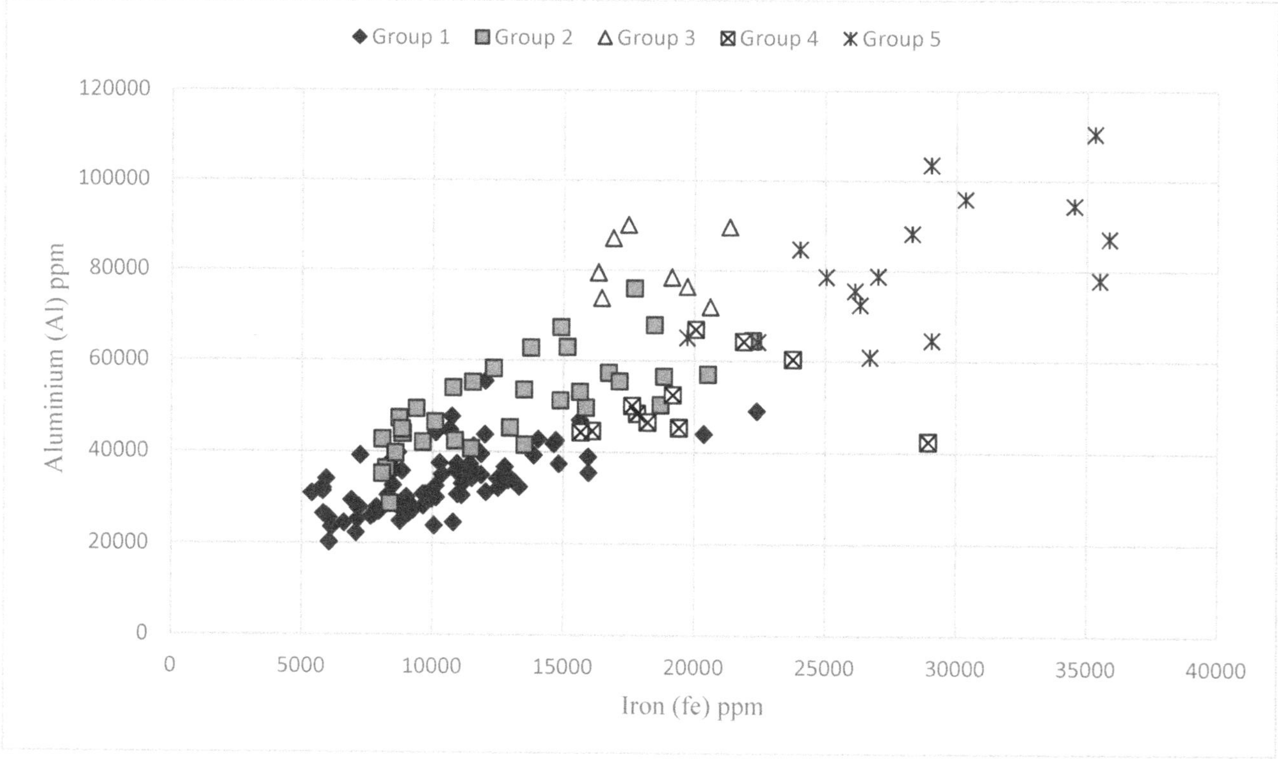

Figure 7-7. Aluminium (Al) against iron (Fe)

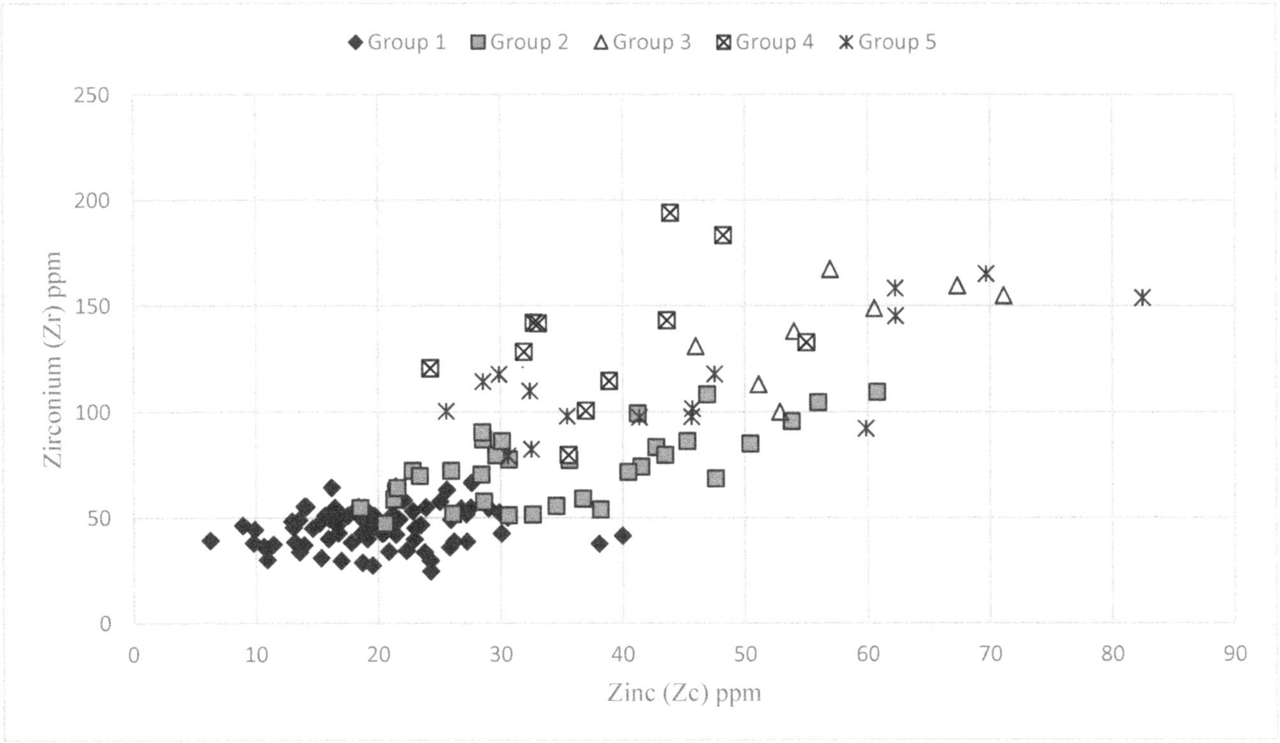

Figure 7-8. Zirconium (Zr) against zinc (Zn)

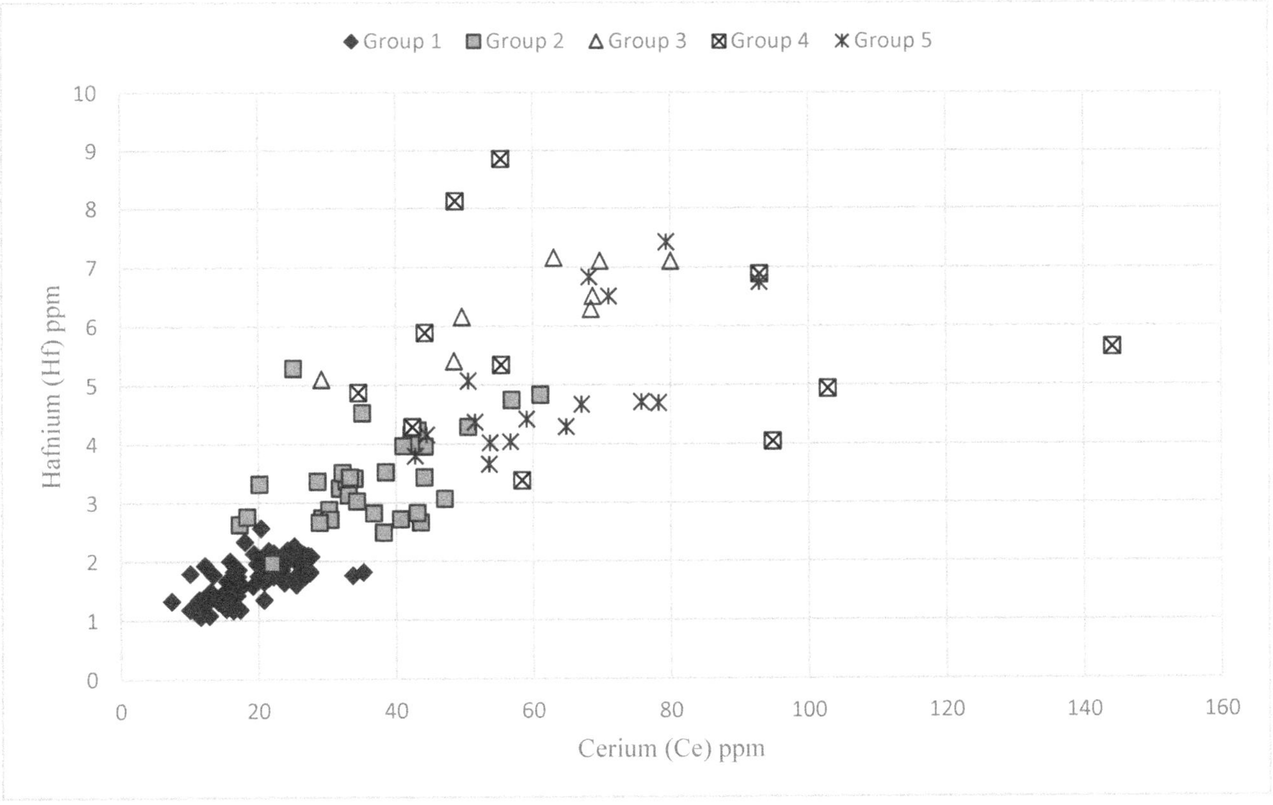

Figure 7-9. Hafnium (Hf) against cerium (Ce)

Table 7-3. No. of Samples within Group 1 within Sites and Ceramic Varieties

Geographical Location	Site Name (n=total chemically analysed)	Red-slipped (Mama)	Unslipped, striated (Yacman)	Xcan chakan	Plain, unslipped (Navula)	Fine, red-slipped (Payil)
North-central area						
	Tepich (n=14)	6	5		1	
	Tecoh (n=17)	4	4		1	
	Telchaquillo (n=19)	4	3	1	4	
	Mayapán (n=39)	12	3		11	
	Tekit (n=15)	3				
	Mama (n=15)	5	2			
	Tipikal (n=16)	4	4			
	Teabo (n=10)					
Eastern area						
	Coba (n=13)	3 (*)			2	
	Culubá (n=15)					
	Chac Mool (n=12)					

(*) Cancún variety

In sum, Group 1 is characterised by having low concentrations in europium (Eu), samarium (Sm), cerium (Ce), iron (Fe), aluminium (Al), hafnium (Hf), lanthanum (La), zirconium (Zr), and zinc (Zn). Within these elements, hafnium (Hf) is the element that best distinguishes Group 1 from the other groups. On the other hand, Group 1 and Group 2 samples are highly calcareous in relation to Groups 3, 4, and 5, with the technological implications that this entails (discussed in section 9.6).

It is possible that an area with raw materials within chemical Group 1 is located within the crater's basin, and that the differences in chemical composition between Group 1 and 2 are related to the age of the rocks. As mentioned in section 4.1, the sedimentation patterns in northern Yucatán are closely linked to the impact the meteorite (Figure 4-1). During the Late Tertiary, while the sedimentation took place within the basin, the area outside (Early Tertiary) the ring may have been exposed and undergoing weathering and karstification made evident by the cenotes. The depleted mineral composition of Group 1, when compared to other groups, may be related to the younger age of the sediments. In addition, a marl (#11, Table 5-4) from within the crater's basin matched the composition of pottery samples within Group1, and the sites in which Group 1 dominate (criterion of abundance) are located within the basin.

7.5 Chemical Group 2 (n=34)

This group is considered the least homogeneous of all the groups (Pierce and Glascock 2015, p.19). Neither principal components nor discriminant analysis completely pulls this group apart from all others. There is always an overlap.

Nevertheless, Group 2 can be distinguished by using the results from these methods to pull Group 2 away from other groups in a group-by-group basis and by comparing the concentrations of pairs of elements (bivariate plots). The comparisons that distinguish Group 2 from other groups are presented next.

Discriminant function #2 distinguish Group 2 from Group 1, as discussed in section 7.3.1. When plotting the variables driving the group separation, Group 2 appears similar to Group 1 but with higher concentrations, as can be observed in Figures 7-6 to 7-9. In these bivariate plots, Group 2 overlaps other groups. It is not the highest or lowest on any of the elements. What better distinguishes Group 2 from Group 1 is the range of concentrations for hafnium (Hf) which shows less overlap than other elements (Figure 7-9).

Bivariate plots show that Group 2 can be distinguished from Group 3 based on antimony (Pierce and Glascock 2015, Figure 7-3) and by Group 3's higher aluminium (Al) concentrations (Figure 7-7). Principal component #2 distinguishes Group 4 from Group 2, with some overlap (Figure 7-2). In this component, caesium (Cs), potassium (K), and rubidium (Rb) pull in the positive direction and accounting for most of the variation. Figure 7-10 shows the low concentration of rubidium of Group 4 separating it from the rest and Group 2.

The overlap between Group 2 and 5 is resolved by the discriminant functions # 3 (Pierce and Glascock 2015). Among the elements driving the separation of groups are europium and iron. Group 2 is clearly separated from Group 5 based on its relatively low iron concentrations, as shown in Figure 7-11. Other bivariate plots presented earlier show that Group 2 is relatively low in many elements in relation to Group 5: europium (Eu) and samarium (Sm)(Figure 7-6), aluminium (Figure 7-7), zirconium (Zr) (Figure 7-8), and cerium (Ce) and hafnium (Hf) (Figure 7-9).In sum, there is no one single element that distinguishes Group 2 from the rest. It is similar in many ways to Group 1, with relatively high calcium and low concentrations for many elements that are higher than Group 1.

7.5.1 Associations between Group 2, Ceramic Typology, and Sites

Group 2 represents a small portion of the analysed samples from Mayapán (only two samples out of 39, #119 and #146 in Table 8-18), Telchaquillo (1 of 19, #172), Mama (1 of 15, #347), Tepich (0 of 14). Table 7-4 shows the number of samples within each location with compositions within Group 2. Tepich, Telchaquillo, and Mayapán are also the sites at which Group 1 is found almost exclusively.

The majority of Group 2 samples are from Tekit, Teabo, Tipikal, and Tecoh. At these sites, some of the Mama, Cancun, Navula, and Yacman, which are considered utilitarian, were found with a composition within Group 2.Group 2 is also found at the eastern sites of Cobá and Culubá.

Table 7-4. Samples within Group 2 within Sites and Ceramic Varieties

Geographical Location	**Site Name (n=total chemically analysed)**	**Total Group 2**
North-central area		
	Tepich (n=14)	0
	Tecoh (n=17)	5
	Telchaquillo (n=19)	1
	Mayapán (n=39)	2
	Tekit (n=15)	6
	Mama (n=15)	1
	Tipikal (n=16)	5
	Teabo (n=10)	5
Eastern area		
	Coba (n=13)	4
	Culubá (n=15)	5
	Chac Mool (n=12)	
Total		34

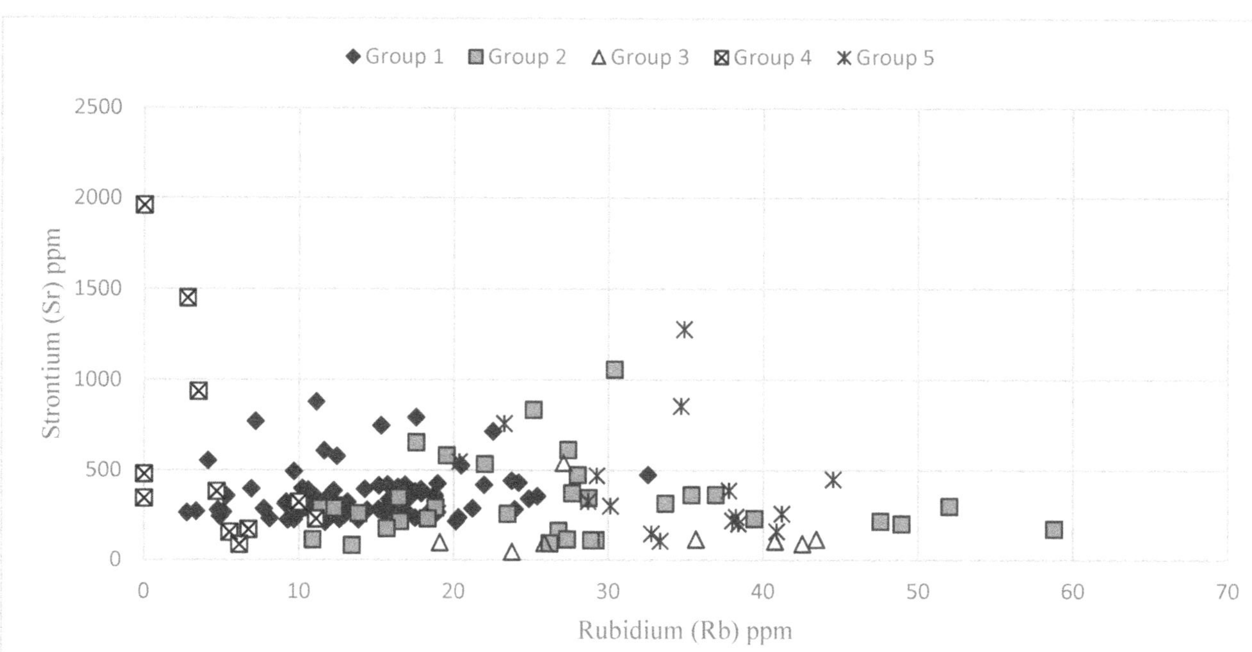

Figure 7-10. Strontium (Sr) against rubidium (Rb)

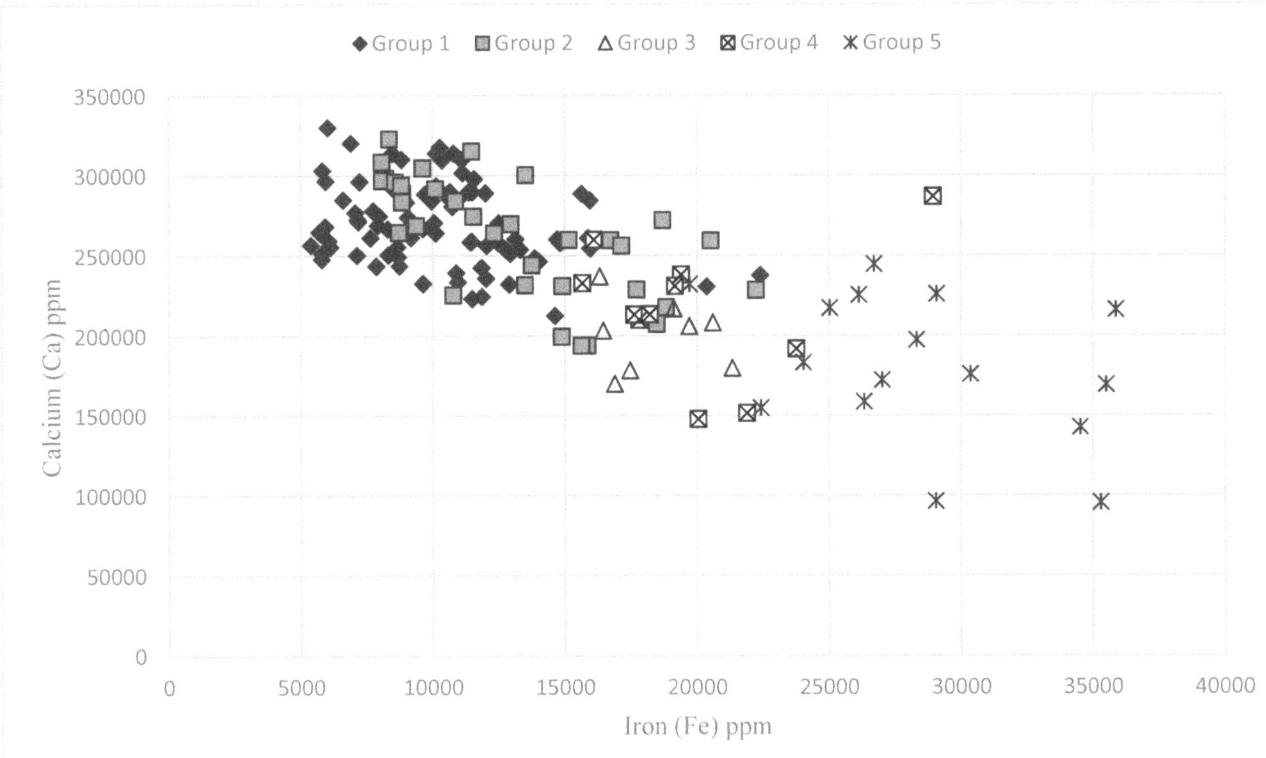

Figure 7-11. Calcium (Ca) against Iron (Fe)

7.5.2 Sources of Group 2 Vessels

None of the raw material selected matched the composition of this group. Euclidean distance calculations between samples in this group and samples in the MURR database for the region provided matches that are too varied and are not useful to determine provenance.

In the north-central area, the samples from Tepich, Telchaquillo, and Mayapán rarely were found to have a composition within Group 2, while most Group 2 samples are from Tecoh and south of Mayapán. These sites coincide with the Early Tertiary area in Figure 4-3. The differences in composition between Group 2 and Group 1 may, then, be related to differences in the age of sedimentation.

7.6 Chemical Group 3 (n=8)

This group is clearly distinguished from Groups 1,2, and 4 based on caesium and from all the groups based on chromium (Pierce and Glascock 2015, Figure 4), although it overlaps with Groups 2 and 5 when visualised in principal component #1 and #2 (Figure 7-2). It is also well differentiated in discriminant analysis (Pierce and Glascock 2015, Figures 7, 8), in particular in discriminant function #2 that separates the overlap between Group 3 and Groups 2 and 5.

Inspection of the main variables driving the separation of groups in this function shows that Group 3 is characterised by its richness in many of the element. It shows relatively high concentrations in several elements including antimony (Sb) (Pierce and Glascock 2015, Figures 3, 4), aluminium (Al) (Figure 7-7), zinc (Zn) (Figure 7-8), zirconium (Zr) (Figure 7-8), hafnium (Hf) (Figure 7-9), iron (Fe) (Figure 7-11). At the same time, it is relatively low in calcium (Figure 7-11).

7.6.1 Associations between Group 3, Ceramic Typology, and Sites

The samples do not show a pattern with regard to ceramic varieties, with slipped, unslipped, coarse and finegrained samples found in this group. In the samples in this study, Group 3 is associated with the site of Culubá: there are eight sherds in this group, of which seven were found at Culubá (Figure 7-5).

7.6.2 Sources of Group 3 Vessels

No raw materials samples matched the composition of this group. Euclidean distance calculations between samples in this group and samples from the region previously analysed in MURR database provided matches that are too varied and not useful to determine provenance.

In the samples in this study, Group 3 is associated with the site of Culubá: there are eight sherds in this group, of which seven were found at Culubá (Figure 7-5). The criteria of abundance point to Culubá as the source area for the raw materials. However, this study does not include other sites from the Culubá area, and the degree of chemical homogeneity of the area is unknown. It could be that raw materials from many other sites within the area fall within Group 3.

7.7 Chemical Group 4 (n=11)

Group 4 differs from the rest of the groups. Particularly, discriminant function #1 totally differentiates this group from all the others (Pierce and Glascock 2015, Figure 7). This group is also clearly distinguished in principal components #2 (Figure 7-2).

The variables most driving principal component # 2 in a positive direction are caesium (Cs), potassium (K), and rubidium (Rb). Group 4 has a relatively low concentration of caesium (Pierce and Glascock 2015, Figure 6) and low concentration of rubidium (Figure 7-10).

7.7.1 Associations between Group 4, Ceramic Typology, and Sites

This group contains all the samples of the eastern, fine-grained, red-slipped Payil (including Palmul, the incised Payil), with one exception from Culubá (#349 in Table 8-18). Conversely, Group 4 only contains samples classified as Payil (or Palmul) with one exception from Tekit (#455 in Table 8-18). Of the 11 samples in this group (Figure 7-5), five samples were found at Chac Mool, five at Mayapán, and one at Tekit (which is not red-slipped but unslipped).

7.7.2 Sources of Group 4 Vessels

Raw materials from the Belize region matched the chemical composition of Group 4. Previous research (Cecil 2012) has shown that Payil, or at least some of Payil, was manufactured using clays from Caye Coco (located in the vicinity of Laguna de On, Figure 7-4). This is demonstrated by sixteen clays collected by Cecil (2012) from around Caye Coco that matched the chemical compositions of a group of red-slipped Payil samples found at Caye Coco and Laguna de On (Cecil 2012, p.23).

The chemical data for Group 4 were compared to a larger dataset (n=1335) kindly provided to this study by Dr. Leslie Cecil (Pierce and Glascock 2015, p.21). Within the larger dataset, there are 34 Payil samples collected from Laguna de On, Belize (Figure 7-4). A comparison based on discriminant analysis fitted 18 of the 34 Cecil's Payil samples within Group 4 (Figure 7-12).

Mahalanobis distance calculation, using both the entire MURR and Cecil's data, was used to test Group 4 against all samples. It found that there are 42 samples in the combined dataset (all regional MURR and Cecil's data) that strongly fit within Group 4. Of these 42 samples, 28 came from Laguna de On (Pierce and Glascock 2015, p.21). The significance of this is higher because Laguna de On comprises only 4.3 per cent of the MURR dataset for the region (Pierce and Glascock 2015, p.21) which shows a strong association between Group 4 and Laguna de On.

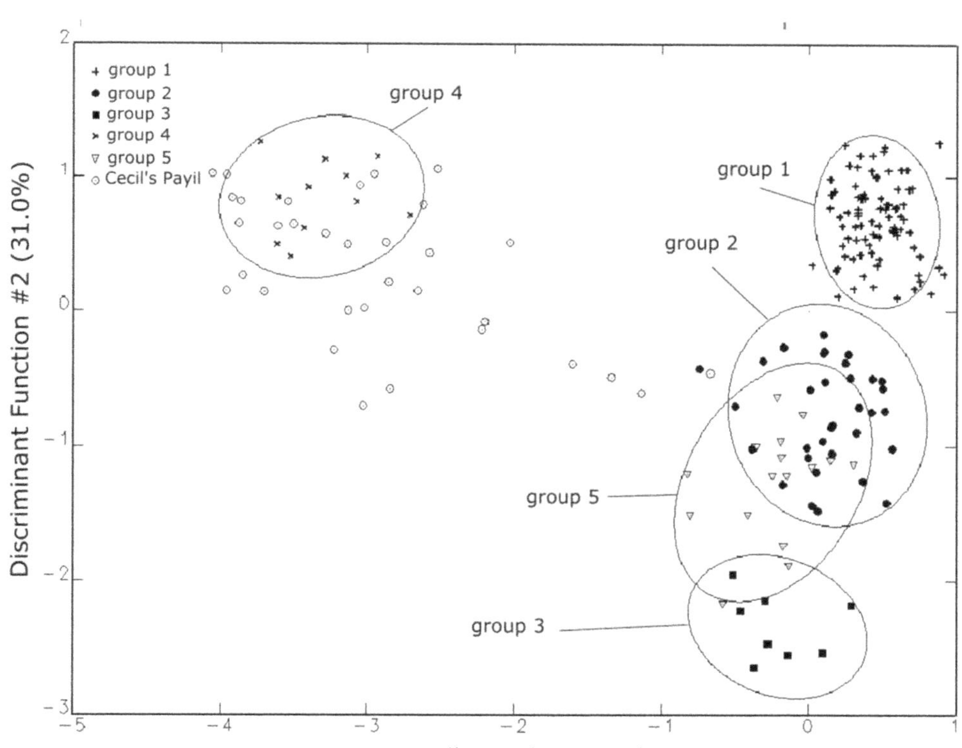

Figure 7-12. Comparison with Cecil's Laguna de On data (Pierce and Glascock 2015, Figure 11)

From these combined lines of evidence, it can be said with a good measure of confidence that Payil samples found at Mayapán (Group 4) are imports from northern Belize, likely from around the Laguna de On region. In the next chapter, petrographic analysis of these samples is used to confirm this result.

7.8 Chemical Group 5 (n=16)

Chemical Group 5 overlaps with Groups 2 and 3 when considering principal components #1 and #2. It is clearly differentiated from Group 2 and 3 in discriminant function #3. In addition, higher concentrations of Iron (Fe) distinguish Group 5 from the rest (Figure 7-11). Other elemental bivariate plots show that Group 5 is relatively high in caesium (Cs) (Pierce and Glascock 2015, Figure 6), aluminium (Figure 7-7), manganese (Mn) (Figure 7-13), and chromium (Cr) (Pierce and Glascock 2015, Figure 4). On the other hand, this group is low in calcium (Ca) concentrations when compared to groups 1 and 2 (Figure 7-11).

7.8.1 Associations between Group 5, Ceramic Typology, and Sites

In regard to the ceramic varieties associated with Group 5, only two of the sixteen samples within Group 5 are red-slipped. Fourteen of sixteen samples within this fabric were taken from Yacman, Navula, and Xcanchakan vessels.

7.8.2 Source of Group 5 Vessels

Vessels with Group 5 composition were found at most north-central sites, with the notable exceptions of Mayapán and Mama. Regarding the eastern sites, they were found at Cobá and Chac Mool. Ten of the samples have Santa Rita Corozal in Belize as most common Euclidian match (Table 7-5).

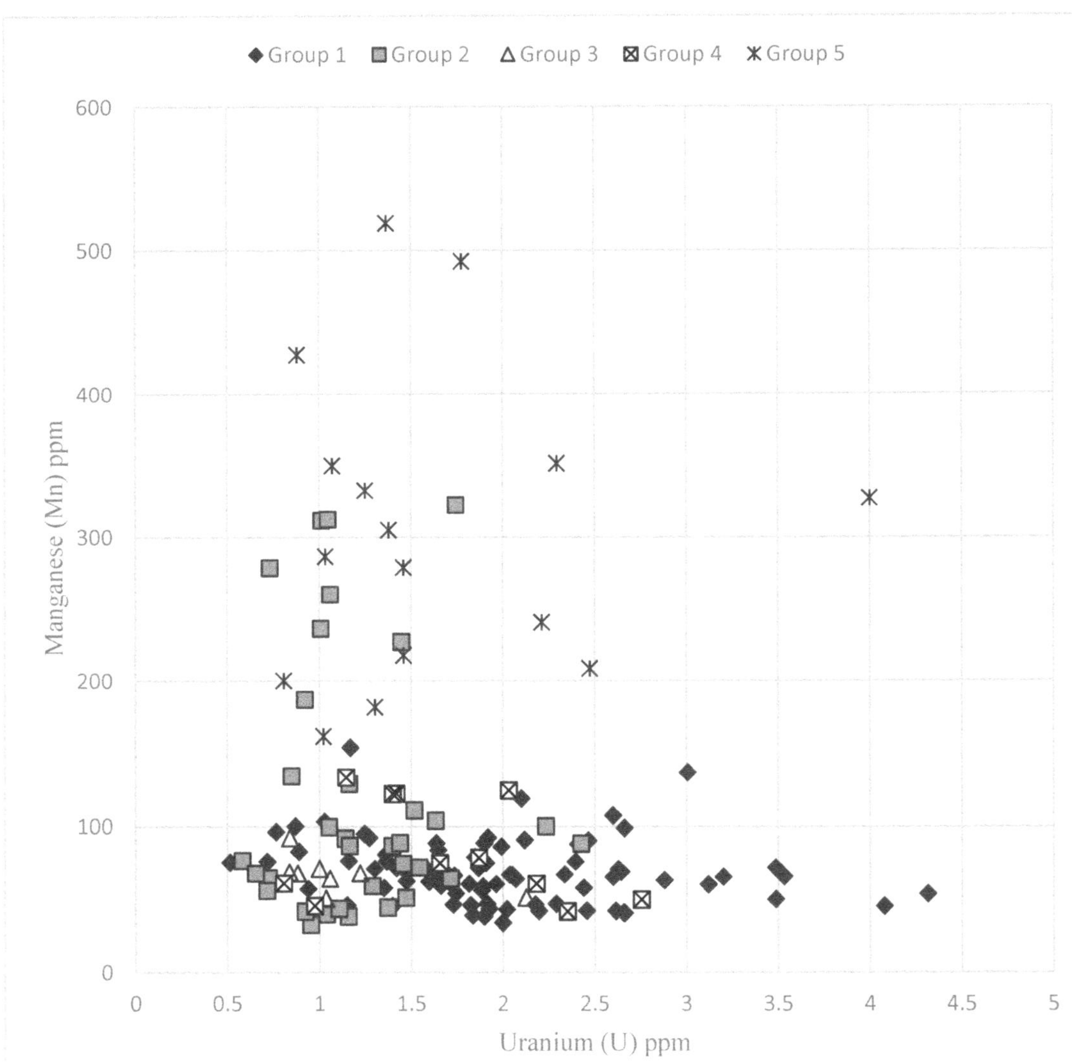

Figure 7-13. Uranium (U) against Manganese (Mn)

Table 7-5. Group 5 Most Common Euclidian Match

Site Name	Ch. Grp	Most common Euclidian match	Variety
Chac Mool	5	Ticul	Xcanchakan
Chac Mool	5	Santa Rita Corozal	Navula
Chac Mool	5	Ticul	Yacman
Chac Mool	5	Ticul	Xcanchakan
Coba	5	Tepakan	Cancun
Teabo	5	various	Yacman
Teabo	5	Santa Rita Corozal	Yacman
Tecoh	5	Santa Rita Corozal	Navula
Tecoh	5	Santa Rita Corozal	Navula
Tekit	5	Santa Rita Corozal	Mama
Tekit	5	various	Yacman
Tekit	5	Santa Rita Corozal	Yacman
Telchaquillo	5	Santa Rita Corozal	Navula
Telchaquillo	5	Santa Rita Corozal	Yacman
Tepich	5	Santa Rita Corozal	Xcanchakan
Tipikal	5	Santa Rita Corozal	Yacman

7.9 Conclusions

As was mentioned at the start of this chapter, the aim of the chemical analysis is to characterise the pottery samples into compositional groups that potentially reflect differences in materials sources, or techniques applied, or both. Raw materials were available from around the north-central sites. When raw materials are available and provide distinctive chemical fingerprints, provenance can potentially be determined. In this chapter, a variety of multivariate statistical analyses was used to characterise the samples. Possible zones of raw materials procurement were also proposed.

Thanks to the variety of methods used the chemical analysis was successful in characterising the samples into five chemical groups that may represent different sources or different ways of making the pottery (techniques). Groups 3, 4, and 5 can be clearly distinguished from each other and the rest using principal components and discriminant function analyses. Group 1 and Group 2 are very similar in that they are depleted of many minerals. They can be differentiated from each other by the degree of depletion of some minerals that is higher in Group 1. Contextual divisions were also identified that strengthen the validity of the groups.

Regarding provenance determination, Group 1, Group 2, and Group 5 comprise the bulk of samples from the north-central sites, including Mayapán. Samples within each one of these groups were found at several sites with no division of chemical composition by site. Zones of raw materials procurement for Group 1 samples could not be determined with certainty owing to the spread of the chemical Group 1 throughout the sites. Arguments were presented for the production of Group 1 pottery samples found at Mayapán at or near that site. The overwhelming majority of samples from Mayapán (after Payil samples are removed) belong to Group 1, and the closest Euclidian match for Mayapán samples in this study is Mayapán. In addition, a marl collected near Mayapán has chemical composition within Group 1, matching Mayapán pottery.

All Group 3 samples (with one exception) were found at Culubá. All Group 4 (with one exception) are Payil samples (fine-grained thought to be elite items). Of the 11 Payil samples in this group (Figure 7-5), five samples were found at Chac Mool and five at Mayapán. Their chemical compositions match the chemical composition of previously analysed clays and pottery from Caye Coco and Laguna de On in Belize (Cecil 2012). Fourteen out of 16 samples in Group 5 are coarse Navula, Yacman, or Xcanchakan from the north-central area, with 10 out of 16 samples in Group 5 pointing to Santa Rita Corozal as the closest Euclidian distance match.

The chemical groups have been useful in characterising the samples. However, Groups 1, 2 and 5 comprise most of the samples and cross cut several sites. Therefore, the groups do not provide the level of detail needed to address the hypotheses and research questions (section 1.3). In the next chapter, another dimension is added to this study with the use of petrographic analysis of thin sections. This analysis focuses on potters' choices related to the selections of raw materials and the processing and construction of the pottery clay, allowing the characterisation of the fabrics based on the techniques or processing applied.

Chapter 8

RESULTS FROM PETROGRAPHIC THIN-SECTION ANALYSIS

This chapter presents the results of petrographic thin-section analysis. The results of the hand-specimen analysis pointed to coarse divisions in the pottery samples mainly based on the inclusions in the fabrics and ceramic types. The inclusions, however, could not be identified with certainty using this type of analysis. The chemical analysis provided significant information about the main chemical divisions in the north-central and the three eastern sites. The chemical groups from the north-central sites, however, are distributed across several sites and are not distinctive of these locations. Hand-specimen and chemical analyses provided important new information about raw material selection and processing. However, they have not provided the level of detail in the variability of the data needed to test the hypotheses proposed in this study (see section 5.5 with hypotheses).

Petrographic analysis using the polarising microscope adds a new dimension to the investigation of the variability in the pottery by focusing on the examination of attributes related to the selections of raw materials and the processing and construction of the pottery clay. Even in a mineralogical and chemically uniform environment, potters may have applied techniques that vary from one group to another. When these variations are repetitive and follow a pattern, a technological group is uncovered. Different technological groups reflect variations in the techniques (way of doing things).

In subsequent chapters, these groups are examined in combination with the information gathered during hand-specimen and chemical analyses of raw material and pottery samples to draw inferences about ceramic techniques for the construction of the vessels and aspects of the organisation of production and exchange. Table 8-1 contains the number of samples included in thin-section analysis within sites and ceramic varieties.

Table 8-1. Prepared Thin Sections by Site and Type of Vessel

	SLIPPED				**UNSLIPPED**				
	MEDIUM and COARSE			**FINE**	**COARSE**				
	MAMA /CANCUN (red, jars and cajetes)		**XCAN CHAKAN (jars)**	**PAYIL (red, bowls and cajetes)**	**NAVULA (plain, jars and cajetes)**	**YACMAN (striated, open-mouthed jars)**	**VISTA ALEGRE (eastern, jar)**	**CHEN MUL (censer)**	**Total**
North-Central									
Tepich	7		1		6	6			20
Tecoh	7				7	5			19
Telchaquillo	7		3		8	5			23
Mayapán	22 (**)		2	5	17	14		1	61
Tekit	5				4	5			14
Mama	11				4	3			18
Tipikal	7				5	10			22
Teabo	4				2	5			11
Eastern									
Coba		6			3				9
Culubá	7	3		1		2	1		14
Chac Mool	1		4	5 (*)	4	4			18
Total	78	9	10	11	60	59	1	1	229

(*) includes Palmul (Payil incised) samples; includes one Tecoh variety sample.

The chapter is structured in three major sections. The first section describes the main types of inclusions found. The second section describes the fabric classes identified in the pottery samples, as well as the associations between these fabrics and contextual information such as sites and ceramic groups or varieties. The last main section describes the petrographic analysis of local raw materials collected.

8.1 Types of Inclusions Found

This section deals with the description of the main particles observed in the thin sections. When observing a thin section using the polarising microscope, two main components can be distinguished. One is a very fine mass, which is the clay portion that used to be plastic before firing, and the second component comprises particles (also called in many other ways, such as aplastics, grains, or inclusions) that appear embedded in this fine mass. The inclusions present in a thin section could be of many types including minerals (which are usually found), clumps of organic materials, shells, or grass.

Calcite and dolomite were the main minerals found. These are both carbonates with similar crystalline structures and optical characteristics. This similarity usually makes their distinction difficult or unreliable. Under the microscope, differentiation was achieved by tinting one-half of the each thin section with the dye alizarin red. This dyes calcite a pink to red colour but not dolomite (Adams and MacKenzie 2001, p.7; Nesse 1986, p.144). In Figure 8-1 (#441, Table 8-18), tinting with alizarin distinguishes dolomite (bottom colourless particles) from calcite (two reddish fragments at left and right of central crystal).

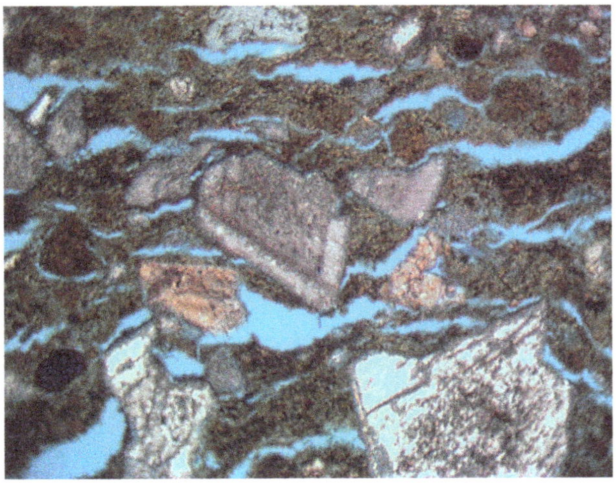

Figure 8-1. Dolomite (colourless or greyish) and calcite (tinted reddish) (#441, pp, width of field 1.1mm)

Calcite and dolomite were present in multiple textural varieties that will be explained in this chapter. Quartz is also present in the thin sections. Quartz in significant amounts was not common, mostly found in Payil sherds from Mayapán and Chac Mool. However, traces of quartz were observed in many thin sections of vessels from the north-central sites.

The detailed description of the main particles found is presented in the rest of this section. Scales or measurements, in particular of crystal and particle size, are used in these descriptions. Their definitions can be found in section 8.2.1.

Most attributes are described as seen under the polarising microscopes. Minerals and other particles, when analysed with the lights of the polarising microscope, exhibit a set of optical characteristics. They are used to identify most particles and characterise the thin section. Although the description and operation of the polarising microscope, as well as the analysis performed with it, are well established in the literature (Philpotts 2003; Nesse 1986), a brief explanation of the main concepts and the most used terms in this research in relation to the polarising microscope are summarised next for ease of reference.

Crystalline materials, based on their crystal structure, could be isotropic or anisotropic. In an *optically isotropic* material, the light travels with the same velocity in all directions, while in an *anisotropic* substance the velocity of light is different in different directions (Nesse 1986, p.10).

These differences occur because the electron clouds of the substance do not have the same vibrations on all directions. When light interacts with these clouds, the velocity of light varies. In contrast to a binocular microscope, the polarising microscope has polarising prisms or filters (Nesse 1986, p.10). One is located under and one above the stage, constraining the light to vibrate in one direction. The upper polarising filter can be moved in and out. In the polarising microscope, the light that exits the lower polar and enters the crystal material is polarised in one direction (*plane polarised or pp*). If plane polarised light enters an anisotropic crystal, it resolves into two light components. If the upper polarising filter is inserted (*crossed polarised or xp*), the filter combines the two components generating *interference colour* (Philpotts 2003, p.9).

These colours are related to attributes that characterise the crystals such as *birefringence,* or the difference between the maximum and minimum refractive index (Philpotts 2003, p.7), also expressed as the range of interference colours observed. The interference colours are dependent on the thickness of the thin section, but this is usually constant and set to 0.030 mm for all the thin sections. If under crossed polars the thin section is rotated, there are positions in which the light is parallel to the polar and no light is transmitted, and the crystals are said to be in *extinction* (Philpotts 2003). If under plane polarised light the stage is rotated and an anisotropic mineral changes colour, it said that the mineral is *pleochroic.*

For the purpose of describing the types found, particles were broadly divided based on their mineralogy into calcite, dolomite, and non-calcareous particles. Table 8-2 contains the list of the observed types of inclusions within these three categories. They are described in the next sections.

Table 8-2. Main Inclusions in Thin Sections

Main Mineral or Inclusion Type	Inclusion Type	Inclusion Code
Calcite	Micritic	CM
	Dark Micrite	CG
	Sparry or sparite	CS
	Skeletal Remain	CH
	Single Crystal	CX
Dolomite		
	Dolospar (sparry)	DS
	Single Crystal	DX
Non-calcareous	Quartz single crystal	QX
	Argillaceous fragments with crystals	MX
	Grog	GROG

8.1.1 *Calcite*

The different textural varieties of calcite found in the samples are described in this section as they appear under the polarising microscope. Some of the characteristics of these textural varieties were introduced and described in chapter six.

8.1.1.1 Micrite

Under plane polarised light, micrite particles (more in section 6.2.1) are sub-translucent (light going through) with a very pale grey to light brown colour. Given micrite crystal size of less than 0.005 mm (Adams and MacKenzie 2001, p.6) that is smaller than the thickness of a thin section (0.030 mm), many crystals are packed vertically through the thin section and the individual crystals cannot be resolved under the polarising microscope. The particle to the top right in Figure 8-2 is translucent micrite having, in this case, a speckled appearance.

8.1.1.2 Dark Micrite

Under the polarising microscope, grey micrite (more in section 6.2.3) appeared opaque or semi-opaque with very limited light passing through, differing from sub-translucent micrite. Similar to micrite, the individual crystals cannot be resolved. The reasons for the dark colour and high levels of opaqueness are varied. One reason could be higher levels of micritisation, with very small crystal sizes. Another possible reason is the presence of organic or mineral matter affecting the colour and translucency of micrite. Figure 8-2, centre, shows a semi-opaque (xp) particle composed of dark grey micrite.

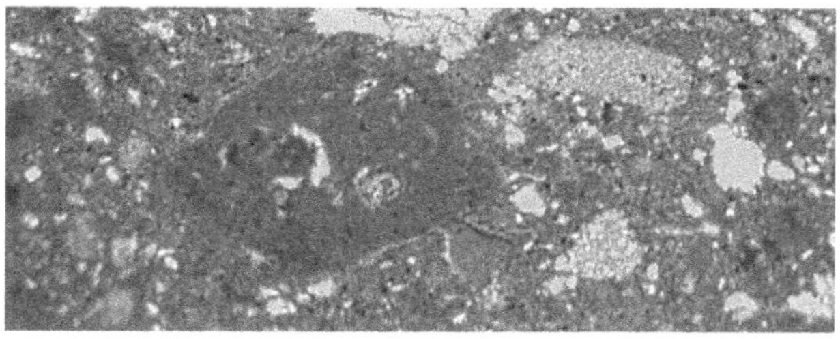

Figure 8-2. Centre: dark micrite or semi-opaque; top, right: translucent micrite (#339, xp; width of field (wof) 1.1 mm).

8.1.1.3 Sparite

As can be recalled (section 6.2.2) the term sparite, spar, or sparry calcite is used for crystals usually over 0.01 mm (10 microns) (Adams and MacKenzie 2001, p.6; Folk 1980, p.156; Greensmith 1989, p.106).

Under the polarising microscope, sparry calcite was usually observed forming bundles of subhedral (subhedral means with a few crystal faces [Philpotts 2003, p.115]) to anhedral (without crystal faces [Philpotts 2003, p.115]) crystals in which the size of each crystal usually ranged between 0.02 mm and 0.04 mm (Figure 8-3a), which is considered finely crystalline sparite (Folk 1980, p.165). Medium crystalline (Folk 1980, p.165) or reaching 0.06 to 0.1 mm and more (Figure 8-3b) was also observed to a lesser degree.

8.1.1.4 Single Crystals

The term single crystal is used to refer to discrete individual crystals and differentiate them from grains composed of an aggregation of crystals, such as the cases mentioned under sparry calcite, or sparite. Figure 8-4 shows single calcite crystals embedded within the matrix.

8.1.1.5 Skeletal Remains

Skeletal remains were easily identified in thin section when part of the original organism could still be discerned, even if it was only a cast or a mould of the original organism. Skeletal remains in various stages of micritisation and sparite infilling were observed in the samples, showing relicts of the original skeletal structure. Figure 8-5 shows small fragments of fossil shells in which the original carbonate (probable aragonite) skeleton was micritised while preserving some of the original animal structures.

A specific type of micritisation that may occur in shells was observed in some of the samples. Some fragments presented a texture that has been called homogeneous calcite (Adams and MacKenzie 2001:37). In homogeneous calcite, the shell structures were micritised

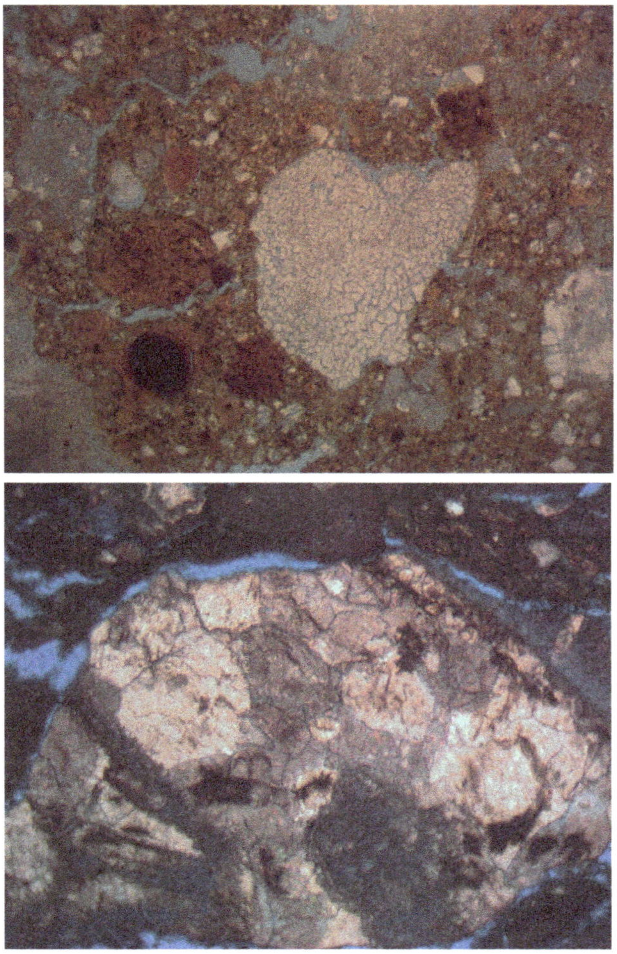

Figure 8-3. Two sparry particles. One (top) has finely (# 165, pp), and the other, (bottom), medium-crystalline (#277, xp) calcite crystals (width of field 1.1 mm).

Figure 8-4. Single calcite crystals, some showing characteristic pastel interference colours (#204, xp, width of field 2.82 mm)

in such a way that the minute crystals have similar orientation (Adams and MacKenzie 2001:37). The whole particle, then, presents interference colours and zoned (moving slowly and not all at once) or shadow extinction, in contrast to regular micrite. It looks almost clear in plane polarised light. Fragments of 'homogeneous' calcite shells can be confused with single calcite crystals.

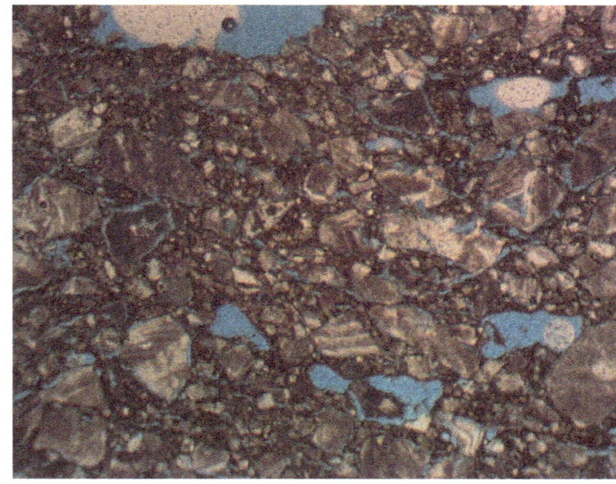

Figure 8-5. Fossil shell fragments showing micritisation (#450; xp; width of field 2.82 mm)

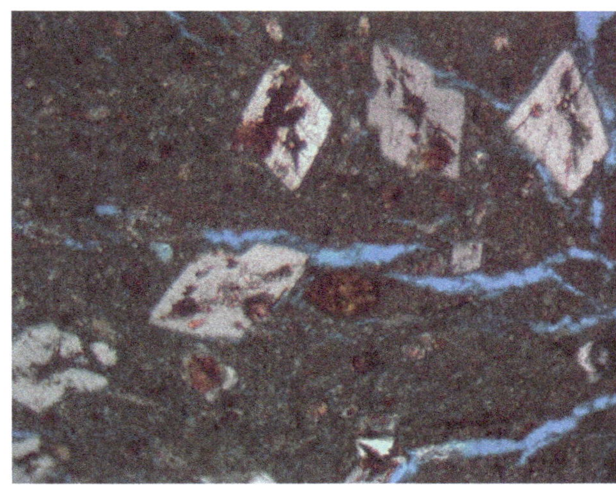

Figure 8-6. Dolomite as single crystals (#339, xp, width of field 1.1 mm)

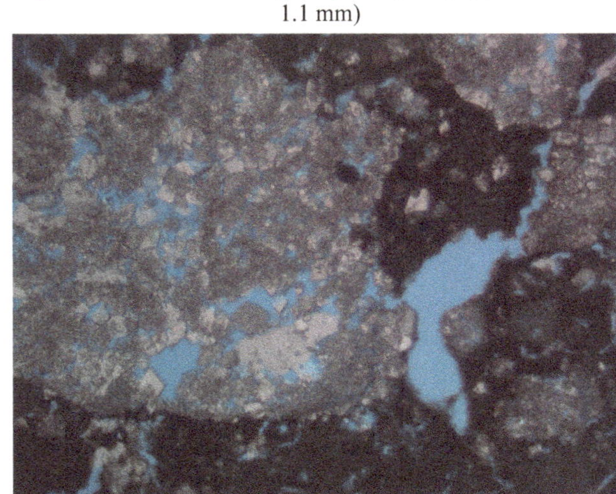

Figure 8-7. Dolospar particle showing rhombohedral dolomite crystals and grain alteration by micrite (#98, xp, width of field 1.1 mm)

8.1.2 Dolomite

Calcite and dolomite are many times found together in limestone, dolomitic limestone, and the rock dolomite (Nesse 1986:144), in this case referring to the rock as opposed to the mineral. These two minerals are both carbonates with similar crystalline structures and optical characteristics in hand specimen and under the polarising microscope. This similarity usually makes their distinction difficult or unreliable in both analytical methods.

Under the microscope, differentiation was achieved by tinting one-half of the each thin section with the dye alizarin red. This dyes calcite a pink to red colour but not dolomite (Adams and MacKenzie 2001, p.7; Nesse 1986, p.144). In Figure 8-1 (#441, Table 8-18), tinting with alizarin distinguishes dolomite (colourless particles at the bottom) from calcite (two reddish fragments at left and right of central crystal).

Dolomite ($CaMg(CO_3)_2$) may form from calcite that undergoes replacement of some of its calcium by magnesium (Adams and MacKenzie 2001, p.132). It also may occur from direct precipitation (Folk1 980, p.98) and as cement or vein filling (Folk 1980, p.98). The magnesium in dolomite may be replaced by manganese or iron, and if iron-rich it is known as ankerite (Adams and MacKenzie 2001, p.132), forming a continuum from dolomite $CaMg(CO_3)_2$ to ferro dolomite $CaFe(CO_3)_2$ (Folk 1980, p.97).

According to Nesse (1986, p.145), ankerite is less common than dolomite. They both have similar optical characteristics under the polarising microscope and, usually, rhombohedral form. Ankerite is often brownish due to iron staining (Folk 1980, p.98).

The carbonate identified in this study, using the polarising microscope, as dolomite, could be ankerite. One sample from Mayapán (#276, Table 8-18) was analysed using X-ray diffraction (XRD). The XRD analysis was performed by PMET Laboratories (Shannon 2016, included in Appendix C for easy of reference). In this sample, petrography had identified dolomite as the main inclusion, while X-ray XRD identified ankerite instead. In this study, minerals having the characteristics of dolomite (or ankerite) are referred to as dolomite, given that they could not be distinguished with the polarising microscope.

Dolomite crystals usually occur as euhedral rhombohedra, contrasting with calcite crystals, and usually do not show the characteristic calcite pastel colours under polarised light. Figure 8-6 shows single crystals of dolomite. Dolomite is usually colourless, cloudy, and may have red or brown iron alteration (Nesse 1986:144) starting at the centre of the crystal (Figure 8-6).

In the samples, dolomite was found as clusters of sparry dolomite and as single crystals. It was not found forming microcrystalline grains. In this study, dolospar refers to bundles of sparry dolomite to indicate the presence of dolomite and differentiate it from calcite spar or sparite.

Figure 8-7 shows one particle of dolospar composed of the clustering of dolomite crystals. Many of the dolospar clusters observed have a micritic alteration, to various degrees (Figure 8-6). In hand specimen, these particles could have been taken for micrite.

Although dolomite crystals were most commonly euhedral (with crystal faces) with their characteristic rhombohedral shape, subhedral crystals that have lost their shape were also found. They can be observed in Figure 8-8.

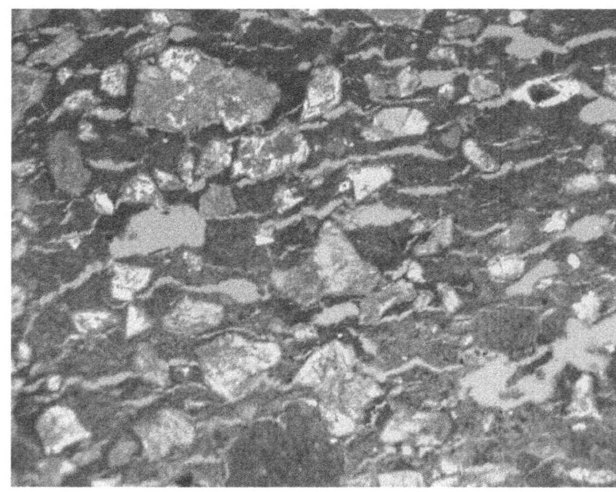

Figure 8-8. Dolomite as single, shapeless crystals (#441, xp, width of field 2.8 mm)

8.1.3 *Non-calcareous Particles*

The main non-calcareous particles observed were quartz, grog, and dark subangular-to-angular argillaceous particles with embedded crystals. In addition to these main or diagnostic inclusions, discrete defined areas that can be distinguished from adjacent areas by an increase or decrease in concentration were also observed (Kemp 1985:32). These units are known as textural concentration features (tcf). In pottery, they may be the result of a variety of processes including, clay mixing, faecal pellets, or concretions (Whitbread 1995:386). Amorphous concentration features (acf) are concentration features that appear opaque under crossed polarised light (Whitbread 1995:386). One example of an amorphous feature is organic matter that may be present. The description of the main non-calcareous inclusions is found next.

8.1.3.1 *Quartz*

Traces of quartz as very fine-to-fine single crystals (0.0625-0.25 mm) were observed in many of the thin sections. As was mentioned earlier, quartz in significant amounts was not common, primarily found only at Mayapán and Chac Mool. Figure 8-9 shows subrounded to subangular quartz grains.

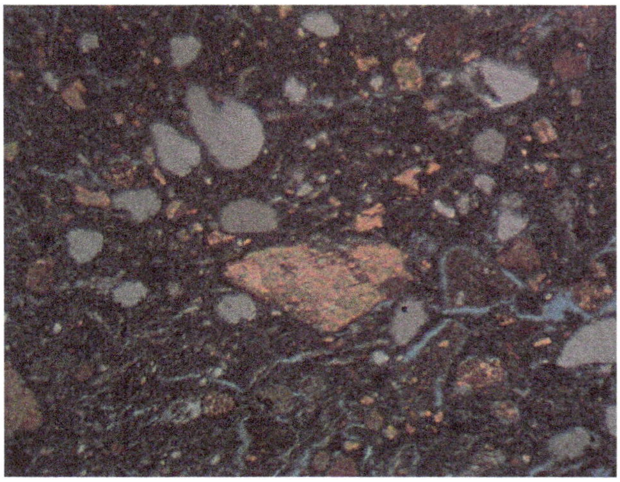

Figure 8-9. Quartz (white) and calcite (tinted pink with alizarin red (#134, xp, width of field 1.1 mm)

8.1.3.2 Grog

Grog (sherd inclusions) was found in four samples (Figure 8-10). There are two types of grog. One consists of unslipped sherds with embedded fine-grained single crystals of calcite (#275, #433, #435, #445), of which one instance can be seen in Figure 8-10(a). The second type consists of fine-grained red-slipped sherds (#445), as seen in Figure 8-10(b).

8.1.3.3 Argillaceous Particles with Embedded Crystals

Angular to subrounded argillaceous particles (Figure 8-10, bottom right) were found in a few sherds. In hand specimens and under the microscope they are mostly dark brown colour. The polarising microscope allowed observing very fine-to-fine calcite crystals embedded in the particles.

These particles are similar to others identified as grog (see previous paragraphs), and they co-occur: grog has been found only in samples having argillaceous inclusions. Therefore, it is possible that argillaceous particles are the result of crushing sherds such as that shown in Figure 8-10 (a).

8.2 Petrographic Fabric Classes

Thin-section analysis was used to refine hand-specimen fabrics by uncovering fine-grained distinctions and reliably identify minerals. For instance, although particles composed of sparry crystals were observed in hand specimens, their mineralogy could not be identified. Thin-section analysis identified them as clusters of either dolomite or calcite crystals.

In addition, thin-section analysis provided much-needed information about the structure of voids and the very fine portions of the clay matrix.

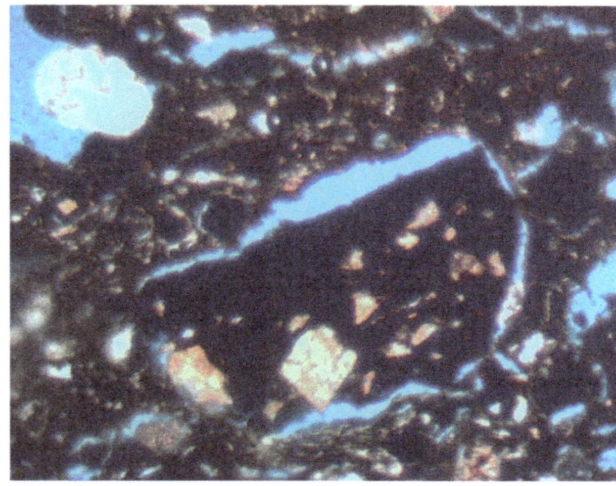

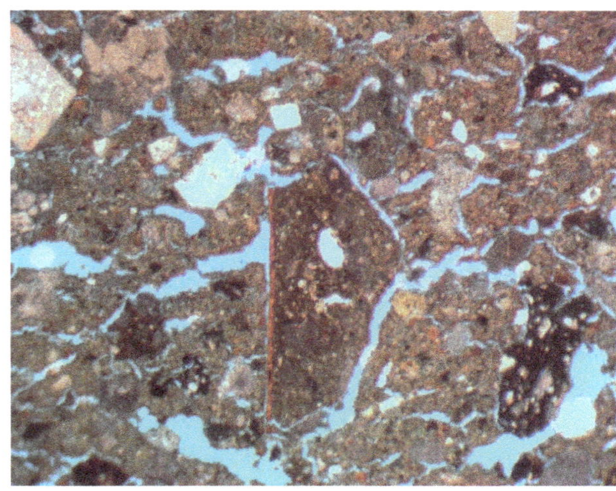

Figure 8-10. (a:top): unslipped sherd with silt size calcite (#275, xp, width of field 1.1 mm); (b:bottom) centre: red-slipped sherd (#445, xp, width of view 2.82 mm); (b) lower right: argillaceous particle with embedded calcite

Therefore, a hand-specimen fabric group yielded several thin-section fabrics based on the identification of the types of inclusion present and other characteristics, such as micromass structure.

The immediate goal of the petrographic analysis is to characterise the samples into fabric classes, in which the individual members of a fabric share a range of properties (Whitbread 1995:372). Briefly, the petrographic analysis of a thin section is such that

1. The starting point of this analysis was the hand-specimen group (chapter 6) to which the thin section belongs.
2. Second, using the polarising microscope, the inclusions were identified and the broad fabric classes further divided usually based on the most abundant or diagnostic coarse inclusions. Each of these types of fabrics (for example, micritic) comprises one or several fabric classes.
3. Next, a set of characteristics was examined following the method for the characterisation into fabric classes devised by Whitbread (1989; 1995, Appendix 3).

Terms that may be unfamiliar and that will be used to describe the thin section are defined next:

Micromass. The micromass (Whitbread 1995:381) is composed of the clay part plus all the material less than 0.010mm (10 microns).

Fabric. In this research, the term fabric refers to the 'arrangement, size, shape, frequency, and composition of components of the ceramic material' (Whitbread, 1995:368) for the coarse and finer portions, including the clay (<0.002 mm). Clay matrix refers to the finer portion of the ceramics that before firing used to be plastic. In contrast to this term, a fabric description includes the description of attributes such as the presence of crystallites, or the relative spatial arrangements of voids and particles in a thin section. In the context of a petrographic analysis, the term fabric is the preferred term for what can also be called paste.

Microstructure. The description of the microstructure includes the description of the shape, arrangement, and abundance of voids in the clay matrix, the description of the packing and preferred orientation of inclusions, and the colour and optical characteristics of the fired clay (Whitbread 1995). The description of the microstructure and micromass follows the approach proposed by Whitbread (1989; 1995, p.380-381).

8.2.1 Scales and Measurements Used for the Description of Inclusions in Thin Section

The scales and measurements used to describe the inclusions observed in thin sections can be found in the literature (Whitbread 1989; 1995), and they are repeated in this section for ease of reference.

(a) Frequency of particles in the coarse portion of the thin section, in this study, is given in relation to the whole thin section. The whole thin section includes voids, and fine and coarse portions of the thin section. The inclusions' frequency was estimated using charts for the visual estimation of percentages of mineral content (Matthew, Woods, and Oliver 1991) and described using the following labels (after Kemp 1985; cited in Whitbread 1995: 379).

Predominant = > 70 per cent
Dominant = 50 – 70 per cent
Frequent = 30 – 50 per cent
Common = 15 – 30 per cent
Few = 5 – 15 per cent
Very few = 2 – 5 per cent
Rare = 0.5 - 2 per cent
Very rare < 0.5

(b) Particle Size was measured using the microscope reticle and described using Wentworth (1922, Table I) scale for the classification of particle sizes:

Silt = 0.04 – 0.0625mm
Very fine sand = 0.0625 – 0.125 mm
Fine sand = 0.0625– 0.25mm
Medium sand = 0.25 – 0.50mm
Coarse sand = 0.50 – 1.0mm
Very coarse sand = > 1.0mm

(c) Angularity was estimated using the scales in Orton et al. (2008: Appendix): rounded, subrounded, subangular, and angular.

(d) Inclusion boundaries refer to the sharpness of the boundary particle/matrix. This boundary can be described using a scale in Kemp (1985; as cited in Whitbread 1986):

Sharp = knife-edge
Clear = (< 0.06mm)
Diffuse = (> 0.06mm), and
Merging = parts of boundary are missing.

(e) Individual crystal size of carbonate rocks was described using Folk's (1980: 165) scale for carbonate rocks:

cryptocrystalline = < 0.001 mm (1 micron)
aphanocrystalline = 0.001 – 0.0039 mm
very finely crystalline = 0.0039 – 0.0156 mm
finely crystalline = 0.0156 – 0.0625 mm
medium crystalline = 0.0625 – 0.25 mm
coarsely crystalline = 0.25 – 1.00 mm
very coarsely crystalline = 1.00 – 4.00 mm
extremely coarsely crystalline = over 4 mm

It is needed to clarify that with sections (b) and (e) two different components are measured. One carbonate particle or carbonate rock fragment can be composed of a single crystal or of a cluster of crystals, and two measurements may be needed to describe them: one measures the size of the overall particle, using (b), and the other the size of the crystals that may be present within a particle, using (e).

8.2.2 Scales and Measurements Used for the Description of the Microstructure and Micromass in Thin Section

Throughout the rest of this and subsequent chapters, the attributes observed in the micromass, as well as the relations between these attributes (structure), are described using the scales and measurements as defined in this section.

(a) Coarse:fine:voids or c:f:v ratio. c:f:v gives the relative proportions (per cent) of the whole thin section area occupied by the coarse particles (c), fine particles (f), and voids (v), following (Whitbread 1995, p.383-384). The coarse to fine boundary is usually 0.01mm which coincides with the micromass boundary; however, this boundary can be set to a different limit (Whitbread 1995,

p.383), and in this study, it is set at 0.0625 mm. Given that this is the silt to sand particle size (section 8.2.1) boundary, for the purpose of calculating the c:f:v ratio, the fine portion includes not only the micromass but also silt that may be present.

(b) The shape of voids in the thin section was described using the scales in Whitbread (1995: Appendix 3):

planar voids = planar in three dimensions
channels = cylindrical in three dimensions
vughs = large and irregular voids
vesicles = regular in shape with smooth surfaces
massive = a microstructure without or with rare voids

(c) Packing of inclusions is based on three variables: grains spacing, fine to coarse distribution, and concentration of crystallites. The spacing is estimated for grains larger than 0.0625 mm. The scale used is from Whitbread (1995):

close-spaced = grains have point of contact
single-spaced = distance between grains is equal to their diameter
double-spaced = distance between grains is double to their mean diameters
open-spaced = distance is more than double

(d) Colour of the micromass under plane and cross-polarised light were described using Munsell Soil-colour Charts (Munsell 2009).

(e) Optical state gives a measure of the micromass' modification or alteration due to firing (Whitbread 2005:382). It is observed by rotating the thin section under crossed polar light (Whitbread 2005:382). The optical state can be described as (Whitbread 1995:382) optically active when it displays interference colours and extinction, or asoptically inactive if no change exist on the optical properties on rotation.

(f) An optically active micromass can be further described by the appearance and shape of the birefringent zones, such as speckled, crystallitic, striated, strial (Whitbread 2005:383).

(g) Optical density describes the opacity of the inclusions relative to the matrix, with those less opaque having low density, more opaque high density, and no difference, neutral density (Whitbread 1986:81).

(h) Porosity abundance was based on the whole thin-section area. It was estimated for each thin section using abundance charts. The scale for the abundance of particles in section 8.2.1 was used to describe porosity.

8.2.3 The Fabric Classes

This section presents the list of the fabric classes identified. They are summarised in Table 8-3 and described in detail in the following sections. To each fabric class, a name and a code were assigned. A fabric class code was assigned because it is particularly useful in some charts and tables where space is limited. In any other case, the fabric name is used.

The fabric class name is descriptive of the main characteristics of the group, in particular the most abundant or diagnostic inclusions (not necessarily the most abundant) present. Earlier in this chapter, Table 8-2 listed the types, codes, and names of these inclusions. More than one inclusion could be included in a name. In the codes given to the inclusions, the first character indicates the main mineral or inclusion type:

D, dolomite;
C, calcite;
Q, quartz; and
M, argillaceous or mud.

The second character gives more information about the particle, usually indicating the crystal structure or texture:

S, sparry;
M, microcrystalline;
X, one single crystal;
H, skeletal remains or homogeneous calcite.

Another important characteristic included in the name of most fabric classes is the overall grain-size in the thin section: coarse (c), medium (m), and fine (f) (see section 6.1). A code for most common colour taken (W=white, B=light reddish brown; and R=red) after refiring (from hand-specimen analysis) is also included. This code is omitted when the fabric does not have a pattern as regard to the refired colour.

The following sections describe the different types of fabrics and fabric classes identified. The sections are presented in terms of the principal types of fabrics, which are mostly related to the diagnostic or main inclusion(s) found in thin section and that groups together multiple fabric classes, for instance, fabrics in which dolospar dominates.

Within each of the main types of inclusions, the individual fabric classes are identified and briefly described. The detailed description of the petrographic characteristics of each fabric can be found in Appendix E. Within each main type of inclusion, and after the descriptions of the fabric classes, the associations of these fabrics with sites and ceramic classes are discussed. Table 8-3 includes the main type of inclusion that groups a set of fabrics, the list of fabric classes, a fabric code mainly for concise displaying in charts and tables, and a very short description of the fabric.

Table 8-3. Petrographic Groups and Fabric Classes

Fabric Class Name	Fabric Code	Main Inclusion and Diagnostic Characteristics
DOLOSPAR (finely crystalline)		
Dolospar-W-Coarse	DS-W-c	finely crystalline sparry dolomite, coarse particle size, silty dolomite matrix, and very porous micromass.
Dolospar-W-Medium	DS-W-m	finely crystalline sparry dolomite, medium particle size, silty dolomite matrix, and usually few voids.
Dolospar-W-Hard	DS-W-mc-h	variant of Dolospar-W-Medium, differing on the frequency of dark brown organic amorphous material and dark micritic inclusions.
Dolospar-B-Coarse	DS-B-c	variant of Dolospar-W-Coarse, differing on its reddish brown colour (xp) and higher frequency of inclusions.
SPARITE (finely crystalline)		
Sparite-W/B-Coarse/Medium	CS-W/B-cm	coarse-to-medium particles made of finely crystalline sparite, usually 0.020 – 0.030 mm; lack of homogeneity across the group in regard to other petrographic attributes.
MICRITE (micro-crystalline)		
Micrite-W-Medium	CM-W-m	medium (mode) micrite grains, usually rare silt, lack of homogeneity across the group in regard to other petrographic attributes.
DARK MICRITE (micro-crystalline)		
DarkMicrite-Micrite-Medium	CG-CM-W-m	medium-grained, opaque or sub-opaque, dark micrite, with low porosity and silt.
DarkMicrite-Micrite-Coarse	CG-CM-W-c	variant of DarkMicrite-Micrite-Medium, with opaque or sub-opaque dark micrite and low silt, differing on its highly porous micromass and coarsely grained fabric.
CALCITE (larger xtals)		
Calcite-Micrite-01-Coarse	CX-CM-01-c	coarse (mode), sub-angular to angular discrete crystals of calcite, with a moderately compacted micromass and very few voids.
Calcite-Micrite-02-Coarse	CX-CM-02-c	variety of Calcite-Micrite-01-Coarse differing in the much higher porosity and large channels.
Sparite-m-W/B-Coarse	CSm-W/B-c	coarse (mode) particles made of mosaics of sparry calcite (sparite) in which each crystal size is medium-crystalline or larger, usually between 0.06 mm to 0.18 mm, and, second, by a very porous matrix.
Calcite-Quartz-B/R-Fine	CX-QZ-B/R-f	fine calcite and quartz single crystals in a compact matrix with rare pores
Calcite-Grog-B/R-Coarse	MX-GROG-B/R-c	grog and coarse angular to irregularly shaped fragments of very dark brown argillaceous particles with embedded silty-to-very fine calcite crystals that in four samples were identified as unslipped sherd fragments. Usually discrete calcite crystals are present.
DOLOMITE (larger xtals)		
Dolomite-B/W-Medium	DX-B/W-m	fine grained (mode) discrete rhombohedral dolomite.
Dolomite-Calcite-B-Coarse	DX-CX-B-c	coarse (mode), subangular to angular, mostly subhedral to anhedral single crystals of dolomite; calcite crystals may be present.
Dolomite-Calcite-Medium-Hard	DX-CX-R/B-c-h	variant of Dolomite-Calcite-B-Coarse with medium-grained single crystals of dolomite and calcite.
BIOCLASTS		
Bioclast-W-Coarse/Medium	CH-W-cm	medium-to-coarse (mode) micritised fossil shells in a moderately compacted and homogeneous across the group fabric.
Bioclast-B-Coarse	CH-B-c	coarse micritised angular fossil shells of varied sizes in a very porous matrix with long channels and fabric that is inhomogeneous across the group.

(*) Inclusion codes: Table 8-2; Colours: W=white, B=light reddish brown, R=red; Texture: c=coarse, m=medium, f=fine

Table 8-4. Dolospar Fabrics: Summary of Main Petrographic Characteristics

Fabric Class	Main Inclusions	Grain Size (mode)	Porosity	Fired Colour under xp
Dolospar-W-Coarse (DS-W-c)	Coarse portion: dolospar -------------------- Fine (<0.0625 mm) portion: dolomite s-xtal	coarse	high	10YR 7/2 light grey; 5Y 7/2, 6/2 very light olive grey; 5YR 6/4,6 Light & reddish yellow; strong brown 7.5YR 5/6; light yellowish brown 10YR 6/4;
Dolospar–B-Coarse (DS-B-c)	Coarse: dolospar -------------------- Fine (<0.0625 mm) portion: dolomite s-xtals	coarse	high	2.5YR 7/3 Light reddish brown; 5YR 5/4 reddish brown; 7.5 YR 6/8 yellowish red
Dolospar–W-Medium (DS-W-m)	Coarse portion: dolospar; dark lumps/dark micrite micrite -------------------- Fine (<0.0625 mm) portion: dolomite s-xtals	medium	moderate	5Y 7/2, 6/2 very light olive grey; 10 YR 6/2 light brownish grey; 10 YR7/3 very pale brown; 10YR 5/4 yellowish brown; 10YR 5/3 brown;
Dolospar-W-Hard (DS-W-mc-h)	Coarse: dolospar brown lumps -------------------- Fine (<0.0625 mm) portion: dolomite s-xtals	medium to coarse	moderate	10 YR 7/3 very pale brown; 5YR 5/4 reddish brown; 10 YR 5/3 brown

s-xtals: single crystal

8.3 Dolospar Fabric Classes

One of the mayor fabric groups that resulted from hand-specimen analysis contains sherds with sparry grains as the main inclusion. The analysis of thin sections taken from some of these samples revealed that sparry crystals are composed of two types of minerals: dolomite (dolospar) and calcite (sparite).

Some samples containing very fine-grained sparite or dolospar had been misclassified as micrite during hand-specimen analysis and were reclassified as sparite and dolospar. Similarly, many samples classified during hand-specimen analysis as having dark micrite as the main inclusion were found in thin-section analysis to have sparry inclusions as the dominant inclusion type. The high abundance of darker inclusions attached to some samples was most probably a trick played by the dark colour on the eye that led to overestimating darker particles during hand-specimen analysis.

This section deals with the group of fabrics in which petrographic analysis showed that sparry grains were composed of dolomite, or dolospar (DS). Dolospar is present in coarse (> 0.5 mm) or medium-grained (0.25 - 0.5 mm) particles. Within these particles, each crystal size usually is finely crystalline (0.015 to 0.060 mm; see scales and measurements in section 7.2.1). Eighty-three of the 226 thin sections have dolospar as the main inclusion,making up a large part of the total samples, representing 36 per cent of the total.

The dolospar fabrics are very homogeneous with regard to the main types of inclusions (dolospar), the secondary inclusions, and the properties of the micromass. The main division between them resides in the grain size. There are coarse (Dolospar-W-Coarse) and medium grained samples (Dolospar-W-Medium), while fine-grained samples are absent.

Some of the medium-grained samples are defined as a variant fabric (Dolospar-W-Hard) because they present a somewhat darker colour micromass and more areas with, apparently, organic material than most medium-grained samples. These differences may represent a variety of reasons that include differences in firing temperature, firing conditions, or differences in clays. Similarly, some of the coarse samples markedly differ on the micromass reddish colour and frequency of inclusions (Dolospar-B-Coarse).

Table 8-4 summarises the main petrographic characteristics of these fabrics. The next four sections contain short descriptions of the dolospar fabrics (see Appendix E for detailed description). Following these descriptions, the associations observed between these fabric classes and contextual information for the samples, such as find spot and ceramic varieties, are examined.

8.3.1 Dolospar-W-Coarse (DS-W-c) Fabric Class

In thin section, Dolospar-W-Coarse fabric class (DS-W-c) is characterised by its finely crystalline sparry dolomite, coarse particle size, silty dolomite matrix, and very porous micromass. A characteristic of this fabric is its homogeneity across the samples in regard to the composition and texture of the inclusions and colour and optical characteristics of the micromass.

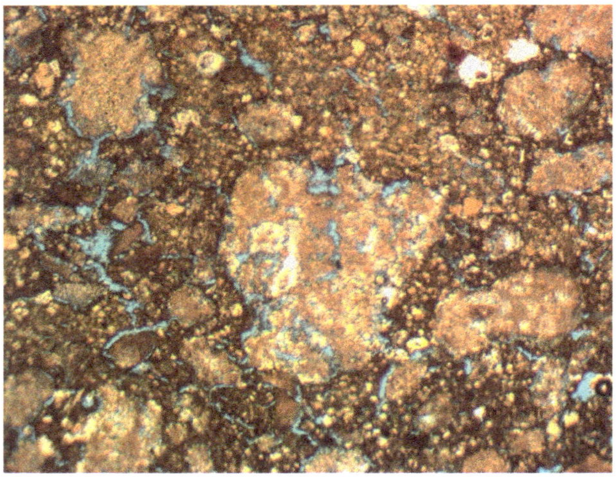

Figure 8-11. Dolospar-W-Coarse, showing coarse dolospar grains and silty dolomite (#405, xp, width of field 2.82 mm)

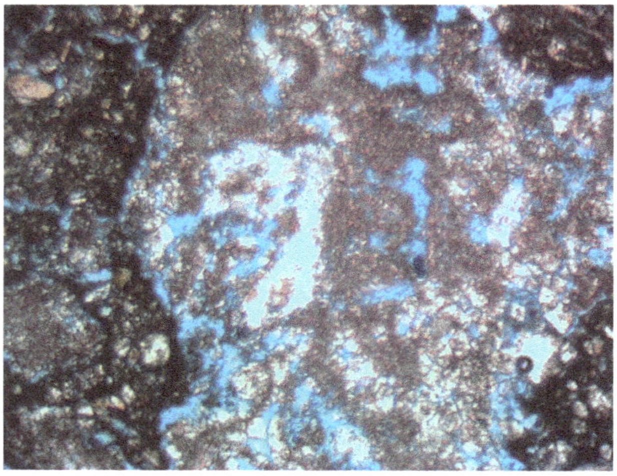

Figure 8-12. Dolospar grain (dolomite clusters) with micritic alteration (#405, pp, width of field 1.1 mm)

Figure 8-11 illustrates a sample having this fabric class showing dolospar grains in a heavily silted matrix, while Figure 8-12 shows a closer look at a dolospar grain with micrite alteration

8.3.2 Dolospar-W-Medium (DS-W-m) Fabric Class

In thin section, Dolospar-W-Medium is characterised by its finely crystalline sparry dolomite, medium particle size, and usually few voids. Most characteristics of this fabric including the types of inclusions and the characteristic of the micromass are homogeneous across the samples. While inhomogeneous bands of darker brown colour, by one wall, may be present (shown in Figure 8-13a), the colour of this fabric within and between thin sections is very homogeneous, mostly of very pale brown (10 YR 7/3) colour (Figure 8-13b).

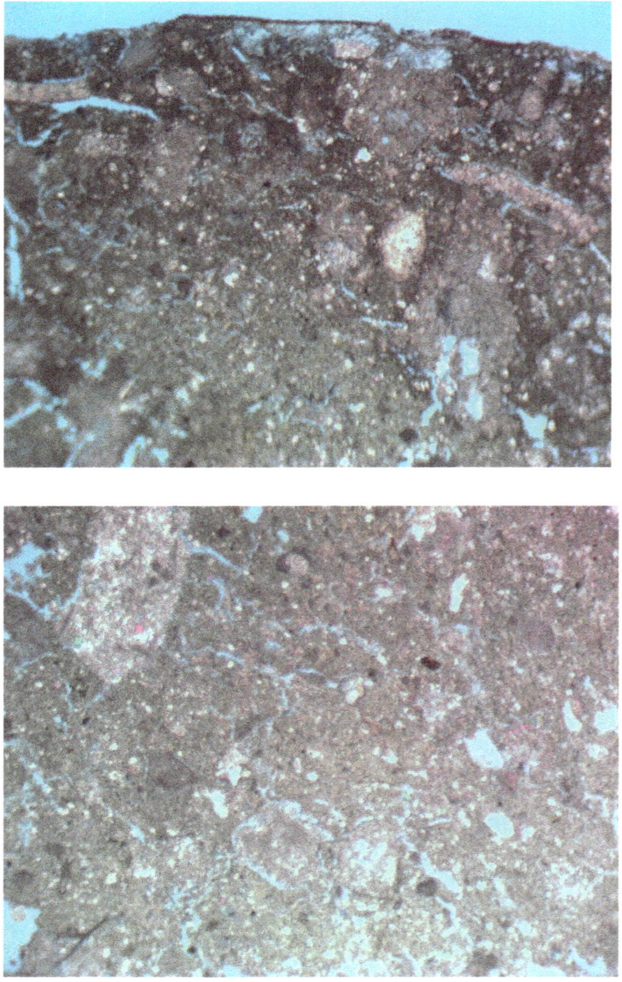

Figure 8-13. Fabric Dolospar-W-Medium: (a: top) mostly darker brown colour, but lighter (see b: bottom) areas are present (#267, xp, width of field 2.82mm)

This fabric differs from Dolospar-W-Coarse (DS-W-c) not only in its grain size but also in the frequency of inclusions and porosity. While the frequency of coarser inclusions in Dolospar-W-Coarse fabric is consistently high (frequent to common), the frequency in Dolospar-W-Medium fabrics could be few. In addition, Dolospar-W-Coarse is more porous than Dolospar-W-Medium and voids separating the particles from the matrix (separation voids) more common than in Dolospar-W-Medium.

8.3.3 Dolospar-W-Hard (DS-W-mc-h) Fabric Class

This fabric is a variant of Dolospar-W-Medium sharing with this fabric the same type and frequency range of dolospar inclusions and largely medium-grained size (but coarse samples were also found). They differ in the darker colour of the micromass, the frequency of dark brown amorphous material that appears to be organic matter, and the frequency of dark micritic inclusions.

The micromass colours are mostly browns (5YR 5/4 reddish brown, 10 YR 5/3 brown, 10YR 6/2 brownish grey) going almost wall to wall, but bands of light brownish grey or light greenish grey areas by one wall may be present, as seen in Figure 8-14b. This difference within the same thin section and the brown micromass colours are indicative of local variation in the firing effects likely due to the presence of organic matter (Whitbread 1995:387), representing amorphous concentrations and depletion features.

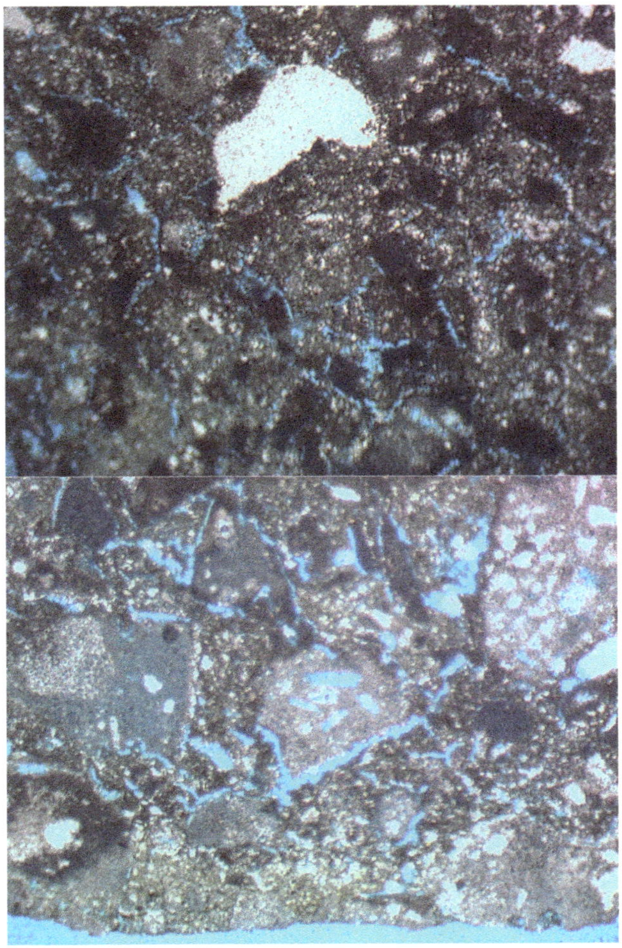

Figure 8-14. Fabric Dolospar-W-Hard. (a: top) brown amorphous concentrations and micromass; (b: bottom) depleted dolospar, brown micromass, light colour band by wall (#342, xp, width of field 2.82mm)

The organic matter in the samples is taking the form of impregnations of the micromass and of nodules that can be observed in Figure 8-14a. In addition, the organic matter also impregnates micrite grains; in areas of brown micromass, the micrite is also medium brown colour, while micrite is translucent in areas with light-colour micromass, reflecting variations in firing temperature and the burning off of the organic material that may be present.

8.3.4 Dolospar-B-Coarse (DS-B-c) Fabric Class

This fabric is a variant of the also coarse Dolospar-W-Coarse. Similar to this fabric, Dolospar-B-Coarse is characterised, under the polarising microscope, by the presence of coarse particles (or grains) made of finely crystalline dolomite (dolospar) as the main inclusion, silty dolomite matrix, and very porous matrix (Figure 8-15).

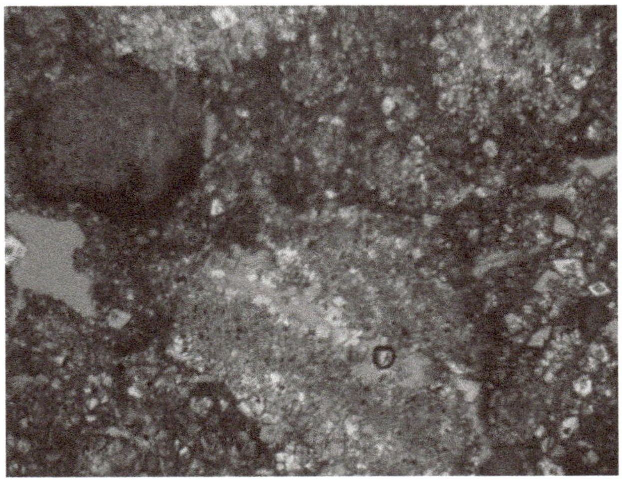

Figure 8-15. Fabric Dolospar-B-Coarse, showing dolospar grains and silty dolomite (#258, xp, width of field 1.1 mm)

Dolospar-B-Coarse differs from Dolospar-W-Coarse fabric in the higher frequency of inclusions and dominant yellow red or reddish brown fired colour (xp), contrasting with the light greenish grey (xp) that dominates in Dolospar-W-Coarse.

8.3.5 Associations between Dolospar Fabrics, Sites, and Ceramic Typology

In this section, the associations between the dolospar samples and other contextual information available for these samples is examined. The chemical variability and the location at which the samples were found, as well as the ceramic variety into which the samples have been classified, are examined as they relate to the fabric classes. The ceramic variety subsumes a series of properties that reflect not only the decorative style of the vessel but also properties that may be related to the intended function of the vessel such as vessel form and presence of slip.

Table 8-5 summarises the contextual information and associations. Each letter (or cell value) in the table represents one thin section characterised within the dolospar fabrics. Each thin section is located within the ceramic typology, fabric class, and site to which corresponds. The two axes of the table are used for the fabric classes and sites, while letter codes are used to indicate the ceramic typology of the sample.

With regard to the chemical composition associations, the dolospar fabrics are largely chemically homogeneous. Chemical variation is indicative of variations in raw materials sources or their processing. Of fifty-six chemically analysed (not all petrographically analysed

Table 8-5. Associations between Dolospar Fabrics and Sites and Ceramic Typology

←----------------------north-central sites--------------------------→←-eastern- sites→

Fabric Texture	Fabric Class	Tepich	Tecoh	Telchaquill o	Mayapán (n=33)	Tekit	Mama	Tipikal	T eabo	Cobá	Chac Mool	Culubá
	DOLOSPAR											
Medium	Dolospar-W-Medium (DS-W-m)		*MM* N	*MMMM* NNN	*MMMMMM* NNNNNN	*MM*		*MM*				
	Dolospar-W-Hard (DS-W-cm-h)				*MMM*		*MMMM* *MMM* Y					
Coarse	Dolospar-B-Coarse (DS-B-c)		YY	NN YY	*M* NN YYYYYYYY			Y				
	Dolospar-W-Coarse (DS-W-c)		*M* N	N Y X	*MMM* NNN YYYY **C**		*M*	*MM* YYYYY YYY				

M = Mama (red-slipped), N = Navula (plain, unslipped), Y = Yacman (striated, unslipped), X = Xcanchakan, and **C** = Chen Mul.

samples are also chemically analysed) dolospar samples, fifty-one (91 per cent) have a composition within Group 1, four are unassigned, and one falls within Group 5.

Concerning dolospar fabrics and associations to sites, a division between the samples from the eastern and north-central sites is evident in Table 8-5. Based on the sampling strategy of this research, sherds from the same type of vessels (for instance, Mama sherds from jars and cajetes) were selected for analysis across all the sites, including north-central and eastern sites. However, dolospar fabrics were not found in any of the samples from the eastern sites (Chac Mool, Culubá, and Cobá), acknowledging that few sites were sampled from the eastern area and that it very well be that dolospar fabrics are present at other eastern sites.

Samples from most north-central sites contain dolospar fabrics, except for Tepich and Teabo, located furthest to the north and the southeast of the area. Based on the samples in this research, and on Table 5-2 with the aerial distances from Mayapán to other north-central sites, the distribution of dolospar fabrics extended 11 km to the north (Tecoh) and 30 km to the south (Tipikal). Dolospar fabrics represent 64 per cent of the analysed thin sections from Mayapán. That is, 39 of 61 thin sections from Mayapán. The proportions are higher if Payil (which is foreign) is not included in the calculation.

Regarding the association between fabrics and ceramic typology, as was mentioned at the beginning of this chapter, the same types of ceramic from both north-central and eastern regions were analysed; however, only Mama, Navula, and Yacman (and a Chen Mul and a Xcanchakan samples) from the north-central sites were found to have dolospar fabrics. Payil and Cancun, both eastern ceramics, were not found to have this type of fabric; nor were Mama, Yacman, or Navula from the eastern sites found with dolospar fabrics.

Among the samples that were found with dolospar fabrics, there is a division between slipped samples (***M***) and the unslipped striated *ollas* (**Y**) into medium and coarse fabrics. Differences in the grain size are associated with other properties of the fabrics including colour and porosity. When found to have a dolospar fabric, the unslipped and striated open-mouthed Yacman jars (**Y**) in 25 out of 26 instances present coarse fabrics. The division may be related to the intended function of the vessels. Yacman jars have an open-mouthed and globular body (*ollas*), and, based on high frequencies of burnt, carbon-coated bottoms (Brown 1999a, p.326) and ethnographic work (Thompson 1958), most likely were used as cooking pots. Slipped vessels, based on ethnographic work (Thompson 1958) involving similar Yucatecan vessels, are not put on the fire. Medium-grained fabrics were preferred by potters using dolospar fabrics for the construction of the red-slipped Mama vessels; however, Mama samples do not show the clear-cut division by grain size observed in the Yacman samples. The unslipped and plain Navula jars and cajetes do not show a grain size pattern.

The Dolospar-W-hard (DS-W-cm-h) fabric is associated with red-slipped Mama samples (10 out of 11 samples). This association, and the fact that this fabric appears to present more organic matter (with firing effects likely due to the presence of organic matter [Whitbread 1995:387]) than other samples with dolospar, may be related to firing at lower temperature or faster in an attempt to preserve the lustre of the slip.

The location at which samples presenting dolospar were found varies with the ceramic typology. Red-slipped Mama and the unslipped, striated Yacman jars were found at many of the north-central sites, whereas the samples of plain unslipped Navula were found, mostly, within the vicinity of Mayapán (Mayapán and Telchaquillo, plus two dolospar sherds at Tecoh). This interpretation may change with more sampling, given that the number of sherds analysed from secondary centres was small, and in particular of the Navula variety.

In summary, among the samples, fragments of sparry dolomite (dolospar) are a common inclusion. Two main dolospar fabrics were defined, differing mainly in the grain size modes: medium and coarse. These fabrics are very homogeneous across sites. Two variant fabrics were also defined for each of these fabrics.

This division by grain size in the dolospar fabrics is highly correlated with the type of vessel. Red-slipped samples (sherds from jars and cajetes) are medium-grained, whereas the unslipped open-mouth striated Yacman samples are coarse, and the plain Navula do not show any pattern.

The samples containing dolospar were found in north-central sites only. At Mayapán, Mama, Yacman, and Navula samples were found with these fabrics comprising two-thirds of the samples from this site. It was also observed that the red-slipped Mama and striated Yacman samples having dolospar inclusions were found at most north-central sites, while Navula samples within this fabric were found at Mayapán and Telchaquillo almost exclusively. However, the number of Navula samples from the lesser sites is small, and this interpretation may change with a larger sample.

8.4 Finely Crystalline Sparite Fabric Classes

This section deals with a group of 20 thin sections in which petrographic analysis found clusters (grains or particles) of finely crystalline sparry calcite, or sparite, as the main inclusion. Samples in which finely crystalline sparite dominate were grouped not in many but in one fabric class, Sparite-W/B-Coarse/Medium.

The main characteristics of this fabric are summarised in Table 8-6. This fabric is not homogeneous in petrographic characteristics but, due to the small number of samples and high variability, it was not possible to separate this fabric further according to the different characteristics.

A more extensive sampling may reveal that many sparite fabrics may have existed. What the samples categorised within this fabric have in common is a basic recipe consisting of the use of medium-to-coarse rock particles compose of finely crystalline sparite as dominant inclusion.

8.4.1 Sparite-W/B-Coarse/Medium (CS-W/B-cm) Fabric Class

This section presents the main characteristics of the sparitic fabric that catches a variety of samples in which sparite dominates (see Appendix E for detailed description). Samples having this fabric class are characterised by the presence of common coarse-to-medium particles composed of clusters of finely crystalline calcite crystals (sparite), with crystal size usually between 0.020 – 0.030 mm (section 8.2.1). Figures 8-16 and 8-17 show examples of the sparite clusters.

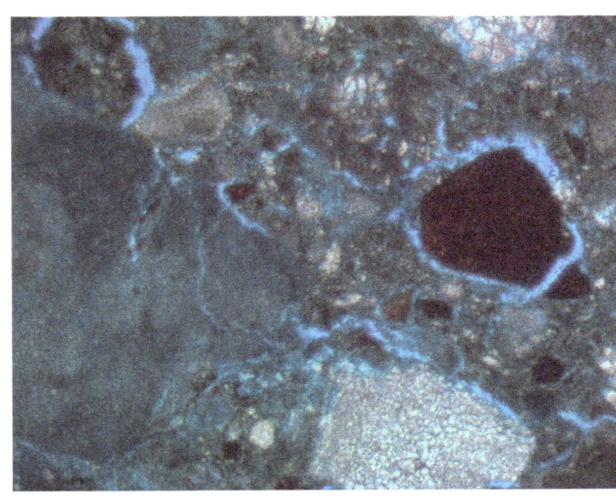

Figure 8-16. Fabric Sparite-W/B-Coarse/Medium (CS-W/B-cm) showing sparite particles (top and bottom) (# 262, xp, width of field 2.82 mm).

Table 8-6. Main Petrographic Characteristics of Finely Crystalline Sparitic Fabrics

Fabric Class	Main Inclusions	Grain Size (mode)	Porosity	Fired Colour (xp)
Sparite-W/B-Coarse/Medium (CS-W/B-cm)	Coarse portion: finely crystalline sparite	Medium to coarse	Varied	light red (7.5 YR 7/6), light-reddish brown (5yr 6/4), dark brown, light greenish grey 7.5YR 7/2 - 7/1, very pale brown (10 YR 8/2)

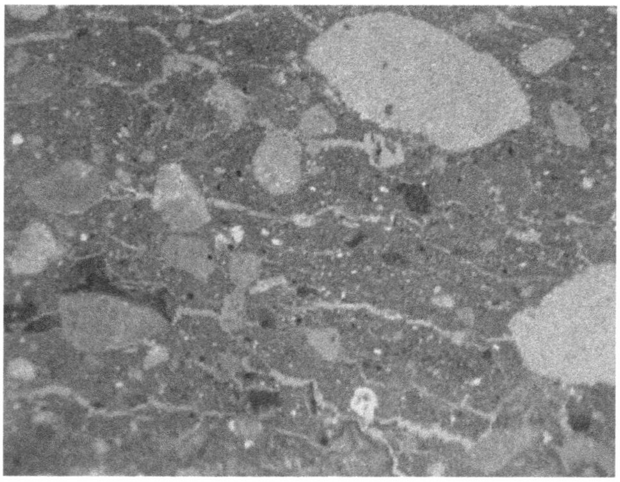

Figure 8-17. Sparite-W/B-Coarse/Medium showing sparite at top and left, and micrite to the right (# 272, pp, wof 2.82 mm)

Other than sharing the main inclusion type, the samples with this fabric are not homogeneous across the group with regard to their petrographic attributes such as porosity, silt frequency, colour of the micromass, or the presence/absence of other coarse (> silt) particles. These coarse particles may include micrite, skeletal remains, red concretions, or discrete dolomite crystals.

8.4.2 Associations between Sparite Fabrics, Sites, and Ceramic Typology

The objective of this section is to examine the associations that exist between the samples with this sparitic fabric and their context related to the find spot and the ceramic type/variety. The main associations are the following:

1. Samples with the sparitic fabric were found at most north-central sites and Cobá.
2. Red-slipped (Mama), cream-slipped (Xcanchakan), and Navula share this fabric.
3. Navula and Yacman jars are both unslipped, but the samples present a division by fabric: Yacman vessels do not have this calcite sparite fabric. Apparently, potters did not select this fabric for the construction of striated jars. Navula vessels sampled do have this fabric.

These associations are illustrated in Table 8-7. Even with the small number of samples, significant information was gathered.

Briefly summarising this section, although technically, sparry samples can be dolomite or calcite, in this study, the term sparite is used when dealing with calcite. The term dolospar is used when dealing with clusters of sparry dolomite. One sparite fabric was defined representing a common basic recipe comprising medium-to-coarse particles composed of finely crystalline sparite in a low to moderately porous micromass. This sparite fabric is not homogeneous in its petrographic characteristics across the sites, and it may comprise multiple origins or producers. This fabric (or better said, this basic recipe) is correlated with the type of vessel: only slipped (Mama and Xcanchakan) and plain Navula samples were found to have this fabric, whereas no samples of the striated open-mouth Yacman, thought to be for cooking, were found to have this fabric.

8.5 Micrite Fabric Class

A major group resulting from hand-specimen analysis has chalky, white micrite as the dominant inclusion. Seventeen thin sections in which micrite inclusions dominate are the topic of this section. Thin sections made from samples taken from this broad hand-specimen fabric showed that a portion of samples classified in hand specimen as dominantly micritic (microcrystalline calcite) contained fine-grained dolomite or sparite instead. They were described in the previous.

One fabric class, Micrite-W-Medium, was defined by grouping the samples having the common characteristic of micrite as the main inclusion. Thin sections classified in this fabric are mostly medium-grained and have little in common with regard to other petrographic characteristics. More extensive sampling may uncover that we are dealing with multiple fabrics, but not enough samples are present in this study to accommodate the number of different combinations of attributes found and separate them into more fabric classes.

Table 8-7. Associations between Sparite Fabric and Sites and Ceramic Typology

← north-central sites →← eastern sites →

Fabric Class	Tepich	Tecoh	Telchaquillo	Mayapán	Tekit	Mama	Tipikal	Teabo	Cobá	Chac Mool	Culubá
SPARITE											
CS-W/B-cm (Grp 2, u/a)	X	*M* N	*M* N X	*MM*	*N*		*MM* *N*	*MMM* *N*	*MMM* *N*		

KEY: *M* = Mama (red-slipped), N = Navula (plain, unslipped), X = Xcanchakan

This section contains a brief description of this micritic fabric class (see Appendix E for a detailed description). The main characteristics of Micrite-W-Medium are summarised in Table 8-8. The section ends with the presentation of the associations between this fabric, sites, and ceramic varieties.

8.5.1 Micrite-W-Medium (CM-W-m) Fabric Class

Samples with this fabric composition are characterised by the presence of medium-size (mode) micrite (sub-translucent under crossed polarised light) grains as the main inclusion and, usually, rare silt. These attributes are illustrated in Figure 8-18.

Other petrographic characteristics are not homogeneous within thin sections and between samples. Micrite-W-Medium displays a wide range of variation on the colours (xp) of the micromass, porosity, presence/absence of iron concretions, or secondary (in addition to micrite) inclusions. For instance, at Mayapán, the porosity of this fabric ranges from low (around 3 per cent of thin sections) to high (around 15 per cent) and varies widely in the absence/presence of bioclasts and silt.

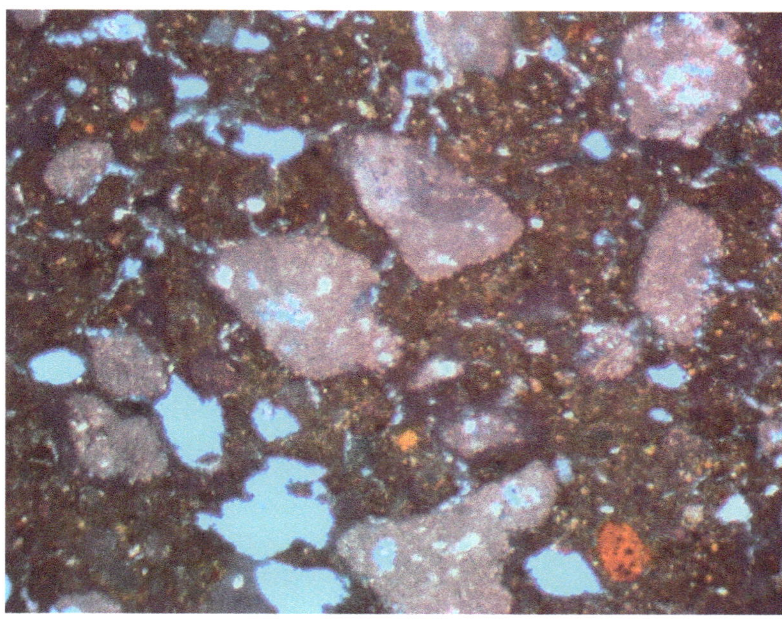

Figure 8-18. Fabric CM-W-m showing micrite particles (# 437, xp, width of field 2.8 mm)

Table 8-8. Main Petrographic Characteristics of Micritic Fabrics

Fabric Class	Main Inclusions	Grain Size (mode)	Porosity	Fired Colour
Micrite-W-Medium (CM-W-m)	coarse portion: micrite; a variety of other inclusions may be present (sparite mainly) -- fine portion: varied	medium	Moderate (mode), with wide variation	Varied

Table 8-9. Associations between Micrite fabric class within Sites and Ceramic Typology

←------------------------- north-central sites --------------------------→←-------eastern sites --------→

Fabric Class	Tepich	Tecoh	Telchaquillo	Mayapán (n=33)	Tekit	Mama	Tipikal	T eabo	Cobá	Chac Mool	Culubá
MICRITE											
CM-W-m (Grp 1, u/a)		YY	*M*	*MMM* NNN X	*M*	*MMM* N	*M*	*M*			

KEY: *M* = Mama variety (red-slipped), N = Navula (plain, unslipped), Y = Yacman (striated, unslipped), X = Xcanchakan

8.5.2 Associations between Micrite Fabrics, Sites, and Ceramic Typology

In this section, the associations between Micrite-W-Medium fabric and other contextual information are examined, in relation to the find spot, chemical group, and the ceramic variety into which the samples have been classified. The Table 8-9 summarises these associations:

1. Samples with this fabric composition were not found at the three eastern sites: Cobá, Chac Mool, and Culubá.
2. Samples were only found at the north-central sites.
3. With two exceptions, vessels sampled having this fabric are slipped (Mama and Xcanchakan) or plain (Navula).

Two Yacman samples, with micrite as dominant inclusion, present a fabric that, nevertheless, differs markedly from the rest in the micritic group. It has common silt and very high porosity and may represent a separate fabric class.

Micrite-W-Medium seems to represent a basic recipe consisting of medium-grained micrite inclusions, but other than that, it is very heterogeneous in its petrographic characteristics. This micritic fabric and the sparitic fabric defined in the previous section (Sparite-W/B-Coarse/Medium) present many similarities: they were both found at multiple north-central sites, they have heterogeneous petrography, they are medium-grained, and they were overwhelmingly found in the red-slipped Mama and plain Navula.

8.6 Medium and Coarsely Crystalline Calcite Fabrics Classes

A main group of fabrics defined during hand-specimen analysis contains semi-translucent to translucent single crystals (XT) detected during that analysis. Samples were taken from these fabrics, and thin sections were prepared.

Their analysis under the polarising light revealed that not only calcite but also dolomite and quartz are present. The petrographic analysis also showed that, in some cases, what looked like single crystals in hand specimens are a variety of other carbonate rock textures including fragments of coarse shells, homogeneous (micritic) calcite (usually skeletal remains), or particles composed of mosaics of sparite or dolospar particles in which each crystal is relatively large (medium-to-coarsely crystalline).

This section deals with those thin sections containing an abundance of few or more (abundance tables in section 8.2.1) single crystals of calcite. This section also includes samples containing few or more particles composed of sparry calcite in which each crystal ranges in size from medium-crystalline to larger crystals, looking like a translucent rocks when observed in hand specimen or low magnification.

The samples can be broadly divided based on particle size into coarse and fine groups of fabrics. Within the coarse group, four fabric classes were formed: Calcite-Micrite-01-Coarse (CX-CM-01-c), Calcite-Micrite-02-Coarse (CX-CM-02-c), Sparite-mc-W/B-Coarse (CSm-W/B-c), and Calcite-Grog-B/R-Coarse (MX-GROG-B/R-c). One crystalline fabric was found to have a fine texture, Calcite-Quartz-B/R-Fine or CX-QZ-B/R-f (and a singleton CX-QZ-B/R-c). A list of the fabrics and a summary of the main characteristics of these five fabrics are found in Table 8-10 (see Appendix E for a detailed description).

8.6.1 Calcite-Micrite-01-Coarse (CX-CM-01-c) Fabric Class

This fabric is characterised by the presence of coarse (mode), discrete crystals of calcite, with a moderately compacted micromass and very few voids. Usually, these crystals are sub-angular to angular. The fabric is not homogeneous across the group, in particular regarding the presence/absence of micrite. Micrite presence could be significant in some thin sections and negligible in others. Figures 8-19 and 8-20 illustrate some of the range of variation in this fabric.

Fabric Calcite-Micrite-01-Coarse has similarities to Calcite C fabric of the Crystalline Calcite-tempered fabric classes described by Linda Howie (2012, p.244-250) for the site of Lamanai, Belize. It differs from this fabric in that in Calcite-Micrite-01-Coarse quartz is absent while Calcite C has a wider range of variation, including a few to rare quartz crystals. Figures 8-19 and 8-20 show the main inclusions found in Calcite-Micrite-01-Coarse fabric.

8.6.2 Calcite-Micrite-02-Coarse (CX-CM-02-c) Fabric Class

Fabric Calcite-Micrite-02-Coarse (CX-CM-02-c) is a variety of the Calcite-Micrite-01-Coarse fabric just described and presents similar attributes of composition: the presence of coarse (mode), sub-angular to angular, discrete crystals of calcite. Micrite or sparite may also be present.

The two fabrics differ in the high percentage of voids and large channels present in Calcite-Micrite-02-Coarse that create a very open porous matrix. Figure 8-21 illustrates Calcite-Micrite-02-Coarse fabric.

Table 8-10. Main Petrographic Characteristics of Calcite-Crystal Fabric Classes

Fabric Class	**Main or Diagnostic Inclusions**	**Grain Size mode**	**Porosity**
Calcite-Micrite-01-Coarse (CX-CM-01-c)	Coarse portion: coarse particles of single calcite crystals micrite	Coarse	Low to Moderate
Calcite-Micrite-02-Coarse (CX-CM-02-c)	Coarse portion: coarse particles of single calcite crystals micrite	Coarse	High
Sparite-m-W/B-Coarse (CSm-W/B-c)	Coarse portion : coarse particles of sparite bundles with crystal size = medium-crystalline (usually 0.07 to 0.18 mm)	Coarse	High
Calcite-Quartz-B/R-Fine (CX-QZ-B/R-f)	Coarse portion: fine (0.1-0.25mm) calcite and quartz single crystals	Fine	Low
Calcite-Grog-B/R-Coarse (MX-GROG-B/R-c)	Coarse portion: dark argillaceous particles, single calcite crystals, grog --- Fine portion (< 0.652 mm) single calcite crystals	Coarse	Varied

KEY s-xtals: single crystals; l-r-b: light reddish brown ; u/a = unassigned; Ch. Grp. = chemical group

8.6.3 Sparite-m-W/B-Coarse (CSm-W/B-c) Fabric Class

This fabric is distinguished by, first, the presence of coarse (mode) particles composed of mosaics of sparry calcite (sparite) in which each crystal size is medium-crystalline or larger, usually between 0.06 mm to 0.18 mm, and, second, by a very porous matrix. This is not a homogeneous fabric across the group, with wide variability related, mainly, to the frequency of the sparite grains, the colour of the micromass, and the presence/absence of micrite.

The difference between this fabric and the two previous fabrics (Calcite-Micrite-01-Coarse and Calcite-Micrite-02-Coarse) is that the two previous are based on single crystals, whereas in this fabric (Sparite-m-W/B-Coarse) mosaics of sparry crystals dominate. To illustrate the appearance of these mosaics, Figure 8-22 shows two views of the grains in this fabric at two different magnifications.

Sparite-m-W/B-Coarse, therefore, could have been considered a variant of the finely crystalline sparite fabric (Sparite-W/B-Coarse/med) described earlier given that both contain sparite clusters, differing only on the size of the individual crystals. This fabric is included with fabrics characterised by the presence of single crystals because in both cases in the hand specimens, they may have looked like translucent or semi-translucent rocks to potters, and potters may have used them indistinctively.

8.6.4 Calcite-Quartz-B/R-Fine (CX-QZ-B/R-f) Fabric Class

This fabric is easily recognised by its very-fine to fine calcite and quartz single crystals in a compact, usually dark in colour, matrix with rare pores. Figures 8-23 and 8-24 illustrate two instances of this fabric.

8.6.5 Calcite-Grog-B/R-Coarse (MX-GROG-B/R-c) Fabric Class

This is a fabric characterised by the presence of coarse angular to irregularly shaped fragments of very dark brown (10YR 2/2) argillaceous particles with embedded silty to very fine calcite crystals. In addition, grog may be present. Furthermore, in four samples, the argillaceous particles were identified as unslipped sherd fragments, or grog.

Usually, the argillaceous particles contain discrete calcite crystals. The argillaceous particles may not be the most abundant inclusion, as shown in Figure 8-25, but they are diagnostic for this fabric.

Given that some of argillaceous particles were identified as grog, they all may be grog, and the subrounded particles such as that at the top left corner of Figure 8-25 (or right bottom in Figure 8-27) may be an indication that the original vessel had a friable paste. One argillaceous inclusion identified as an unslipped sherd fragment is illustrated in Figure 8-26. In addition, finely grained red-slipped sherd fragments were found in one sample, shown in Figure 8-27.

Figure 8-19. Fabric Calcite-Micrite-01-Coarse showing single calcite crystals (#354, xp, wof 2.8 mm)

Figure 8-20. Fabric Calcite-Micrite-01-Coarse showing single calcite crystals (#204, xp, wof 2.8 mm)

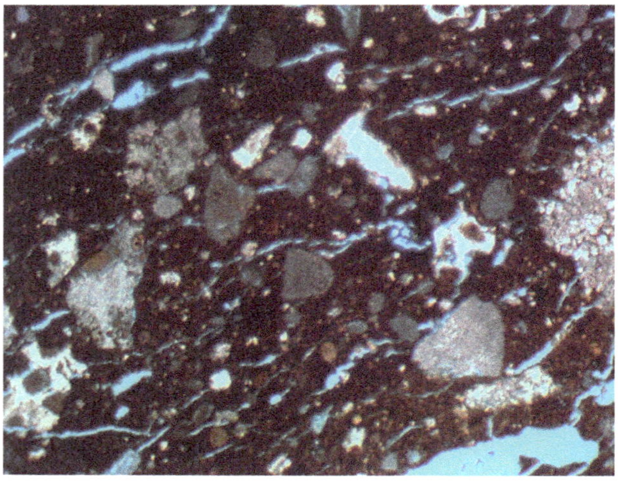

Figure 8-21. Fabric Calcite-Micrite-02-Coarse (CX-CM-02-c) showing calcite crystals with spot alteration and open porous matrix (#191, xp, wof 2.82 mm)

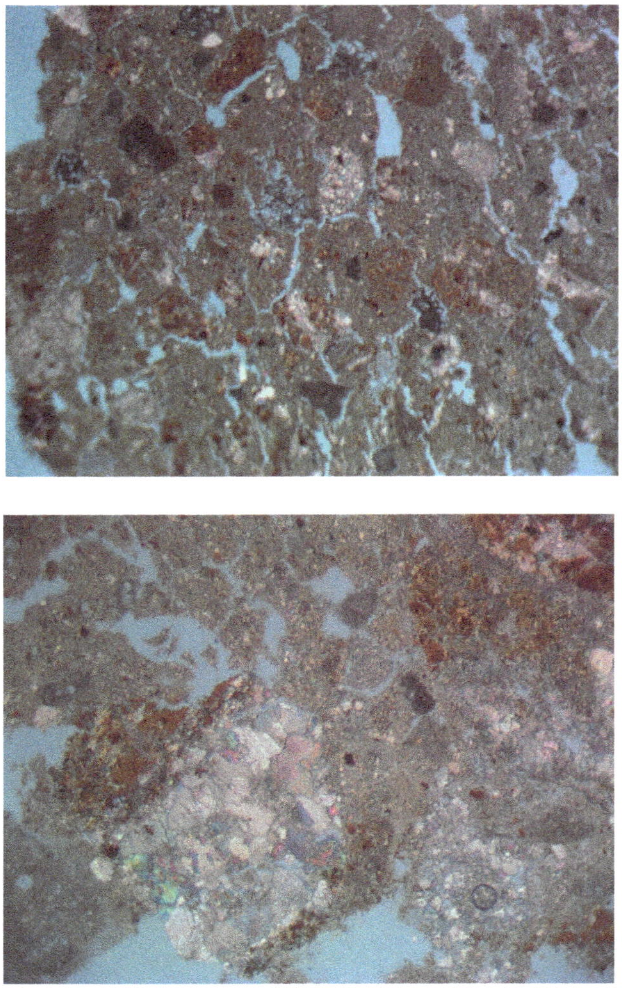

Figure 8-22. Fabric Sparite-m-W/B-Coarse showing two views of: (a:top) medium-crystalline sparite (# 335, xp, wof 2.82 mm) and (b:bottom) medium crystalline sparry calcite (#335, xp, wof 1.1 mm)

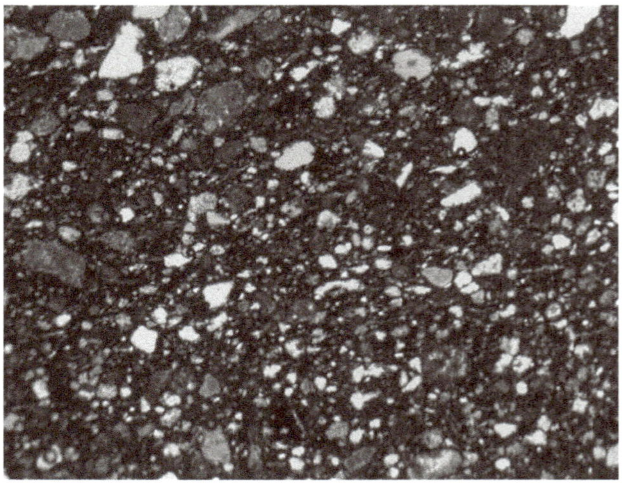

Figure 8-23. Fabric Class Calcite-Quartz-B/R-Fine showing calcite and quartz (#131, pp, wof 2.82 mm)

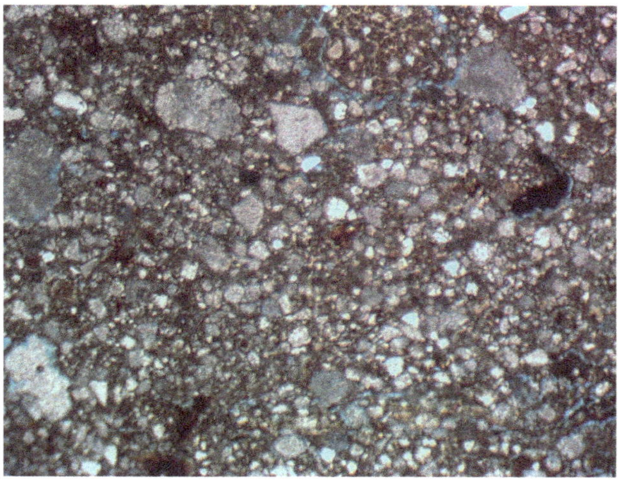

Figure 8-24. Fabric Class CX-QZ-R/B-f showing calcite and scatter quartz (white); (# 382, xp, wof 2.82 mm)

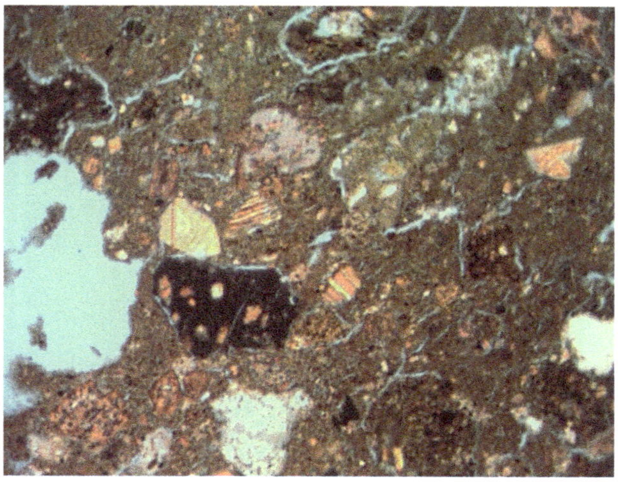

Figure 8-25. Fabric Calcite-Grog-B/R-Coarse, stained with alizarin red, showing calcite crystals (pink) and dark argillaceous particles with embedded calcite crystals (#275, xp, wof 2.82 mm)

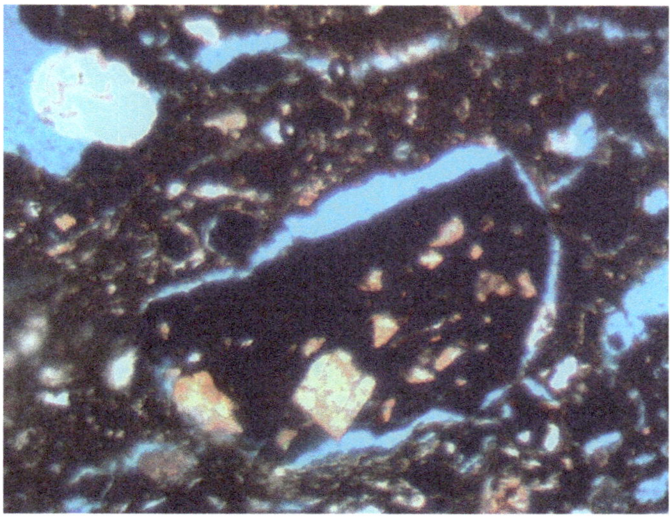

Figure 8-26. Fabric Calcite-Grog-B/R-Coarse showing fragment of unslipped sherd (#275, xp, width of field 1.1 mm)

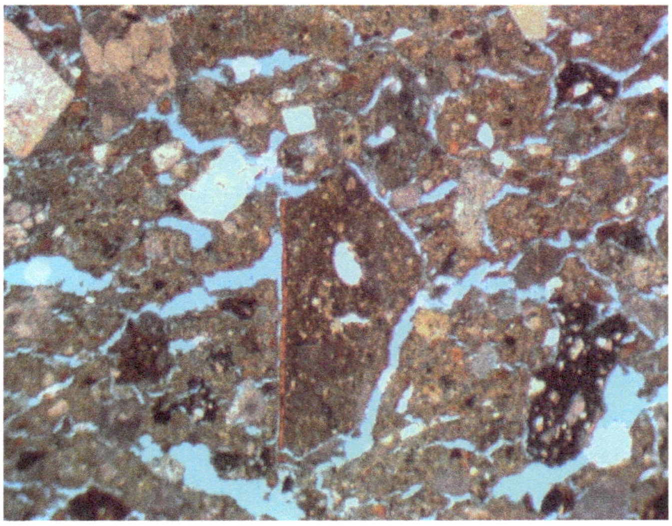

Figure 8-27. Fabric Calcite-Grog-B/R-Coarse showing red-slipped sherd (centre) and dark argillaceous particles with embedded crystals (#445; width of view 2.82 mm, xp)

8.7 Medium and Coarsely Crystalline Dolomite Fabrics Classes

Coarse to fine discrete crystals of dolomite were also found in the samples. It is likely that potters were not able to differentiate them from calcite crystals with the same texture. Three fabrics were defined having single dolomite crystals as the dominant inclusions:

Dolomite-W/B-Medium (DX-W/B-m),
Dolomite-Calcite-B-Coarse (DX-CX-B-c), and
Dolomite-Calcite-Medium-Hard (DX-CX-m-h).

They are briefly described in this section (more detail in Appendix E). Their main characteristics are listed in Table 8-11.

8.7.1 Dolomite-W/B-Medium (DX-B/W-m) Fabric Class

This is a very distinctive fabric characterised by fine-grained (mode) discrete rhombohedral dolomite crystals, in a largely compact matrix, as shown in Figure 8-28 and Figure 8-29. Dolomite is mostly euhedral to subhedral. Many dolomite crystals present alteration to red iron oxides, particularly, at the crystals' centre. There are five sherds within this fabric class.

8.7.2 Dolomite-Calcite-B-Coarse (DX-CX-B-c) Fabric Class

This fabric is characterised by the presence of coarse (mode), subangular to angular, mostly subhedral to anhedral single crystals of dolomite. Calcite crystals were also present in three of the four samples characterised within this fabric. The main difference with Dolomite-W/B-Medium is in the size (mode) of each dolomite crystal and the presence of discrete calcite crystals.

This is not a homogeneous fabric class (Figure 8-30, 8-31 and 8-32). It is defined with a wide range of variation. It is possible that Dolomite-Calcite-B-Coarse comprises two fabric classes having in common the presence of coarse dolomite and differing in the frequency of inclusions. Mayapán samples contain very few dolomite particles (Figure 8-30), whereas Tekit (Figure 8-31) samples have a higher frequency of inclusions. This fabric class was originally identified and described by Brown (1999a, p.352) who associated it with new ceramic types that he called Calcite-tempered Unslipped Type and Striated Calcite-tempered Unslipped Type.

8.7.3 Dolomite-Calcite-Medium-Hard (DX-CX-m-h) Fabric Class

This fabric is a medium-grained variant of the previous fabric Dolomite-Calcite-B-Coarse. Similar to Dolomite-Calcite-B-Coarse, this fabric is characterised by the presence of dolomite and single calcite crystals. The micromass is also moderately to highly porous. The main difference from Dolomite-Calcite-B-Coarse is in the dominant medium-grained dolomite particles. This fabric was found in two thin sections from Culubá (Figure 8-33).

8.7.4 Singleton Fabrics

This section contains the description of three samples in the crystalline fabrics that could not be categorised within any of the fabrics described earlier. A fabric code was given to these samples.

One shred (CX-DXsilt-DS-W-c, # 291) found at Tepich contains few fine to medium calcite crystals. It differs from the previous crystalline fabrics in the presence of silt-size dolomite, probable dolospar (very altered to micrite), and coarse spar and micrite.

Another sample (CX-CM-01-f, #135) contains calcite crystals and micrite. It differs from Calcite-Micrite-01-Coarse (a fabric class that also contains calcite crystals and may contain micrite) in its fine-grained and low porosity. The only sample within this fabric was found at Mayapán and classified as Payil.

The third sample (CX-QZ-B-c, #375) was found at Chac Mool and contains significant quantities of coarse calcite crystals and quartz. It differs from the fabric Calcite-Quartz-B/R-Fine, found in the red-slipped Payil (eastern Tases) samples, in the coarse texture of the fabric. This singleton was an unslipped sample classified as Navula

Table 8-11. Main Characteristics of Dolomite Fabrics

Fabric Class	Main Inclusions	Grain Size (mode)	Porosity
Dolomite-W/B-Medium (DX-B/W-m)	Coarse portion: fine dolomite s-xtals	Fine	Low
Dolomite-Calcite-B-Coarse (DX-CX-B-c)	Coarse portion: coarse dolomite and medium calcite s-xtals	Coarse	Varied
Dolomite-Calcite-Medium-Hard (DX-CX-m-h)	Coarse portion: medium grained dolomite and calcite s-xtals	Medium	Medium to high

KEY s-xtals: single crystals

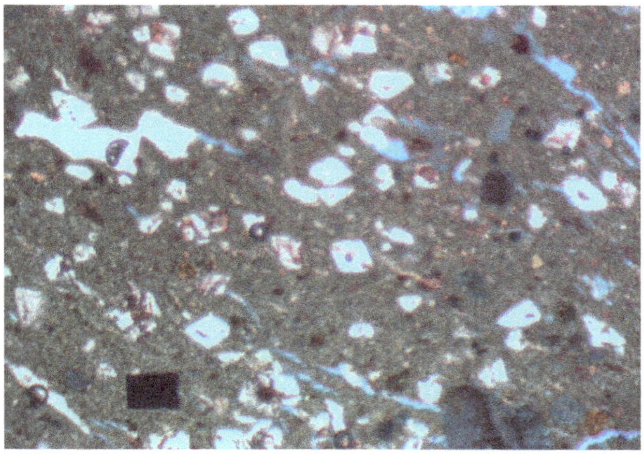

Figure 8-28. Fabric Dolomite-W/B-Medium with single crystals of dolomite (# 339, xp, width of view 2.82 mm)

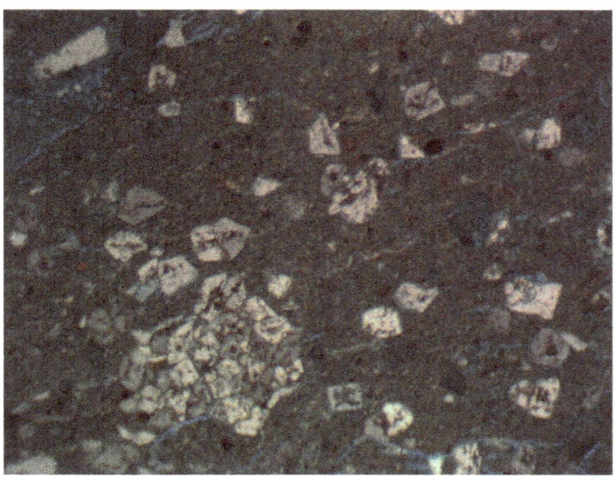

Figure 8-29. Fabric Dolomite-W/B-Medium with dolomite (# 237, xp, width of view 2.82 mm)

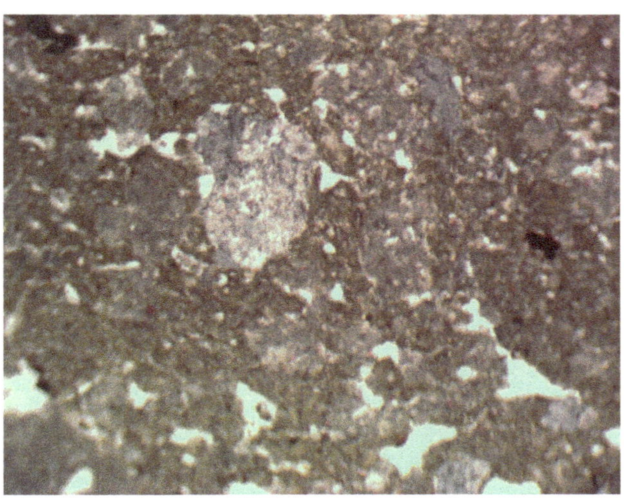

Figure 8-30. Dolomite-Calcite-B-Coarse showing anhedral dolomite (#60, xp, with of field 2.8 mm)

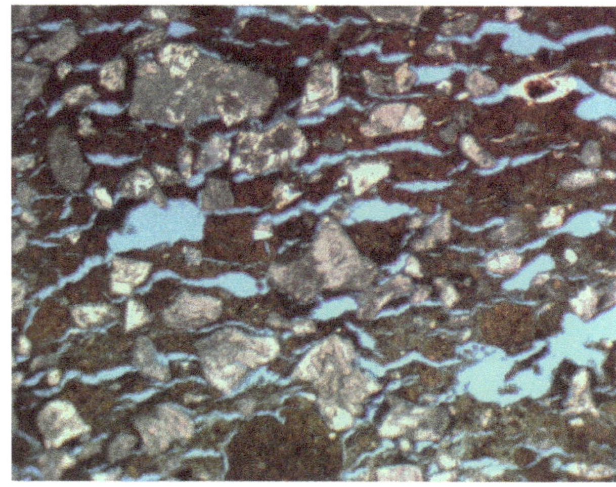

Figure 8-31. Fabric class Dolomite-Calcite-B-Coarse (DX-CX-B-c) showing subangular anhedral dolomite (#441, xp, width of field 2.8)

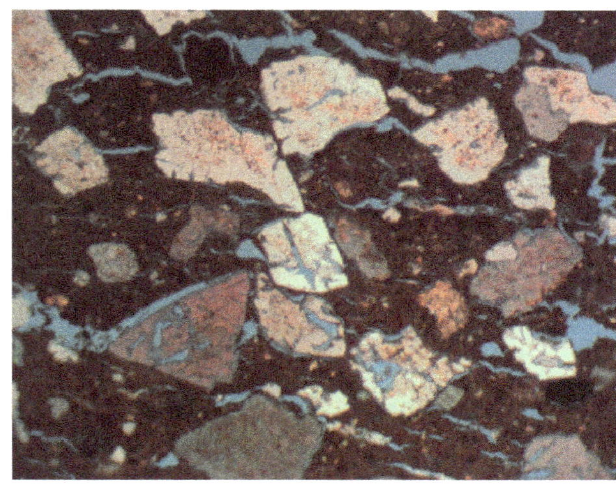

Figure 8-32. Fabric Dolomite-Calcite-B-Coarse showing subhedral dolomite (top white, with pink alizarin red stripes) and calcite (bottom, tinted with alizaring red) (#442, xp, width of field 2.82 mm

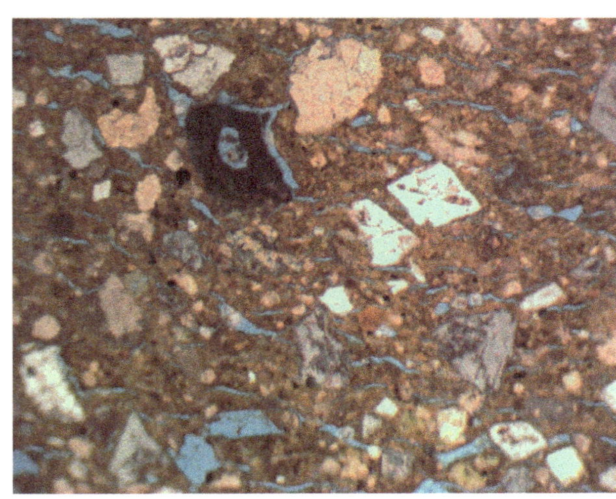

Figure 8-33. Fabric class Dolomite-Calcite-Medium-Hard (#320, xp, width of field 2.82 mm)

8.7.5 Associations between Single-crystal Fabrics, Sites, and Ceramic Typology

This section presents the associations that exist between the crystalline fabric classes, described in the previous sections, and contextual information such as find spot, surface finish, or vessels form (subsumed in the ceramic variety). Table 8-12 presents the fabrics with fine to very coarse crystals (calcite, dolomite, or quartz) arranged by their grain size to better show patterns in the data. Even with the small number of samples, patterns in the data can be inferred from this table.

Fabric classes with coarse crystals when found at the north-central sites are associated with unslipped jars, in particular striated Yacman jars. Of the total of 20 samples, 19 are unslipped jars. Of these, 12 are Yacman *ollas*, most likely used for cooking.

On the other hand, these types of coarse crystalline fabrics classes when found at Cobá, Chac Mool or Culubá are associated with all utilitarian ceramic varieties: Mama, Yacman, Navula, and Xcanchakan. In addition, it can be observed from Table 8-12 and Table 8-19 at the end of this chapter that the coarse crystalline fabric classes found at the eastern and north-central sites are different and do not overlap.

Not all the crystalline fabrics from eastern and north-central sites are coarse. A fine (mode) grained dolomite fabric (Dolomite-B/W-Medium) was found at the north-central sites of Tecoh and Tekit, and fabric Calcite-Quartz-B/R-Fine containing fine calcite and quartz was found in Payil samples from Mayapán and Chac Mool.

Table 8-12. Associations between Single-Crystal Fabrics and Sites and Ceramic Typology

←-------------------North-central sites -------------------→←-------Easter sites-------→

Fabric Class	Tepich	Tecoh	Telchaquillo	Mayapán	Tekit	Mama	Tipikal	T eabo	Cobá	Chac Mool	Culubá
Fine Particle Size											
Calcite-Quartz-B/R-Fine (CX-QZ-B/R-f)				P P P P						P P P P P	
Dolomite-B/W-Medium (DX-B/W-m)		*M M* N			*M* Y						
Coarse Particle Size											
Calcite-Quartz-B-Coarse (CX-QZ-B-c)										N	
Dolomite-Calcite-Medium-Hard (DX-CX-m-h)											*M M*
Calcite-Micrite-01-Coarse (CX-CM-01-c)				X			N		*M M M* N	XXXX	*M M M M M M*
Calcite-Micrite-02-Coarse (CX-CM-02-c)										N Y Y Y	*M* Y Y V
Sparite-mXtal-W/B-Coarse (CSm-W/B-c)	Y Y	N N N	Y			Y	N	Y Y	N		
Calcite-Grog-B/R-Coarse (MX-GROG-B/R-c)			Y		*M*		Y	Y Y			
Dolomite-Calcite-B-Coarse (DX-CX-B-c)				N N	Y Y						

KEY: *M* = Mama (or Cancun) variety (slipped), N = Navula (plain), Y = Yacman (striated), P = Payil . X = Xcanchakan; uk = unknown; u/a = unassigned

Table 8-13. Main Characteristics of Skeletal Remains Fabrics

Fabric Class	**Main Inclusions**	**Grain Size (mode)**	**Porosity**
Bioclasts-W-Coarse/Medium (CH-W-cm)	Coarse portion: micritised skeletal remains and homogeneous calcite Fine portion: no silt	Medium to coarse	Low to Medium (1-7 per cent)
Bioclasts-B-Coarse (CH-B-c)	Coarse portion : micritised skeletal remains and homogeneous calcite Fine portion: varied silt from few to common	Coarse	High (10-20 per cent)

8.8 Fabric Classes with Skeletal Remains

Fossilised fragments of maritime faunal skeletal remains (CH) were commonly found in some of the samples. These remains had undergone processes such as micritisation that created micro to medium-crystalline infilled fossils. This section deals with those thin sections in which skeletal remains are the dominant inclusion. Two fabric classes were found, CH-W-cm and CH-B-c. Their main characteristics are summarised in Table 8-13.

8.8.1 Bioclast-W-Medium (CH-W-cm) Fabric Class

This fabric is characterised by the presence of micritised shell fragments (see Figure 8-34) in which most fit the description of 'homogeneous calcite' (section 8.1.1), which can lead to their classification as single crystals.

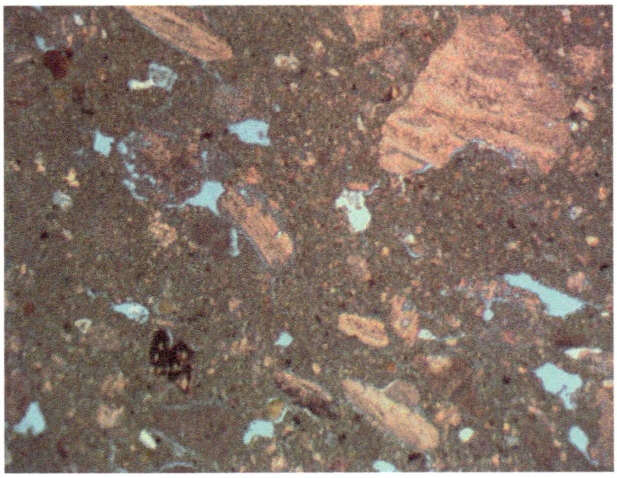

Figure 8-34. Fabric class Bioclast-W-Coarse/Medium (CH-W-cm) showing skeletal fragments (thin section is tinted) (#284, xp, width of field 2.8 mm)

8.8.2 Bioclast-B-Coarse (CH-B-c) Fabric Class

This is a coarse fabric, with abundant angular, recognisable shell remains. This fabric differs from the skeletal remains fabric just described, or Bioclast-W-Coarse/Medium, not only on the fired and refired colours but also in the higher porosity, long channels, and the colour under xp and pp.

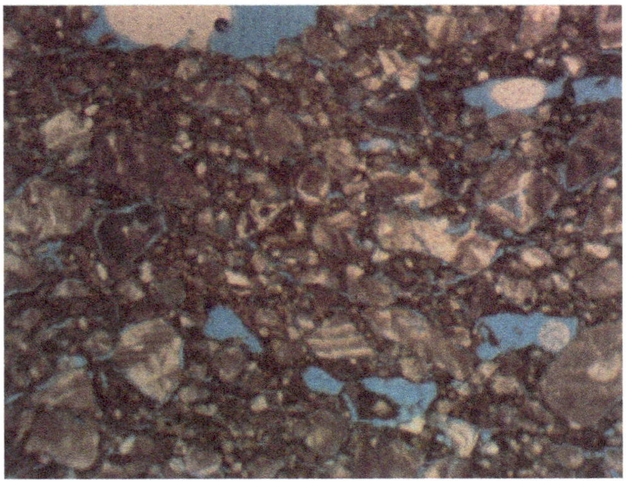

Figure 8-35. Fabric Class Bioclast-B-Coarse (#450, xp, width of field 2.82mm)

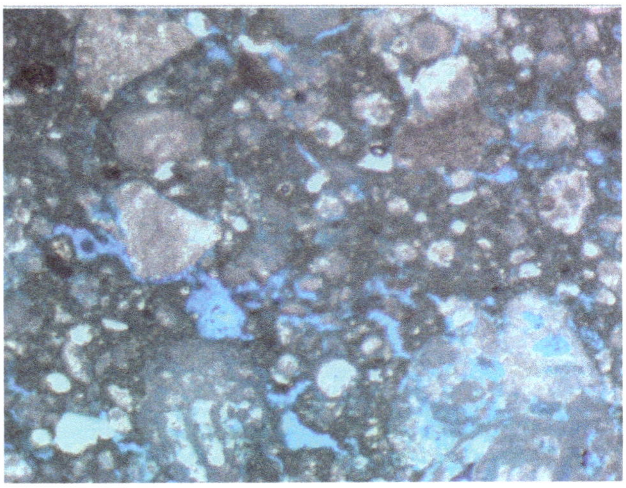

Figure 8-36. Fabric class Bioclast-B-Coarse (#409, xp, width of field 2.82 mm)

This fabric is not homogeneous as illustrated in Figures 8-35 and 8-36. In Figure 8-35, the inclusions are largely homogeneous micrite and shell remains are recognisable, whereas the inclusions in Figure 8-36 are a mixture of biomicritic rocks, micritised shell fragments, and bio-sparite fragments

Table 8-14. Associations between Bioclast Fabric Classes and Sites and Ceramic Typology

	←-----North-central sites							-----→	←---Easter sites--→		
Fabric Class	Tepich	Tecoh	Telchaquillo	Mayapán	Tekit	Mama	Tipikal	Teabo	Cobá	Chac Mool	Culubá
BIOCLASTS											
Bioclast-W-Coarse/Medium (CH-W-cm)	*MMMM MMM* N Y			N							
Bioclast-B-Coarse (CH-B-c)					N N N Y	N N N	N	N			

KEY: *M* = Mama variety (slipped), N = Navula (plain), Y = Yacman (striated)

8.8.3 Singleton Fabrics

One single sherd defines fabric CH-R-f. The fabric is represented by one Palmul sherd from Culubá characterised with fine-grain, predominant presence of fine angular and sub-angular shell and calcite homogeneous (likely shells) fragments, and a relatively compacted paste.

8.8.4 Associations between Bioclast Fabrics, Sites, and Ceramic Typology

Table 8-14 helps to visualise the distribution of the two fabrics in which the skeletal remains dominate. The table clearly shows that the two fabrics have different geographical scope. Fabric class Bioclast-W-Coarse/Medium, or CH-W-cm, was found at Tepich almost exclusively.

The north-central sites in Table 8-14 are presented ordered according to their location, in a north to south direction. Tepich is the farthest site to the north of Mayapán in this study, and the fabric classes found at Tepich do not overlap, for the most part, with the fabrics found at the other north-central sites.

These fabrics appear correlated with the surface finish (or ceramic type), although the number of samples characterised within these fabrics is small, and this interpretation can change with more extensive sampling. Fabric class Bioclast-B-Coarse, or CH-B-c, was found in unslipped samples only, mostly plain Navula jars (but one cajete).

8.9 Dark Micrite Fabric Classes

Dark Micrite fabrics comprise a group of samples containing dark brown or grey, semi-opaque particles of micrite as main inclusions. A significant portion of hand specimens has as a main characteristic the presence of dark grey, brown, or black particles. In many instances during thin-section analysis, the dark micrite hand-specimen fabric was re-classified under a variety of different fabric classes. Apparently, during hand-specimen analysis dark particles played a trick on the eye leading to overestimating their contribution to the total of inclusions.

This group is composed of two fabric classes: a medium-grained DarkMicrite-Micrite-Medium (CG-CS-W/B-m) class and a coarsely grained (CG-CM-W-c) class. Table 8-15 summarises the main characteristics of these fabrics.

8.9.1 DarkMicrite-Micrite-Coarse (CG-CM-W-c) Fabric Class

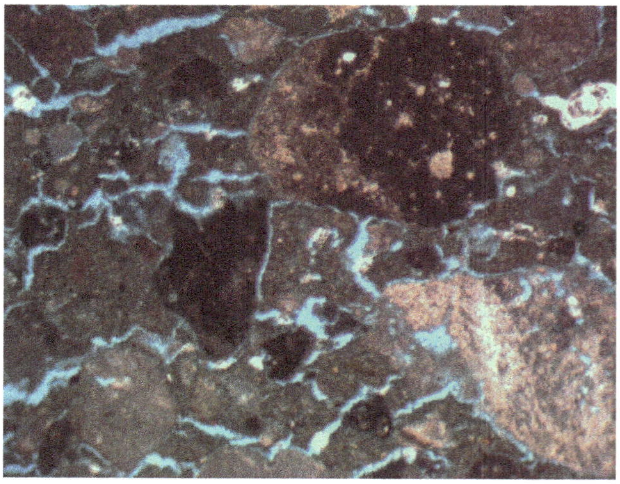

Figure 8-37. Fabric DarkMicrite-Micrite-Coarse (CG-CM-W-c) showing coarse dark micrite and very porous micromass (#146, xp, width of field 2.82 mm)

Samples with DarkMicrite-Micrite-Coarse fabric are characterised by the presence of opaque or sub-opaque dark micrite, low silt, highly porous micromass, and coarsely grained fabric, as illustrated in Figure 8-37.

Table 8-15. Main Petrographic Characteristics of Dark Micrite Fabrics

Fabric Class	**Grain Size**	**Main Inclusions**	**Porosity**	**Fired Colour (xp)**
DarkMicrite-Micrite-Medium (CG-CM-W-m)	coarse	COARSE portion: dark brown micrite	Low	Pale brown, red
DarkMicrite-Micrite-Coarse (CG-CM-W-c)	coarse	COARSE portion: dark brown micrite (xp, pp)	High	Light olive, very light grey

Table 8-16. Associations between Dark Micrite Fabric Classes and Sites and Ceramic Typology

←---------------------North-central sites -----------------------→←-- Easter sites -→

Fabric Class	**Tepich**	**Tecoh**	**Telchaquillo**	**Mayapán**	**Tekit**	**Mama**	**Tipikal**	**T eabo**	**Cobá**	**Chac Mool**	**Culubá**
DARK MICRITE											
DarkMicrite-Micrite-Medium (CG-CM-W-m)	N N N N N			*M*							
DarkMicrite-Micrite-Coarse (CG-CM-W-c)	**Y Y**	**Y**	*M*	**Y Y**	**Y**						

KEY: *M* = Mama variety (slipped), N = Navula (plain), **Y** = Yacman (striated), uk = unknown

8.9.2 DarkMicrite-Micrite-Medium Fabric Class

Samples with this fabric present opaque or sub-opaque, medium-grained, dark micrite. The fabric has low porosity and silt. DarkMicrite-Micrite-Medium differs from the DarkMicrite-Micrite-Coarse fabric just described in the texture and porosity with DarkMicrite-Micrite-Coarse being coarser and more porous.

8.9.3 Associations between Dark Micrite Fabrics, Sites, and Ceramic Typology

In this section, the associations between the dark micrite fabrics and contextual information are examined. Each entry in Table 8-16 represents one thin section and shows the site or find spot, as well as the ceramic variety and the fabric class to which it was assigned.

It can be observed from the table that these fabrics have a limited geographical scope. They were found within the northern portion of the north-central area from Tepich to Tekit.

The table also shows that there is a correlation between the fabrics and the types of vessels. Dark micrite fabrics are associated with unslipped jars and cajetes. The coarser fabric is associated with open-mouth Yacman jars, cooking pots most likely, while the medium-grained fabric is, apparently, associated with plain unslipped (Navula) samples.

8.10 Petrographic Analysis of Collected Clays and Marls

This section reports the results of the thin-section analysis of raw materials from the north-central area. As can be recalled from section 5.4, no clay deposits were located within the vicinity of Mayapán (one potential clay mine within Mayapán had its entrance plugged in with natural beehives). The closest mine was located 20 km from this site in the Chapab area. Samples (Table 5-4) of red soils, off-white to yellow marl sascab, off-white to yellow marl sascab, off-white to yellow rock sascab, light brown marl sascab, and a light grey clay were collected from various locations (see map in Figure 5-4).

In subsequent chapters, these results are used to assess the range of options available to potters regarding raw materials. They are compared to the pottery samples to investigate the techniques that might have been involved in changing them into ceramics.

Marls (plastic sascab) comprise most of the petrographically analysed raw materials. They contain carbonate rock fragments in a calcareous clay or marl. First, after manually removing the larger pebbles, briquettes were prepared from these marls. They were fired at 700°C for 40-45 minutes. Thin sections were prepared and examined under the polarising microscope. Table 8-17 contains the results of the petrographic analysis of these raw materials.

Table 8-17. Results of Petrographic Analysis of Local Marls and Clay

Raw Material #1 (CGS187)
I) Microstructure (section 8.2.2)
Porosity: no porosity
IIa) Groundmass: homogeneity and optical characteristic (section 8.2.2)
Micromass optically inactive. Fired (original) colour in xp, pp: light greenish grey GLEY1 5G 7/1 (xp, pp);.
IIb) Groundmass: coarser inclusions (section 8.2.1)
Coarser Inclusions (>0.0625 mm) (frequency ranges of whole thin section). No inclusions
IIc) Groundmass: finer inclusions (section 8.2.1)
Fine Portion (<0.0625 mm) (frequencies are of the fine portion only) No inclusions
III) Textural Concentration Features
Very rare yellow red subrounded concretions
Raw Material #2 (CGS188)
I) Microstructure (section 8.2.2)
Porosity: no porosity
IIa) Groundmass: homogeneity and optical characteristic (section 8.2.2)
Homogeneity : homogeneous colour and optical activity Optically inactive micromass Micromass Fired (original) colour in xp, pp: r: light greenish grey GLEY1 5G 7/1 (xp, pp).
IIb) Groundmass: coarser inclusions (section 8.2.1)
Coarser Inclusions (>0.0625 mm) (frequency ranges of whole thin section). **Common** SPARITE (finely crystalline), very coarse particle size, subrounded to rounded. (NO = Single dolomite crystals, dolospar, single calcite crystals, quartz, micrite)
IIc) Groundmass: finer inclusions (section 8.2.1)
Fine Portion (<0.0625 mm) (frequencies are of the fine portion only) No silt size particles
III) Textural and amorphous Concentration Features
Very rare red concretions Very rare medium brown opaque (xp) irregularly shaped units
Raw Material #4 (CGS189)
I) Microstructure (section 8.2.2)
Porosity: no porosity
IIa) Groundmass: homogeneity and optical characteristic (section 8.2.2)
Homogeneity : homogeneous colour and optical activity Optically activity: inactive micromass Micromass fired (original) colour in xp, pp: light greenish grey GLEY1 5G 7/1 (xp, pp)
IIb) Groundmass: coarser inclusions (section 8.2.1)
Coarser Inclusions (>0.0625 mm) (frequency ranges of whole thin section). **Few** SPARITE (finely crystalline), with very coarse (mode) particle size ranging from fine to very coarse. **Few** BROWN MICRITE (NO = Single dolomite crystals, dolospar, single calcite crystals, quartz, micrite)
IIc) Groundmass: finer inclusions (section 8.2.1)
Fine Portion (<0.0625 mm) (frequencies are of the fine portion only) No silt size particles
III) Textural and Amorphous Concentration Features
Red concretions, fine size, rounded, count of 5 Amorphous irregularly shaped concentrations of brown colour
Raw Material #11 (CGS193)
I) Microstructure (section 8.2.2)
Porosity: five per cent channels
IIa) Groundmass: homogeneity and optical characteristic (section 8.2.2)
Homogeneity : homogeneous colour and optical activity Optically activity: optically inactive Micromass fired (original) colour in xp, pp: Reddish yellow 10 YR 7/8 pp,xp and pale brown yellow 10YR 8/8;
IIb) Groundmass: coarser inclusions (section 8.2.1)
Coarser Inclusions (>0.0625 mm) (frequency ranges of whole thin section).

Few finely crystalline DOLOSPAR, grain size very coarse (mode) but ranging from fine to very coarse **Few** finely crystalline SPARITE, coarse (mode) grain size. **Few** BROWN MICRITE, semi-opaque **Rare** BIOCLAST **Very rare** QUARTZ (NO = dolomite or calcite single crystals, micrite)
IIc) Groundmass: finer inclusions (section 8.2.1)
Fine Portion (<0.0625 mm) (frequencies are of the fine portion only) **Very rare** DOLOMITE
III) Textural and Amorphous Concentration Features
Few red iron concretions, rounded Amorphous brown irregularly shaped clay impregnations and nodules
Raw Material #19a (CGS196)
I) Microstructure (section 8.2.2)
Porosity: No porosity
IIa) Groundmass: homogeneity and optical characteristic (section 8.2.2)
Homogeneity : homogeneous colour and optical activity Optically activity: optically inactive Micromass fired (original) colour in xp, pp: very pale grey
IIb) Groundmass: coarser inclusions (section 8.2.1)
Coarser Inclusions (>0.0625 mm) (frequency ranges of whole thin section). **Few** finely crystalline SPARITE, fine to medium particle size, subangular, very finely crystalline **Very Few** DARK MICRITE **Very rare** SINGLE CALCITE CRYSTAL (counted 1) (NO Micrite, Quartz, Red concretions, Shells)
IIc) Groundmass: finer inclusions (section 8.2.1)
Fine Portion (<0.0625 mm) (frequencies are of the fine portion only) **Very rare** calcite
III) Textural and Amorphous Concentration Features

The particle sizes of the carbonate inclusions in the raw materials do not match the pottery samples. The pottery samples from the north-central area do not include un-tempered, fine-grained, or very coarsely grained samples. Contrastingly, these types of grains were found in these marls. One thin section does not have coarse (larger than silt) inclusions (#1). Samples #2 and #4 contain very coarse particles of finely crystalline sparite. Very coarse particles are rare in the north-central samples. Sample #11 includes a mixture in which coarse dolospar and sparite comprise the main inclusions.

In the next chapter, these issues are examined further, and the raw materials and pottery samples are compared in depth. They are also examined in terms of the possible ceramic techniques that could have been used to produce the vessels sampled in this study. The ceramic techniques examined in this study include firing techniques appropriate for such calcareous raw materials.

8.11 List of Samples

Table 8-18 contains the list of pottery samples. The table includes the ANID or id used in the chemical analysis, the fabric class code assigned to the sample, the chemical group (if the sample was selected for chemical analysis), the ceramic variety, and the most common Euclidian distance match calculated as part of the chemical analysis.

Table 8-18. List of Samples in this Study

#	Site	Variety	ANID	Ch. Grp	Fabric Class (thin section)	Most common Euclid match	Vessel Shape	(*)	Sample ID
1	Mayapan	Mama			CM-W-m		Cajete	F	MY-BROWN- 87
2	Mayapan	Mama			DZ-B-c		Cajete	F	MY-BROWN- 87
3	Mayapan	Mama					cajete	A	MY-BROWN- 87
4	Mayapan	Mama					Cajete	F	MY-BROWN- 87
5	Mayapan	Mama					tinaja	A	MY-BROWN- 87
6	Mayapan	Mama			CG-CM-W-m		tinaja	F	MY-BROWN- 87
7	Mayapan	Mama					tinaja	A	MY-BROWN- 87
8	Mayapan	Mama					cajete	F	MY-BROWN- 87
9	Mayapan	Navula					Olla	A	MY-BROWN- 87
10	Mayapan	Navula					olla	F	MY-BROWN- 87
15	Mayapan	Yacman			DZ-B-c		olla	G	MY-BROWN- 87
16	Mayapan	Yacman			DZ-B-c		olla	G	MY-BROWN- 87
17	Mayapan	Tecoh					cajete	A	MY-BROWN-87
18	Mayapan	Mama			DZ-W-c		cajete	G	MY-BROWN-87
23	Mayapan	Mama					tinaja	A	MY-BROWN-89
24	Mayapan	Mama					tinaja	F	MY-BROWN- 89
25	Mayapan	Mama			DZ-W-m		tinaja	F	MY-BROWN- 89
26	Mayapan	Mama					tinaja	A	MY-BROWN- 89
27	Mayapan	Mama			CS-W/B-cm		tinaja	G	MY-BROWN- 89
28	Mayapan	Mama					tinaja	F	MY-BROWN-89
32	Mayapan	Navula			DZ-W-c		olla	G	MY-BROWN- 89
34	Mayapan	Yacman					Olla	A	MY-BROWN- 89
35	Mayapan	Yacman			DZ-W-c		olla	A	MY-BROWN- 89
38	Mayapan	Polbox					tinaja	C	MY-BROWN- 89
39	Mayapan	Yacman					Olla	F	MY-BROWN- 97
42	Mayapan	Yacman					Olla	G	MY-BROWN- 96
43	Mayapan	Yacman			DZ-B-c		Olla	G	MY-BROWN- 96
44	Mayapan	Navula					Olla	A	MY-BROWN- 91
45	Mayapan	Yacman					Olla	G	MY-BROWN- 96
48	Mayapan	Chen Mul			DZ-W-c		censer	G	MY-BROWN- 91
54	Mayapan	Papacal					tinaja	F	MY-BROWN- 86
58	Mayapan	Yacman					Olla	F	MY-BROWN- 87
59	Mayapan	Navula			DX-CX-B-c		Olla	C	MY-BROWN- 87
60	Mayapan	Navula			DX-CX-B-c		Olla	C	MY-BROWN- 87
62	Mayapan	Kukula					tinaja	A	MY-BROWN- 50
64	Mayapan	Kukula					tinaja	A	MY-BROWN- 50
65	Mayapan	Kukula					tinaja	A	MY-BROWN- 50
68	Mayapan	Tecoh					tinaja	F	MY-BROWN- 50
70	Mayapan	Polbox					tinaja	H	MY-BROWN- 50
72	Mayapan	Xcanchakan					tinaja	F	MY-BROWN- 50
74	Mayapan	Moyos					velero	G	MY-BROWN-35
76	Mayapan	Unslipped-Exterior					tinaja	A	MY-BROWN- 35
78	Mayapan	Unslipped-Exterior					tinaja	A	MY-BROWN- 35
79	Mayapan	Polbox					Tinaja	A	MY-BROWN- 35
80	Mayapan	Sulche					Tinaja	A	MY-BROWN- 35
82	Mayapan	Yacman			CG-CM-W-c		Olla	A	MY-BROWN- 35
85	Mayapan	Xcanchakan					tinaja	G	MY-BROWN- 58
87	Mayapan	Mama					tinaja	F	MY-BROWN- 58
88	Mayapan	Kukula					tinaja	F	MY-BROWN- 51
89	Mayapan	Yacman					Olla	C	MY-BROWN- 110
90	Mayapan	Yacman					Olla	C	MY-BROWN- 110
91	Mayapan	Payil					tinaja	F	MY-BROWN- 110
95	Mayapan	Papacal					tinaja	F	MY-BROWN- 110
98	Mayapan	Yacman	CGS065	1	DZ-W-c	Mayapan	Olla		MY-BROWN- 110
101	Mayapan	Navula	CGS070	1	DZ-B-c	Mayapan	cajete	F	MP980702-1
102	Mayapan	Navula	CGS071	u/a	CM-W-m	Mayapan	Olla	F	MP980702-1
103	Mayapan	Navula					cylinder	F	MP981506-2

#	Site	Variety	ANID	Ch. Grp	Fabric Class (thin section)	Most common Euclid match	Vessel Shape	(*)	Sample ID
104	Mayapan	Navula	CGS072	u/a	CM-W-m	Mayapan	Olla	F	MP989701
105	Mayapan	Navula					Cajete	F	MP980792
106	Mayapan	Navula	CGS073	1	CH-W-cm	Mayapan	Cajete	A	MP980752
107	Mayapan	Navula					Cajete	F	MP980731
109	Mayapan	Navula	CGS074	1	DZ-W-m	Mayapan	Cajete	A	MP980752
110	Mayapan	Navula	CGS075	1	DZ-W-m	Mayapan	Cajete	F	MP980691
111	Mayapan	Mama					tinaja	G	MP980701
112	Mayapan	Mama					Cajete	F	MP970526-1
113	Mayapan	Mama	CGS052	1	DZ-W-c	Mayapan	Cajete	G	MP9770506
114	Mayapan	Mama					Tecom ate	G	MP982014-1
115	Mayapan	Mama			DZ-W-m		Cajete	G	MP982065-3
117	Mayapan	Mama					cajete	F	MP982811-1
119	Mayapan	Mama	CGS053	2	QZ-MX-B-f	Ticul	Cajete	A	MP981505-1
121	Mayapan	Unslipped-Exterior					tinaja	F	MP982926-2
129	Mayapan	Payil	CGS083	4		Laguna de On	tinaja	O	MP980793
130	Mayapan	Payil	CGS084	4		Laguna de On	tinaja	C	MP980793
131	Mayapan	Payil	CGS085	4	CX-QZ-B/R-f	Laguna de On	tinaja	C	MP980793
132	Mayapan	Payil					Cajete	C	MP981740-1
133	Mayapan	Payil	CGS086	4	CX-QZ-B/R-f	Laguna de On	tinaja	C	MP982004-2
134	Mayapan	Payil	CGS087	u/a	CX-QZ-B/R-f	S. Rita Corozal	tinaja	C	MP982041-1
135	Mayapan	Payil	CGS088	u/a	CX-CM-01-f	various	tinaja	C	MP982501
136	Mayapan	Payil	CGS089	4	CX-QZ-B/R-f	Laguna de On	tinaja	C	MP980752
137	Mayapan	Payil					Cajete	C	MP982006-2
141	Mayapan	Polbox					Tinaja	A	MP981527-1
142	Mayapan	Polbox					Cajete	A	MP982969-2
146	Mayapan	Yacman	CGS066	2	CG-CM-W-c	Mayapan	Olla	C	MP980731
147	Mayapan	Yacman			DZ-B-c		Olla	G	MP980752
148	Mayapan	Yacman	CGS067	1	DZ-B-c	Mayapan	Olla	G	MP970526-1
149	Mayapan	Yacman	CGS068	1	DZ-B-c	Mayapan	Olla	G	MP980792
150	Mayapan	Yacman	CGS069	u/a	DZ-B-c	Mayapan	Olla	G	MP980792
151	Mayapan	Yacman			DZ-B-c		Olla	G	MP982547-2
152	Mayapan	Yacman			DZ-W-c		Olla	G	MP982565-1
153	Telchaquillo	Yacman	CGS038	1	DZ-B-c	Mayapan	Olla	F	TQ97 0970
154	Telchaquillo	Yacman	CGS039	5	MX-CX-B/R-c	S. Rita Corozal	Olla	C	TQ97 0092-1
155	Telchaquillo	Mama	CGS033	1	CG-CM-W-c	Mayapan	Tinaja	A	TQ97 0727-1
156	Telchaquillo	Mama					Cajete	F	TQ97 0988
157	Telchaquillo	Mama					cajete	A	TQ97 1007-1
158	Telchaquillo	Mama	CGS034	1	DZ-W-m	Mayapan	tinaja	F	TQ97 0970-2
159	Telchaquillo	Mama					tinaja	F	TQ97 0912-1
160	Telchaquillo	Mama	CGS035	1	DZ-W-m	Mayapan	tinaja	F	TQ97 0130-1
161	Telchaquillo	Mama	CGS036	u/a	CM-W-m	Mayapan	Cajete	F	TQ971203-2
162	Telchaquillo	Navula	CGS043	1	DZ-W-c	Mayapan	Olla	F	TQ97 0454-3
163	Telchaquillo	Navula	CGS044	u/a	CS-W/B-cm	Mayapan	Olla	G	TQ97 0728-1
164	Telchaquillo	Navula			DZ-B-c		Olla	G	TQ97 0898
165	Telchaquillo	Navula	CGS045	5	DZ-B-c	S. Rita Corozal	olla	G	TQ 970727
166	Telchaquillo	Navula					olla	G	TQ97 0743
167	Telchaquillo	Navula					olla	G	TQ97 1207
168	Telchaquillo	Xcanchakan					tinaja	A	TQ97 3024-2
169	Telchaquillo	Xcanchakan	CGS050	1	DZ-W-c	Mayapan	tinaja	F	TQ971007-1
170	Telchaquillo	Xcanchakan					tinaja	F	TQ971208-5
171	Telchaquillo	Xcanchakan			Untemprd		tinaja	F	TQ970727
172	Telchaquillo	Xcanchakan	CGS051	2	CS-W/B-cm	Ticul	tinaja	A	TQ970454
173	Telchaquillo	Xcanchakan					tinaja	F	TTQ 97 0783
175	Telchaquillo	Yacman	CGS040	u/a	DZ-B-c	various	Olla	F	TQ97 0791
176	Telchaquillo	Yacman	CGS041	1	CSm-W/B-c	Mayapan	Olla	F	TQ97 1584-2
178	Culuba	Vista Alegre						C	CLBE E11-14 10-E IV
179	Culuba	Vista Alegre						C	CLB 01 11Kloc 13-E III
180	Culuba	Vista Alegre						C	CLB01 9KK 12-II II
181	Culuba	Vista Alegre	CGS154	u/a	CX-CM-02-c	Tipuj		G	CLB01 11Klo 9-C II

#	Site	Variety	ANID	Ch. Grp	Fabric Class (thin section)	Most common Euclid match	Vessel Shape	(*)	Sample ID
182	Culuba	Vista Alegre						C	CLB01 11KLo 18-k II
183	Culuba	Vista Alegre						C	CLB02 9K1C 12-HH II
184	Culuba	Vista Alegre						C	CLB01 9K1C 12-GG
185	Culuba	Vista Alegre						C	CLB01 AK19 5-S I
188	Culuba	Yacman	CGS158	3		various	Olla	A	CLB00 PM 3M I
189	Culuba	Yacman					Olla	A	CLB00 9KK 16-29 I
190	Culuba	Yacman					Olla	A	CLB00 9KK 16-FF I
191	Culuba	Yacman	CGS159	3	CX-CM-02-c	Caye Coco	Olla	C	CLB00 RM 29-R I
200	Culuba	Yacman	CGS160	3	CX-CM-02-c	Ticul	Olla	C	CLB00 9K1C 16-FF I
201	Culuba	Yacman					Olla	C	CLB00 PM 2K I
202	Culuba	Yacman	CGS161	3		Laguna de On	Olla	G	CLB00 9K1C 16-FF I
203	Culuba	Mama	CGS147	2	CX-CM-01-c	Mayapan	tinaja	C	CLB00 9KK 17CC I
204	Culuba	Mama	CGS148	2	CX-CM-01-c	Mayapan	tinaja	C	CLB00 9KIC 16EG I
205	Culuba	Mama					tinaja	A	CLB00 9KIC 16EG I
206	Culuba	Mama			CX-CM-02-c		tinaja	C	CLB00 9KIC 16EE I
207	Culuba	Mama	CGS149	u/a	CX-CM-01-c	various	tinaja	A	CLB00 PM 32-1 II
208	Culuba	Mama	CGS150	2	CX-CM-01-c	various	tinaja	C	CLB00 9KIC 16FF I
209	Culuba	Mama	CGS151	2	CX-CM-01-c	various	tinaja	C	CLB00 PM 18-J I`
210	Culuba	Mama	CGS152	3	CX-CM-01-c	Laguna de On	tinaja	A	CLB00 9KK 16-FF
211	Culuba	Mama					tinaja	A	CLB00 PM 18-J I
212	Culuba	Mama					tinaja	C	CLB00 PM 32-0 I
213	Culuba	Cancun					tinaja	C	CLB00 PM 28 I
214	Culuba	Cancun	CGS155	2		Mayapan	tinaja	H	CLB00 PM 26-K II
215	Culuba	Cancun			CX-CM-01-c		tinaja	H	CLB00 PM 35-A I
216	Culuba	Cancun			DX-CX-m-h		tinaja	C	CLB00 PM 35-K
217	Culuba	Cancun	CGS156	u/a		Mayapan	tinaja	H	CLB00 PM 35-K I
218	Culuba	Cancun					tinaja	C	CLB00 PM 35-J I
219	Culuba	Cancun					tinaja	C	CLB00 PM CT-1 I
229	Culuba	Navula					Olla	C	CLB00 PM 34-0 I
230	Culuba	Navula					Olla	C	CLB00 PM 36-M I
231	Culuba	Navula					Olla	G	CLB00 PM 18-J
232	Culuba	Navula					Olla	G	CLB00 PM 34-0 I
233	Culuba	Navula					Olla	C	CLB00 PM 36-M I
234	Culuba	Navula					Olla	C	CLB00 PM 32-I I
235	Tecoh	Mama	CGS016	1	DZ-W-m	Mayapan	tinaja	F	TC 97 0844-1
236	Tecoh	Mama	CGS017	u/a	CS-W/B-cm	various	tinaja	A	TC 97 10750-1
237	Tecoh	Mama	CGS015	2	DX-B/W-m	Mama	tinaja	A	TC 97 0130
238	Tecoh	Yacman			DZ-B-c		Olla	G	TC970728-1
239	Tecoh	Yacman	CGS029	1	DZ-B-c	Mayapan	Olla	G	TC970728-1
241	Mayapan	Yacman			DZ-W-c		Olla	G	MP
242	Mayapan	Navula	CGS076	1	CM-W-m	Mayapan	Olla	A	MP 98 0739
243	Mayapan	Navula	CGS077	1	DZ-W-c	Mayapan	Olla	G	MP 98 1513-3
244	Mayapan	Navula	CGS078	1	DZ-W-c	Mayapan	Olla	G	MP98 0739 D#91
246	Mayapan	Navula			DZ-W-m		Cajete	G	MP 98 1679-1
247	Mayapan	Navula					Olla	F	
248	Mayapan	Mama	CGS054	1	DZ-W-m	Mayapan	tinaja	F	MP980752
249	Mayapan	Mama	CGS055	1	DZ-W-c	Mayapan	tinaja	F	MP9822065-3
250	Mayapan	Mama					tinaja	O	MP9800641
251	Mayapan	Mama					tinaja	A	MP98 0641
253	Mayapan	Navula	CGS079	1	DZ-W-m	Mayapan	Cajete	G	MP 98 0792
254	Mayapan	Navula					Cajete	G	MP
255	Mayapan	Navula					Cajete	G	
256	Mayapan	Navula	CGS080	1	DZ-W-m	Mayapan	Cajete	G	MP 98 0770-1
257	Mayapan	Navula	CGS081	1	DZ-W-m	Mayapan	Cajete	G	MP 930960
258	Mayapan	Navula	CGS082	1	DZ-B-c	Mayapan	Olla	G	MO 98 2065-4
259	Mayapan	Mama	CGS056	1	CM-W-m	Mayapan	tinaja	F	MP 981S08-2
260	Mayapan	Mama	CGS057	1	DZ-W-m	Mayapan	tinaja	F	MP98 0752
261	Tipikal	Mama	CGS131	2	CS-W/B-cm	Laguna de On	Cajete	A	TI99 1087-1
262	Tipikal	Mama	CGS132	1	CS-W/B-cm	Mayapan	Cajete	A	TI 94 1107-1
263	Tipikal	Mama	CGS133	1	DZ-W-m	Mayapan	Cajete	F	TI 99 1125-1
264	Tipikal	Mama					Cajete	A	TI 99 99 1125-1

#	Site	Variety	ANID	Ch. Grp	Fabric Class (thin section)	Most common Euclid match	Vessel Shape	(*)	Sample ID
265	Tipikal	Mama					tinaja	F	T1 99 3000-1
266	Tipikal	Mama	CGS134	1	DZ-W-c	Mayapan	tinaja	A	Ti 99 1124-1
267	Tipikal	Mama	CGS135	1	DZ-W-m	Mayapan	tinaja	A	TI 99 11-24-1
268	Tipikal	Mama			DZ-W-c		tinaja	G	TI 99 1104-1
269	Tipikal	Mama	CGS136	u/a	CM-W-m	Mayapan	tinaja	F	TI 99 1106-1
270	Tipikal	Yacman			DZ-W-c		Olla	G	TI 99 1107-1
271	Tipikal	Yacman	CGS137	1	DZ-W-c	Mayapan	Olla	G	TI 99 1106-1
272	Tipikal	Yacman	CGS138	1	DZ-W-c	Mayapan	Olla	G	TI 99 1107-1
273	Tipikal	Yacman			DZ-W-c		Olla	G	TI 99 1144-1
274	Tipikal	Yacman			DZ-W-c		Olla	G	TI 99 1144-1
275	Tipikal	Yacman	CGS139	5	MX-CX-B/R-c	S. Rita Corozal	Olla	C	TI 99 1087-1
276	Tipikal	Navula					Olla	F	TI 99 1087-1
277	Tipikal	Navula	CGS142	2	CSm-W/B-c	Gruta de Chac	Olla	G	TI 99 1929-1
278	Tipikal	Navula	CGS143	u/a	CS-DZ-W-c	Mayapan	Olla	G	TI 97 1111-1
279	Tipikal	Navula	CGS144	2	CH-B-c	various	Olla	C	TI 99 1509-2
280	Tipikal	Navula					Olla	G	TI 99 1530-1
281	Tipikal	Navula	CGS145	2	CX-CM-01-c	various	Olla	C	TI 49 1530-1
282	Tepich	Mama	CGS001	1	CH-W-cm	Mayapan	Cajete	F	TP 97 0849
283	Tepich	Mama	CGS002	1	CH-W-cm	Mayapan	Cajete	F	TP 47 ??92
284	Tepich	Mama	CGS003	1	CH-W-cm	Mayapan	Cajete	F	TP 97 1032
285	Tepich	Mama	CGS004	1	CH-W-cm	Mayapan	Cajete	F	TP 42 0794
286	Tepich	Mama	CGS005	1	CH-W-cm	Mayapan	Cajete	F	TP 97 3046-1
288	Tepich	Yacman	CGS009	1	CG-CM-W-c	Mayapan	Olla	A	TP 97 1634-1
289	Tepich	Yacman	CGS010	1	CSm-W/B-c	Mayapan	Olla	A	TP 43 3107
290	Tepich	Yacman	CGS011	u/a	CSm-W/B-c	Mayapan	Olla	A	TP 97 0849
291	Tepich	Yacman	CGS012	1	CX-DXsilt-W-c	Mayapan	Olla	F	TP 97 2245
300	Tepich	Yacman	CGS013	1	CG-CM-W-c	Mayapan	Olla	A	TP 97 1034-3
301	Tepich	Yacman	CGS014	1	CH-W-cm	Mayapan	Olla	F	TP 97 3111-1
302	Tepich	Navula			CG-CM-W-m		Cajete	A	TP 97 1928
303	Tepich	Navula			CG-CM-W-m		Cajete	A	TP 97 1634-1
304	Tepich	Navula	CGS007	1	CH-W-cm	Mayapan	Olla	F	TP 97 2231-1
305	Tepich	Navula			CG-CM-W-m		Olla	A	TP 97 5353``
306	Tepich	Navula			CG-CM-W-m		Cajete	A	TP 97 2123
307	Tepich	Navula			CG-CM-W-m		Cajete	A	TP 97 1888-1
308	Tepich	Mama					tinaja	A	TP 97 2271
309	Tepich	Mama	CGS006	1	CH-W-cm	Mayapan	tinaja	A	TP 97 1753
310	Tepich	Mama					tinaja	A	TP 97 0776
311	Tepich	Mama			CH-W-cm		tinaja	F	TP 97 0790
312	Telchaquillo	Yacman	CGS042	1	DZ-W-c	Mayapan	Olla	F	TQ97 0702
313	Telchaquillo	Navula	CGS046	1	DZ-W-m	Mayapan	Olla	F	TQ 970727
314	Telchaquillo	Navula	CGS047	1	DZ-B-c	Mayapan	Olla	F	TQ 0702- 3A
315	Telchaquillo	Navula	CGS048	u/a	DZ-W-m	Mayapan	Olla	F	TQ 97 1016 - 1
316	Telchaquillo	Mama					Cajete	F	TQ 97 08824-1
317	Telchaquillo	Mama			DZ-W-m		Cajete	F	TQ 971220
318	Telchaquillo	Mama			DZ-W-m		Cajete	F	TQ 97 0743
319	Telchaquillo	Mama	CGS037	1	CS-W/B-cm	Mayapan	tinaja	F	TQ 47 0454 - 2
320	Culuba	Cancun	CGS157	3	DX-CX-m-h	Ticul	Cajete	C	CL 800 PM 27-J
321	Culuba	Cancun					Cajete	F	CJB 00 PM 26-J I D
322	Mayapan	Mama	CGS058	1	DZ-W-mc-h	Mayapan	Cajete	F	MP 930506-1
323	Mayapan	Mama	CGS059	1	DZ-W-m	Mayapan	Cajete	F	MP 97 0526-1
324	Mayapan	Mama			CS-W/B-cm		Cajete	A	MY 96 3375
325	Mayapan	Mama					Cajete	A	MY 96 3351
326	Mayapan	Xcanchakan					tinaja	F	MP 98
327	Mayapan	Xcanchakan			CM-W-m		tinaja	F	MP 98
328	Mayapan	Xcanchakan	CGS090	u/a	CX-CM-01-c	Ticul	tinaja	C	MP 9886-1
329	Mayapan	Xcanchakan					tinaja	F	MP 98
330	Mayapan	Xcanchakan					tinaja	A	MP 98
331	Mayapan	Xcanchakan					tinaja	F	MP 98
332	Tecoh	Yacman	CGS030	1	CM-W-m	Mayapan	Olla	G	TC 97 0728-1
333	Tecoh	Yacman	CGS031	1	CM-W-m	Mayapan	Olla	F	TC 97 0728-1
334	Tecoh	Yacman	CGS032	1	CG-CM-W-c	Mayapan	Olla	F	TC 97 0750-1

#	Site	Variety	ANID	Ch. Grp	Fabric Class (thin section)	Most common Euclid match	Vessel Shape	(*)	Sample ID
335	Tecoh	Navula	CGS022	2	CSm-W/B-c	various	Olla	A	TC 97 0844
336	Tecoh	Navula	CGS023	u/a	DZ-W-c	Mayapan	Olla	F	TC 97 0750
337	Tecoh	Navula					Olla	A	TC 97 0106-1
338	Tecoh	Mama					tinaja	A	TC 97 0844
339	Tecoh	Mama	CGS018	2	DX-B/W-m	Mama	tinaja	A	TC 97 0130
340	Mama	Mama	CGS106	u/a	CM-W-m	various	tinaja	A	MM 98 est.H DESM
341	Mama	Mama	CGS107	1	DZ-W-mc-h	Mayapan	tinaja	F	MM 98 1395 - 1
342	Mama	Mama	CGS108	1	DZ-W-mc-h	Kiuic	tinaja	F	MM 98 1395-1
343	Mama	Mama	CGS109	1	DZ-W-mc-h	Mayapan	tinaja	F	MM 98 1395 - 1
344	Mama	Mama	CGS110	u/a	CM-W-m	Mayapan	tinaja	A	MM 98 0946 - 1
345	Mama	Mama			CM-W-m		Cajete	A	MM 98 0891 -
346	Mama	Yacman	CGS113	1	CSm-W/B-c	Kiuic	Olla	A	MM 98 0866-2
347	Mama	Yacman	CGS114	2	CH-YC-W-c-h	Ticul	Olla	F	MM 98 2991- 1
348	Culuba	Palmul					tinaja	C	CLB 99 1M 34.P I O
349	Culuba	Palmul	CGS153	3	CH-R-f	Ticul	tinaja	C	CLB 00 PM 40-M
350	Mama	Yacman	CGS115	1	DZ-W-mc-h	Mayapan	Olla	F	MM98 2991-1
351	Mama	Yacman	CGS116	u/a		various	Olla	F	MM 98 0866
352	Mama	Navula	CGS117	u/a	CH-B-c	various	Olla	F	MM 98 2471
354	Coba	Cancun	CGS162	2	CX-CM-01-c	Laguna de On	tinaja	F	
355	Coba	Cancun	CGS164	u/a	CS-W/B-cm	Tepakan	tinaja	C	
356	Coba	Cancun	CGS165	1		Mayapan	cajete	C	
357	Coba	Cancun	CGS167	1	CS-W/B-cm	Mayapan	cajete	A	
358	Coba	Cancun	CGS166	u/a		Akil	tinaja	G	
359	Coba	Cancun	CGS170	3		Ticul	tinaja	A	
360	Coba	Cancun	CGS169	2	CS-W/B-cm	Laguna de On	tinaja	C	
361	Coba	Navula	CGS173	1		Mayapan	Olla	G	
362	Coba	Navula	CGS174	2	CSm-W/B-c	Kiuic	Olla	G	
363	Coba	Navula					Olla	G	
364	Coba	Navula	CGS171	2	CS-W/B-cm	various	Olla	A	
365	Coba	Navula	CGS172	1	CX-CM-01-c	Mayapan	Olla	F	
366	Chac Mool	Yacman	CGS181	5	CX-CM-02-c	Ticul	Olla	G	
367	Chac Mool	Yacman	CGS182	u/a	CX-CM-02-c	various	Olla	C	
368	Chac Mool	Yacman			Magnt-MX-R-f		Olla	O	
369	Chac Mool	Mama	CGS186	u/a	CX-QZ-B/R-f	S. Rita Corozal	Bowl	C	
370	Chac Mool	Xcanchakan	CGS184	5	CX-CM-01-c	Ticul	tinaja	C	
371	Chac Mool	Xcanchakan	CGS185	5	CX-CM-01-c	Ticul	tinaja	C	
372	Chac Mool	Xcanchakan			CX-CM-01-c		tinaja	C	
373	Chac Mool	Xcanchakan			CX-CM-01-c		tinaja	C	
374	Chac Mool	Navula	CGS183	5	GR-CM-R-m	S. Rita Corozal	Olla	F	
375	Chac Mool	Navula			CX-QZ-B-c		Olla	C	
376	Chac Mool	Payil	CGS175	4	CX-QZ-B/R-f	Laguna de On	Tinaja	C	
377	Chac Mool	Payil	CGS176	4	CX-QZ-B/R-f	Laguna de On	Tinaja	C	
378	Chac Mool	Navula			CX-CM-02-c		Olla	C	
379	Chac Mool	Palmul	CGS180	4	CH-QZ-B-f	Laguna de On	Bowl	C	
380	Chac Mool	Palmul	CGS177	4		Laguna de On	Bowl	C	
381	Chac Mool	Palmul	CGS178	u/a	CX-QZ-B/R-f	S. Rita Corozal	Bowl	C	
382	Chac Mool	Palmul	CGS179	4	CX-QZ-B/R-f	Laguna de On	Bowl	C	
384	Chac Mool	Navula			QZ-R-m		Olla	O	
385	Coba	Cancun	CGS163	1	CX-CM-01-c	Mayapan	tinaja	F	
386	Coba	Navula					Olla	G	
387	Chac Mool	Yacman			CX-CM-02-c				
388	Coba	Cancun	CGS168	5	CX-CM-01-c	Tepakan	tinaja	A	
389	Tepich	Xcanchakan	CGS008	5	CS-W/B-cm	S. Rita Corozal	tinaja		TP 97 2230-1
390	Tecoh	Navula	CGS024	2	CS-W/B-cm	Ticul	Olla	F	TC 97 0106-1
391	Tecoh	Navula	CGS025	5	CSm-W/B-c	S. Rita Corozal	Olla	F	TC97 0764-1
392	Tecoh	Navula	CGS026	2	DX-B/W-m	Mama	Olla	G	TC97 0106-1
393	Tecoh	Navula	CGS027	5	CSm-W/B-c	S. Rita Corozal	Olla	G	TC 97 0106
394	Tecoh	Navula	CGS028	1	DZ-W-m	Mayapan	Olla	F	TC97 0844
395	Tecoh	Mama	CGS019	1	DZ-W-m	Mayapan	Tinaja	F	TC 970844-1
396	Tecoh	Mama	CGS020	1	DZ-W-c	Mayapan	Tinaja	G	TC 97 0852
397	Tecoh	Mama	CGS021	1	Untemprd	Mayapan	Tinaja	G	TC 97 0844

#	Site	Variety	ANID	Ch. Grp	Fabric Class (thin section)	Most common Euclid match	Vessel Shape	(*)	Sample ID
398	Telchaquillo	Navula	CGS049	1	DZ-W-m	Mayapan	Olla	G	TQ 970727-1
399	Tipikal	Navula					Olla	G	TI 99 1163-1
400	Tipikal	Navula					Cajete	G	TI 99110S-1
401	Tipikal	Navula					Olla	G	TI 991123-1
402	Tipikal	Navula					Olla	G	TI 991123-1
403	Tipikal	Navula	CGS146	2	CS-W/B-cm	Kiuic	Olla	G	TI 99 1171-2
404	Tipikal	Yacman			DZ-W-c		Olla	G	TI99 1144-1
405	Tipikal	Yacman	CGS140	1	DZ-W-c	Mayapan	Olla	G	TI99 1146-2
406	Tipikal	Yacman			DZ-W-c		Olla	G	TI99 1610-2
407	Tipikal	Yacman	CGS141	1	DZ-B-c	Mayapan	Olla	G	TI99 1146-2
408	Mama	Navula					olla	C	MM98 008-1
409	Mama	Navula	CGS118	u/a	CH-B-c	Mayapan	Olla	G	MM98 0887-1
410	Mama	Navula	CGS119	u/a	CH-B-c	Kiuic	Cajete	F	MM98 1714-1
411	Mama	Navula	CGS120	u/a	CM-W-m	Mayapan	Cajete	G	MM98 0986-1
412	Mama	Mama			DZ-W-mc-h		Tinaja	F	MM98 1395-1
413	Mama	Mama			DZ-W-c		Tinaja	G	MM98 1472-1
414	Mama	Mama			DZ-W-mc-h		Tinaja	F	MM98 1345-1
415	Mama	Mama	CGS111	1	DZ-W-mc-h	Mayapan	Tinaja	G	MM98 1395-1
416	Mama	Mama	CGS112	1	DZ-W-mc-h	Mayapan	Tinaja	G	MM98 1395-1
417	Teabo	Mama					Tinaja	A	TB97 1854-1
418	Teabo	Mama	CGS121	u/a	CS-W/B-cm	Mayapan	Tinaja	F	TB98 0421-1
419	Teabo	Mama	CGS122	2	CS-W/B-cm	various	Tinaja	A	TB98 1068-1
420	Teabo	Mama	CGS123	u/a	CM-W-m	Mayapan	Tinaja	F	TB98 1726-1
421	Teabo	Mama	CGS124	2	CS-W/B-cm	various	Tinaja	F	TB98 0821-1
422	Teabo	Mama					Tinaja	A	TB98 0890-1
423	Teabo	Mama					Cajete	A	TB98 1096-1
424	Teabo	Mama					Cajete	A	TB98 1033-1
425	Teabo	Navula					Olla	A	TB98 0884-1
426	Teabo	Navula	CGS130	2	CS-W/B-cm	Mayapan	Olla	A	TB98 0847-1
427	Teabo	Navula					Olla	G	TB98 0643-1
428	Teabo	Navula			CH-B-c		Olla	G	TB98 0793-1
429	Teabo	Navula					Olla	G	TB98 0821-1
430	Teabo	Yacman	CGS125	u/a	Tcf-MX-CX-B-c	Chac	Olla	F	TB98 0972-1
431	Teabo	Yacman	CGS126	2	CSm-W/B-c	various	Olla	G	TB98 1097-1
432	Teabo	Yacman	CGS127	2	CSm-W/B-c	various	Olla	G	TB98 1097-1
433	Teabo	Yacman	CGS128	5	MX-CX-B/R-c	various	Olla	A	TB98 0903-1
434	Teabo	Yacman					Olla	C	TB98 1012-1
435	Teabo	Yacman	CGS129	5	MX-CX-B/R-c	S. Rita Corozal	Olla	F	TB98 0793-1
436	Mayapan	Mama	CGS060	1	DZ-W-mc-h	Mayapan	Tinaja	F	Pemy 09 8553
437	Mayapan	Mama	CGS061	1	CM-W-m	Mayapan	Cajete	F	Pemy 09 8553
438	Mayapan	Mama	CGS062	1	DZ-W-mc-h	Mayapan	Cajete	A	Pemy 09 8553
439	Mayapan	Mama	CGS063	1	DZ-W-m	Mayapan	Tinaja	F	Pemy 09 8553
440	Tekit	Yacman	CGS096	2	DX-B/W-m	Mayapan	Olla	A	TK980203-1
441	Tekit	Yacman	CGS097	5	DX-CX-B-c	S. Rita Corozal	Olla	C	TK980239
442	Tekit	Yacman	CGS098	5	DX-CX-B-c	various	Olla	C	TK80239
443	Tekit	Yacman	CGS099	u/a	CH-B-c	Ticul	Olla	C	TK98 0199-1
444	Tekit	Yacman	CGS100	2	CG-CM-W-c	Tepekan	Olla	A	TK980237-1
445	Tekit	Mama	CGS091	5	MX-CX-B/R-c	S. Rita Corozal	Tinaja	A	TK99 0258
446	Tekit	Mama	CGS092	1	CM-W-m	Mayapan	Tinaja	F	TK98 1850-1
447	Tekit	Mama	CGS093	2	DX-B/W-m	Ticul	Tinaja	A	TK98 0014
448	Tekit	Mama	CGS094	1	DZ-W-m	Mayapan	Tinaja	F	TK98 0117-1
449	Tekit	Mama	CGS095	1	DZ-W-m	Mayapan	Cajete	G	TK98 0151
450	Tekit	Navula	CGS101	2	CH-B-c	Mayapan	Olla	G	TK98 0055
451	Mayapan	Mama	CGS064	1	DZ-W-c	Mayapan	Cajete		MP 981505-1
452	Tekit	Navula	CGS102	2	CS-W/B-cm	Kiuic	Olla	G	TK 98 0213
453	Tekit	Navula	CGS103	u/a	CH-B-c	Kiuic	Olla		TK 98 0157
454	Tekit	Navula	CGS104	2	CH-B-c	Sayil	Olla	A	TK 98 2253
9107	Mayapan	Yacman					Olla	A	MY-BROWN- 110
9113	Mayapan	Mama					tinaja	F	MY-BROWN- 87
9114	Mayapan	Mama					tinaja	F	MY-BROWN- 87
9115	Mayapan	Mama					tinaja	F	MY-BROWN- 87

#	Site	Variety	ANID	Ch. Grp	Fabric Class (thin section)	Most common Euclid match	Vessel Shape	(*)	Sample ID
9116	Mayapan	Mama					tinaja	A	MY-BROWN- 87
9117	Mayapan	Mama					tinaja	A	MY-BROWN- 87
9118	Mayapan	Mama					tinaja	A	MY-BROWN- 87
9119	Mayapan	Mama					tinaja	F	MY-BROWN- 87
9120	Mayapan	Mama					cajete	A	MY-BROWN- 110
9121	Mayapan	Mama					tinaja	F	MY-BROWN- 110
9122	Mayapan	Mama					tinaja	F	MY-BROWN- 110
9123	Mayapan	Mama					tinaja	F	MY-BROWN- 110
9124	Mayapan	Mama					tinaja	F	MY-BROWN- 110
9125	Mayapan	Mama					tinaja	F	MY-BROWN- 110
9127	Mayapan	Yacman					Olla	C	MY-BROWN- 89
9130	Mayapan	Chen Mul					censer	A	032195
9132	Mayapan	Tecoh					tinaja	A	963998
9133	Mayapan	Tecoh			DZ-W-m		tinaja	C	963998
9136	Mayapan	Mama					tinaja	G	063616-3
9137	Mayapan	Mama					tinaja	F	063616-3
9150	Mayapan	Mama					Bowl		MP982047-2
9151	Mayapan	Mama					Cajete		MP980506-4
9152	Mayapan	Mama					Cajete		MP98
9153	Cenote Pixyah	Mama					tinaja		
9154	Cenote Pixyah	Mama					tinaja		

(*) Main hand-specimen grouping: A=dark particles; F=micrite; G=sparite; C=single crystals

8.12 Summary

Table 8-19 combines the tables presented earlier in this chapter for each type of fabric. Associations between these types of fabrics, sites, and ceramic varieties can be observed in this table. Two blocks of fabrics have been highlighted in the table. One block contains, in pink, the fabric classes in which most north-central red-slipped (Mama) samples were found. The second contains, in grey, the fabrics in which most of the unslipped, striated samples (Yacman) were found.

Table 8-19 shows that:

1. The samples from the north-central sites show a clear division by grain size between the utilitarian ceramic varieties of red-slipped Mama (pink shaded area) and the striated open-mouth Yacman (pale grey shaded area) samples: red-slipped samples are largely medium-grained whereas the striated open-mouth samples are coarse.
2. The division by grain size between Mama and Yacman that was observed at the north-central sites was not observed in the Mama and Yacman samples from Cobá, Culubá, and Chac Mool.
3. Instead, a partition by grain size was observed in the studied samples from the eastern sites that divides elite, or Payil, and utilitarian (Mama, Cancun, Navula, Yacman, and Xcanchakan) samples into fine and coarse respectively.
4. At the north-central sites, most Mama samples fall within micro or finely crystalline calcite or dolomite fabrics. This is to say, micrite, sparite, or dolospar fabrics. In hand specimens, these three types of inclusions are difficult to impossible to differentiate, many times requiring the polarising microscope to distinguish them.
5. At the north-central sites, most Yacman samples were found within the dolospar, dark micrite, and discrete crystal fabrics.

With the exception of the dolospar fabrics, Mama and Yacman are not usually found sharing the same fabric. The fabrics of Mama and Yacman overlap in one type of fabric, dolospar, although they differentiate in their grain size. On the other hand, Mama and Navula co-occur.

In the next chapter, the information from this chapter about fabric classes and their distribution is combined with the geological (chapter three), hand-specimen (chapter six), chemical (chapter seven), and ethnographic information to draw inferences about the ceramic techniques applied to raw materials in producing the vessels sampled in this study. Aspects of production organisation will also be examined.

Table 8-19. Distribution of the Fabric Classes within Sites and Ceramic Varieties

	Fabric Class	Chemcl Group	Tepich	Tecoh	Telchaquillo	Mayapán	Tekit	Mama	Tipikal	Teabo	Cobá	Chac Mool	Culubá
MEDIUM GRAINED	**DARK MICRITE**												
	CG-CM-Wm	1,2,uk	NNNN N			*M*							
	BIOCLASTS												
	CH-W-cm	1	*MMM MMM M* N Y			N							
	MEDIUM-CRYSTALLINE DOLOMITE												
	DX-B/W-m	2		*M M* N			*M* Y						
	FINELY CRYSTALLINE SPARITE												
	CS-W/B-cm	2, u/a	X	*M* N	*M* N X	*MM*	N		*MM* N	*MMM* N	*MMM* N		
	MICRITE												
	CM-W-m	1, u/a		YY	*M*	*MMM* NNN X	*M*	*MMM* N	*M*	*M*			
	FINELY CRYSTALLINE DOLOSPAR												
	DS-W-m	1		*MM* N	*MMMM* NNN	*MMMMMM* NNNNNN	*MM*		*MM*				
	DS-W-mc-h	1				*MMM*		*MMM MMM M* Y					
COARSELY GRAINED	DS-B-c	1		YY	NN YY	*M* NN YYYYYYYY			Y				
	DS-W-c	1		*M* N	N Y X	*MMM* NNN YYYY C		*M*	*MM* YYYY YYYY				
	DARK MICRITE												
	CG-CM-W-c	1,2	YY	Y	*M*	YY	Y						
	MEDIUM AND COARSELY CRYSTALLINE CALCITE												
	CSm-W/B-c	1,2 u/a	YY	N N N	Y			Y	N	YY	N		
	MX-CX-GROG-B/R-c	5			Y		*M*		Y	YY			
	DX-CX-B-c	uk,5				NN	YY						
	CX-QZ-B-c	uk										N	
	DX-CX-m-h	3											*MM*
	CX-CM-01-c	1,23,5 u/a				X			N		*MMM* N	XXXX	*MMM M M M*
	CX-CM-02-c	3,5, uk										N YYY	*M* YY V
	SKELETAL REMAINS												
	CH-B-c	2, u/a				NNN Y	NNN	N	N				
	FINE CALCITE AND QUARTZ												
FINE	CX-QZ-B/R-f	4, u/a				PPPP						PPPP P	

KEY: *M* = Mama variety (slipped), N = Navula (plain), Y = Yacman (striated), P = Payil, X = Xcanchakan and C= Chen Mul

Chapter 9

LATE POSTCLASSIC CERAMIC PRODUCTION TECHNOLOGY

In this chapter, patterns in the data are sought and inferences drawn aiming to reconstruct two main aspects of ceramic production and organisation: one, the ceramic techniques applied at the different stages of raw material processing and pottery manufacture to produce the vessels sampled in this study; and second, the variation of the fabric classes indicative of pottery groups. In contrast to previous chapters, in which the properties and attributes of the pottery clay, samples, and raw materials were analysed forming petrographic (chapter six and eight) and chemical (chapter seven) groups, in this chapter, a more holistic approach is taken to interpret patterns in the analytical results with reference to geological, ethnographic, and raw materials experimental replication.

Reconstruction of production technologies is presented in terms of the geographical location of north-central sites and sites located toward the east of the peninsula (Cobá, Chac Mool, and Culubá), mainly because the north-central and eastern samples overlap little in their petrographic characteristics, and because raw materials were collected only from the north-central area. Therefore, the investigation of techniques for eastern samples is restricted to the examination of their fabric classes.

In this reconstruction, the aim is to link what has been learned from the hand-specimen, chemical, and thin-section analyses (chapter six, chapter seven, and chapter eight) to the raw materials within the area, and to techniques and processes that might have been applied in manufacturing the vessels sampled in this study. A combination of analytical methods, including examination of, and experimentation with, raw materials complemented with ethnographic data, was used for comparing the characteristics of the samples to raw materials.

The samples included in this study and analysed through these methods do not include complete or partial vessels; in addition, there are no known production locations. Production techniques, defined by Rye (1981, p.4) as the patterned and repetitive series of potters' actions that produced the attributes observed in the ceramic material, were inferred using aspects of the chaine opératoire, as described in the methodology chapter (sections 5.1, 5.3), and informed by contextual information, such as geological and ethnographic sources.

Within the north-central area, reconstruction of production technologies is presented in terms of the different types of vessels included in this study. They are, mainly, red-slipped jars and cajetes commonly found in this area (Mama variety), plain and unslipped jars and cajetes (Navula variety), and striated unslipped (Yacman variety) cooking pots.

There are two reasons for this. One is that different varieties imply differences in attributes such as surface finish (red-slipped, unslipped, or cream-slipped), decoration (plain, striated, or incised), paste texture (fine, or non-fine), and, sometimes, form; attributes may be related to the intended function of a vessel. The second reason is that patterns have emerged in which variations in technology (such as in grain size) are associated with the ceramic varieties, as observed in Table 8-19.

One such pattern is presented by a division in the ceramic varieties as illustrated in Figure 9-1. This figure shows that at north-central sites, medium-grained fabrics (such as DS-W-m, or Dolospar-W-Medium, in Figure 9-1a) dominate in red-slipped Mama samples (solid bars more pronounced in Figure 9-1a), while coarse fabrics (such as CG-CM-W-c, or DarkMicrite-Micrite-Medium, in Figure 9-1b) dominate in unslipped striated Yacman jars (brick pattern is more pronounced in Figure 9-2b than in 9-1a).

For each main ceramic variety in this study, Mama (red-slipped), Navula (plain, unslipped), and Yacman (striated, unslipped), the different stages of ceramic production (including raw material collection and processing, construction of the pottery fabrics, and forming) are analysed, provided that information is available. The objectives are to:

1. Demonstrate that raw materials from the north-central area have the potential of producing the samples analysed in this study and contribute to an assessment of the range of available choices, as opposed to the choices that may have been made by potters.
2. Propose possible techniques that may have been applied to local raw materials to produce the vessels sampled.

3. Draw inferences about the number of potters' groups that manufactured Mama, Navula, and Yacman.

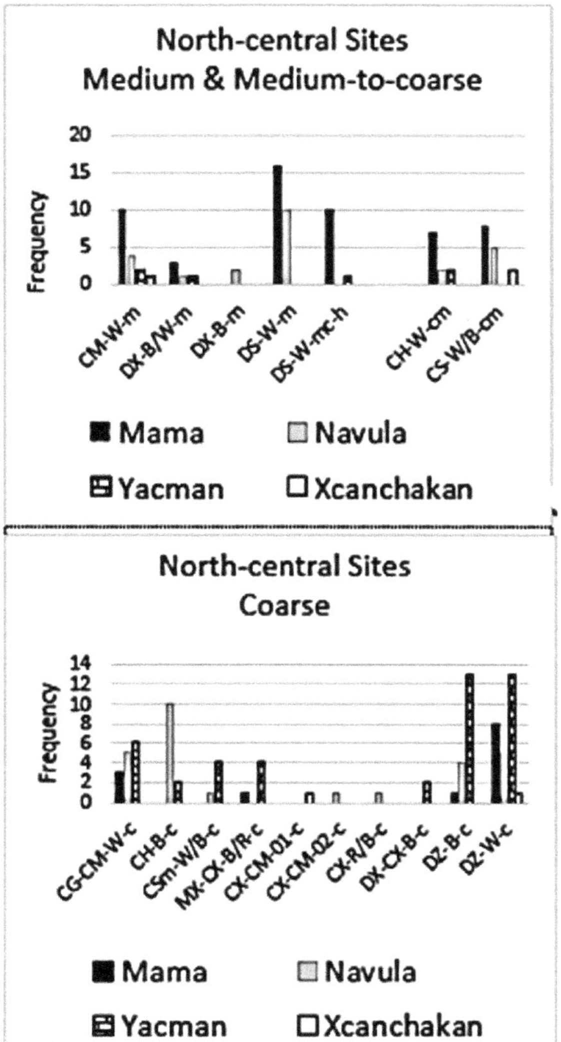

Figure 9-1. Main Medium (top) and coarse (bottom) fabrics of the north-central area

With these objectives in mind, local raw materials were collected from around the north-central area, at localities shown on the map in Figures 5-2 and 5-4, for comparison with the pottery fabrics. The materials collected are described in section 5.4 and include red soils, marl sascab, non-plastic sascab, and light grey clay. As can be recalled (section 4.3), the term sascab is the local name for the pockets of unconsolidated carbonate rocks (such as limestone or dolostone) that are so common in northern Yucatán (Perry et al. 2002).

The sections in this chapter dealing with the reconstruction of technologies are presented in the following order:

1. Reconstructing north-central red-slipped Mama, a variety common at north-central sites (section 9.1)
2. Reconstructing north-central plain, unslipped Navula, a variety common at north-central sites (section 9.2)
3. Reconstructing north-central striated, unslipped Yacman, a ceramic variety common at north-central sites (section 9.3)
4. Firing of the samples (section 9.4)

The firing of the vessels is discussed only once at the end of the reconstruction, encompassing all the ceramic varieties, because the information gathered is not specific to a given ceramic variety. Inferences about the number of Mama, Navula, and Yacman (for which there is more information) potters that the different fabric classes may represent are included in the presentation of the fabric classes found in these varieties.

9.1 Reconstructing Production: North-central Red-slipped Mama

During the Late Postclassic, the red-slipped Mama group and the closely related, thinner, finer-grained, red-slipped Payil group of the eastern sites were the predominant slipped ceramics over most of northern Yucatán and, in many centres, the only type of slipped ceramics. One characteristic of Mama samples from the north-central sites is the homogeneity in their external characteristics of composition, including the general appearance of the inclusions, grain size, and even colour of the paste. In a hand specimen, most red-slipped jars and cajetes are characterised by frequent to common white and chalky or finely crystalline carbonate inclusions, with a medium (mode) grain size, and light fired colours, usually light pinkish grey, light reddish brown, or light red (Figure 9-2).

9.1.1 Raw Materials Selection for Mama

9.1.1.1 Clays

In this section, the local geology and materials from the north-central area are examined with the perspective of their potential to produce ceramics such as red-slipped Mama samples. A variety of clay deposits have been documented (section 4.4). In this section, these clays are examined from a pottery-making perspective.

Kaolinite-montmorillonite has been used by modern potters as the main pottery clay (*K'at*) in the vicinity of Ticul (Arnold and Bohor 1977; Isphording and Wilson 1974), which is located 28 aerial kilometres south of Mayapán. As the name kaolinite-montmorillonite indicates, clay deposits can comprise more than one type of clay, and the properties of this combination may be different from the attributes of the individual components. Ticul potters consider that the properties of their clay (kaolinite-montmorillonite) are superior to any other clays in Yucatán for the construction of large vessels (Arnold 2008, p.155).

Figure 9-2. Sample from Mama vessel, showing its colours and textures.

Figure 9-3. Sample bowl made with clay#11, scale in cm (Table 5-4)

In northern Yucatán, this kaolinite-montmorillonite combination has been found only in a few localities, which include the vicinity of Ticul, Tepekán, and Becal (Schultz et al. 1971; Isphording and Wilson 1974; Arnold 2008). The distance of these quarries to Mayapán makes them an unlikely source of Mayapán pottery clay.

Kaolinite deposits have been reported using XRD (Morales Valderrama 2005, Figure 2) of raw pottery clays from towns in northern Yucatán at Yaxcabá, Akil, and Ticul.

A review of the literature showed that palygorskite deposits have been found in several northern Yucatán locations, with many mapped in Figure 4-8. All of the

Figure 9-4. Small bowls prepared with the different clay samples (#1, #2, #4, #11, #16, #18), with red-soil applied to one

Figure 9-5. Briquettes made with marl sascab (#2) and red soil (#3, Table 5-4)

palygorskite deposit locations are outside of the crater, but this could be the result of studies concentrated around modern pottery towns.The closest mines to Mayapán are located approximately 21 km to its south and south-west, near the current towns of Chapab, Mama, and Sacalum (Figures 4-8, 5-4). Morales Valderrama's (2005, Figure 2) XRD analysis of ethnographic raw clays and tempers reports that palygorskite is present in the pottery clay (*k'at*) of modern potters at Uayma and Mama (Figure 4-8). At Ticul, palygorskite (*sacalum*) is one ingredient mixed with carbonate rocks (calcite, dolomite) to prepare temper, which, according to potters, enhances the strength of the pottery clay (Arnold 1971, 1972, 2008).

In general, the number of pottery clay deposits documented near Mayapán is limited. A cenote from which clay may have been collected (according to local sources) was inaccessible because its entrance was blocked by natural beehives.

The suitability of some of the sascab samples to be successfully fired to ceramic was investigated. Its suitability was assessed through a test of plasticity and a successful firing. The list of raw materials samples and their characteristics, including plasticity, can be found in Table 5-4.

A test of plasticity performed on sascab samples collected from the north-central area (wetting the material, making a coil, and bending into two circles; see Table 5-4 for results) showed that there are two types of sascab: a plastic sascab or marl, and a non-plastic, largely carbonated-rock sascab.

The term marl refers to a friable carbonate rock with clay. This term is used in this study to refer to the sascab samples that pass the test of plasticity. The plasticity of marl sascab samples allowed for the construction of small vessels (Figures 9-3 and 9-4) and briquettes (Figure 9-5).

To test whether marl sascab samples could be fired to produce a ceramic vessel, briquettes were made (Figure 9-5), and experimental firings were performed (more details on firing experiments in section 9.4) with marls and a clay collected from around the north-central sites. Various combinations of conditions were attempted to achieve basic characteristics of a successful firing, which included losing plasticity by becoming a ceramic as demonstrated by not re-hydrating and recovering plasticity, by keeping its form, and by not breaking easily under a bending tension (see section 9.4). Two briquettes met these characteristics under the firing conditions: raw material #1, and #11 (Table 5-4), which are both marls. Sample #11 has a chemical composition within Group 1 that matches most pottery samples from Mayapán.

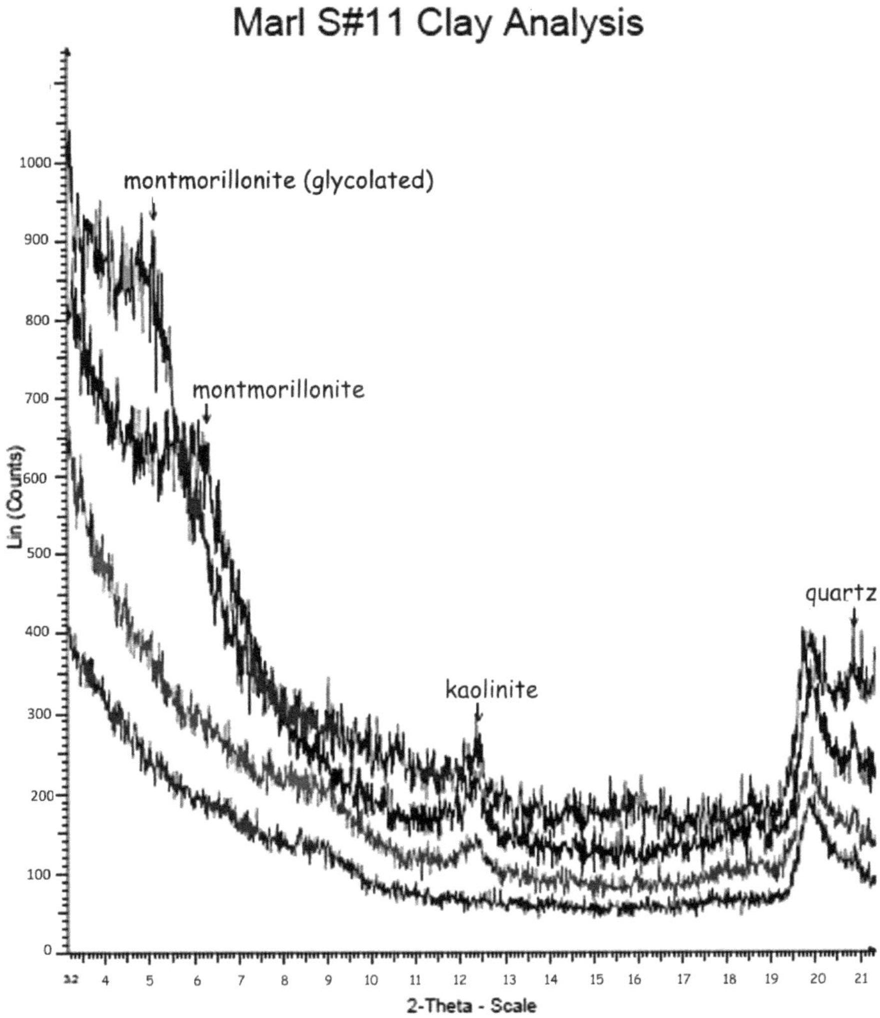

Figure 9-6. Results X-ray powder analysis of the clay fraction of raw material sample #11 (Shannon 2007, Figure 4)

In sum, the number one objective outlined at the start of this section was met, which showed that some of the marl sascab (#1 and #11) samples are suitable for pottery making. The characteristics of these samples when fired (see section 9-4) match the characteristics of most north-central pottery in their light colour and largely soft paste. This shows that suitable materials for pottery production occur in the vicinity of Mayapán, and, given that marl sascab may be ubiquitous, probably also occurs in the vicinity of the other north-central sites.

Raw material sample # 11 was tested further, given that it matches the chemistry of pottery samples (chemical Group 1) and, in addition, is suitable as pottery clay. This sample is composed of a clay fraction and coarser particles. Petrography (section 8.10) showed that the particles are very coarse sparry dolomite (dolospar) and calcite (sparite). Neither petrography nor chemical analysis can detect the type of clay present.

The clay portion of the marl sample #11 was analysed by PMET Laboratories (Shannon 2007, included in Appendix A) using x-ray powder diffraction (XRD) to determine the types of clay(s) present. XRD found that the main clay is montmorillonite, and kaolinite is a minor component (Figure 9-6). Arnold (2008, p.155) has reported widespread deposits of montmorillonite as marl pockets or as clay deposits around Ticul; however, potters consider this clay of inferior quality (than kaolinite-montmorillonite) because large vessels break.

There are indications in a sample from Mayapán that kaolinite or clays with kaolinite were used to prepare pottery clay. In general, depending on firing time and temperature, the clays decompose into other compounds, and it is usually impossible to determine the composition of the original clay(s) using mineralogical or chemical analyses of the fired pottery. In the case of kaolinite, after its decomposition, a mica phase appears that is stable at 800°C (see details in section 9.4.4). The clay analysis of one pottery sample (Navula) from Mayapán, (Shannon 2016 in Appendix C) determined the presence of precisely a mica phase. In addition, the chemical compositions of the marl sample #11 (the Mayapán marl in which montmorillonite/kaolinite was found) and most of the Mayapán pottery fall within the chemical Group 1.

Having examined the composition of the fine portion, the next step in this reconstruction is to examine the composition of coarse inclusions in terms of the raw materials from the area. It was demonstrated that there are marls close to the north-central sites that are suitable for pottery making. The marl sascab samples, however, contain very coarse carbonate fragments and hardly any silt or fine particles. Further processing of the raw materials collected from the area would have been necessary to construct pottery clay similar in grain size and frequency of inclusions to the fabrics of the pottery samples.

9.1.1.2 Coarser Inclusions

In this section, the types of inclusions found in the red-slipped samples during the petrographic analysis are compared to particles found in the collected raw materials, particularly in non-plastic sascab. The objective is to examine and propose the types of raw materials that match the particles observed in the pottery samples.

In Figure 9-7 the fabric classes discussed in chapter eight were grouped by the main types of fabrics or inclusions (a type of fabric refers to all the fabrics with a common main or diagnostic inclusion, for instance, bioclasts) to give a broader picture of the types of inclusions found in the red-slipped (Mama) samples of the north-central area.

Figure 9-7 presents the percentage that Mama samples within a type of fabric (for instance, all micritic or dolospar types of fabrics) represents of the total number of analyzed north-central Mama samples in thin section (n=69). For instance, in Figure 9-7, 16 per cent of Mama thin sections are characterised as micritic type of fabrics and 50 per cent as dolospar fabrics. From Figure 9-7, it can be observed that:

1. At the north-central sites, the bulk of red-slipped Mama samples have three types of fabrics: (a) finely crystalline dolospar (50 per cent), (b) micrite (16 per cent), and (c) finely crystalline sparite (12 per cent), acknowledging that more samples were taken from Mayapán (21 Mama thin sections, or 30 per cent).
2. At Mayapán, dolospar (sparry dolomite as opposed to sparite that is used to refer to sparry calcite) inclusions dominate the samples.
3. Samples in which the main inclusions are single crystals of calcite, dark micrite, or single dolomite crystals are rare in the Mama samples from the north-central sites. Equally rare are fabrics without inclusions and singletons (ungrouped samples).

The division in the composition of the coarser inclusions of most red-slipped samples from the north-central areas into dolospar, micrite, or sparite corresponds with the composition of sascab (see section 6.10). Figure 9-8 shows an example of what is considered by local residents (Don Fernando Flores, personal communication, Feb 2015) a fine-grained sascab or *cuut* (raw material #17, Table 5-4). Figure 9-9 portrays a *sascabera,* or a pocket in the carbonate layer containing sascab. There are two types of sascab, a marl and a non-plastic, mostly carbonate-rock sascab (section 6.6).

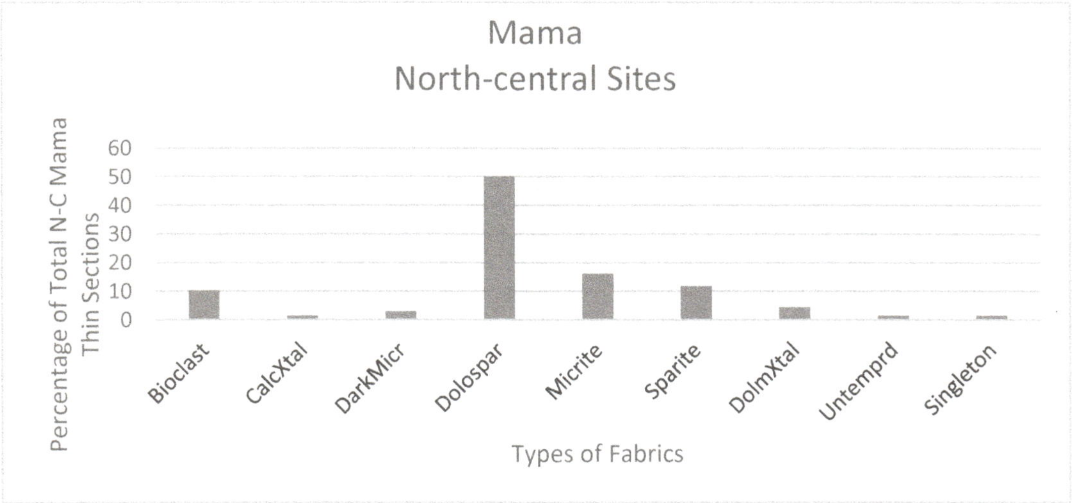

Figure 9-7. Percentage that a fabric type represents of the total Mama (n=69) N-C thin sections; KEY: N-C=north-central;CalcXtal=calcite, single crystals; DarkMicr =dark micrite;DolmXtal=dolomite, single crystals

Figure 9-8. Carbonate-rock sascab (# 17, Table 6-12)

Figure 9-9. Sascabera (sascab pocket or layer) located by the road Mama to Chumayel (#2, Table 5-4)

Figure 9-10. (a:top) Use of *mecapal* by novice priests carrying materials to the temple (Mendoza Codex, Fol. 62r); (b:bottom) complex *mecapal* to transport water-carrying vessels (Thompson 1958, Figure 28 e)

In sum, the bulk of the inclusions in red-slipped vessels sampled from north-central sites are particles of dolospar, micrite, and sparite. With the unaided eye or low magnification (10X), they are difficult to impossible to set apart because of their small crystal size, likely accounting for the observed superficial homogeneity in composition.

9.1.2 Raw Materials Transportation

The Maya people did not use beasts of burden or rolling carts to transport heavy loads. At the time of Thompson's (1958) ethnographic work in Yucatán, depending on the distance, clays and tempers were carried to the potters' work area by *mecapal* or tumpline. Figure 9-10a from the Codex Mendoza (Aztec) illustrates their use. *Mecapals* can be made very complex, such as in Figure 9-10b for the transport of bulky and heavy items such as ceramic jars. It is not known how far Mayapán potters travelled to transport clays. Arnold (2005b) has provided, based on ethnographic work at 117 sites, a maximum distance of around seven km that, approximately, 90 per cent of potters would have travelled on foot while carrying raw materials.

9.1.3 Raw Materials Processing for the Production of Mama

As discussed earlier, it is probable, as shown by firing experiments (section 9.4), that part of the production of vessels used local marl sascab for the plastic portion of the clay body. It was also discussed that the marls collected contain very coarse carbonate fragments and hardly any silt or fine particles. Very coarse particles are not present in the pottery samples. If marls were used, further processing of the raw materials would have been necessary. In this section, the raw materials and pottery samples are compared, aiming to draw inferences about the ceramic technologies to which the raw materials were subjected to prepare the pottery clay.

9.1.3.1 The Use of Temper

Coarser inclusions found in the pottery clay could be naturally occurring or could have been added by the potters, in which case are known as temper (Shepard 1964, p.518; Whitbread 1995, p.374). It is likely that the inclusions found in the north-central samples are temper. A hand-specimen analysis in which samples of non-plastic (carbonate rock) sascab were compared to inclusions in the pottery using the binocular microscope (section 6.10) showed that the examined samples of sascab are composed of sparry particles, which are consistent with what is observed in many pottery samples.

More telling are the divisions observed in the grain size and frequency of the inclusions in pottery, indicating that the inclusions are not naturally occurring and that a selective adding of inclusions dependent on the type of vessels being made took place. This is demonstrated by comparing the red-slipped Mama samples, which are medium-grained (mode), with the striated, unslipped Yacman open-mouth *ollas* that are coarsely grained (section 8.3.1, section 8.3.2, and summary in 8.11).

Ethnographic work in Yucatán has documented the use of sascab as temper. It has been used traditionally for the construction of pots that would not be put on the fire (Thompson 1958, Morales Valderrama 2005; Rendón 1947, p.117). At the time of Thompson's (1958) work, communities specialised in production of cooking versus non-cooking vessels, and the preferential use of sascab was mostly related to a specialisation in non-cooking pots.

9.1.3.2 Processing the Clay

There are indications in Mama pottery samples that the grain size of the inclusions resulted from the crushing and sieving of raw materials. The truncation of the upper size range of the particles in the pottery samples is an indication that sieving may have occurred (Whitbread 1995, p.398). For instance, in Mama thin sections, the most frequent grain size (mode) is medium, while the maximum particle size is coarse. As presented in section 9.3, the mode for Yacman is coarse with a maximum of around 1.5 mm (very coarse). A division of ceramic varieties by grain size is observed: medium-grained in red-slipped Mama and coarsely grained in Yacman ollas. This division points to the selective addition of particles by potters, the addition of temper, depending on the type of vessel.

Crushing of raw materials to reduce them to the desired size is a likely scenario. Analysed raw materials (marls and non-plastic) hardly contain silt to medium particles. Most particles are very coarse (> one mm) to small pebbles, and they do not correspond to the size and frequency of particles within the fabrics of the pottery samples, acknowledging that potters may have obtained their raw materials from deposits different from the locations that were sampled.

Rendón (1947), in her ethnographic work in some Yucatán towns, describes the processing of clay in which crushing and sifting are part of the process. The clay, or *k'at*, is dried under the sun just enough to start crumbling under pressure (also see Arnold 2008). It is then put in a container called *ch'oy* that is the same Maya name given to the wooden containers from previous times (Rendón 1947). Enough water is added to wet the clay, which is then kneaded. Temper is added after kneading. In modern times, sascab was taken from the sascabera and ground. Rendón observed that sascab was first processed by whacking it with a good thick club (Rendón 1948a).

Rendón tells us that to control the grain size for ordinary pots, a coarse sifter is used. For finer ceramics, several sifters (Figure 9-11) may be used; each of finer aperture than the previous until the desired grain-size is achieved.

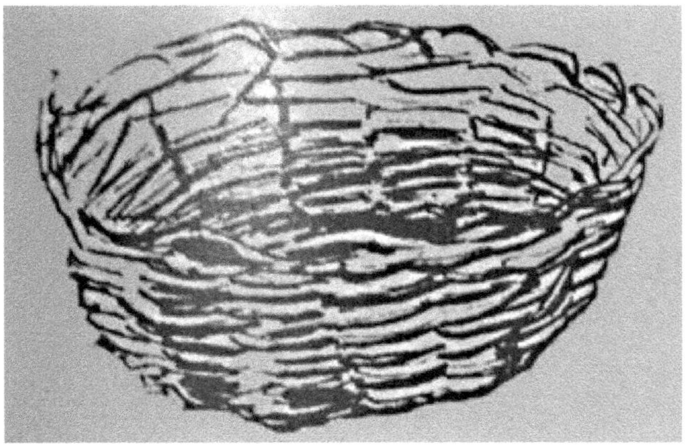

Figure 9-11. Basket-type and palm-rib sifters from Yucatán used by potters to sift temper (from Thompson 1958, Figures 16g and 16h)

In sum, the divisions of grain size and frequency between ceramic varieties indicate that the coarser particles found in Mama samples were probably added by potters. The grain size of the coarser particles observed in the samples is most likely the result of crushing and sifting. Similarities between raw materials and the inclusions in the pottery samples indicate that micritic to finely crystalline carbonate inclusions in the red-slipped Mama samples were likely collected from sascab pockets, locally called sascaberas. Most likely, potters could not distinguish whether sascab consisted of fragments of micrite, or finely crystalline calcite (sparite) or dolomite (dolospar). With the unaided eye or very low magnification (10X) these carbonate varieties are difficult to impossible to set apart due to their small crystal size.

9.1.4 The Fabric Classes of Red-slipped Mama Samples

In this section, the fabric classes of red-slipped Mama samples and the sites in which they were found are presented. Different fabric classes are, in part, the result of differences in the environment as well as in the processes of pottery clay construction discussed earlier, including the selection of raw materials (clays and tempers) and the processing of the materials. One fabric class represents a common technology for the construction of the pottery clay, and when one fabric class is found at multiple sites, they share a common technology or way of doing things.

Figure 9-12 breaks down into fabric classes the broad categories of inclusions illustrated in Figure 9-7 (Mama samples from the north-central sites) and shows on a site by site basis the frequencies of the different fabric classes found at each site. From the sampling restrictions and sampling strategies faced in this study (described in section 5.3), it follows that the frequencies illustrated in Figure 9-12 do not represent the true proportions of these fabrics at the sites. They do, however, represent the variability of the samples.

Figure 9-12 shows that, at the north-central sites, the red-slipped Mama samples were found within multiple fabric classes. The codes used to represent the inclusions were presented in the previous chapter (Table 8-2) and repeated for ease of reference in Table 9-1. Figure 9-12 shows that there were three main types of fabrics found at most of the north-central sites: dolospar (DS-W-m or Dolospar-W-Medium, DS-W-mc-h or Dolospar-W-Hard), micritic (CM-W-m or Micrite-W-Medium), and sparitic (CS-W/B-m or Sparite-W/B-Medium) fabrics. The rest of the fabrics were found at one or two sites.

One type of fabric found at multiple sites represents a significant portion of the red-slipped samples, the very-fine to finely crystalline clusters of dolomite, or dolospar fabrics. Although Figure 9-12 shows this type of fabric under the general name of Dolospar (solid orange), it comprises a list of fabrics, shown in Table 9-2. Most Mama samples have medium-grained fabrics, but some Coarse Mama samples were also found.

Table 9-1. List of Inclusions, Codes, and Short Description

Inclusion code	Short Description
CM	(calcite) micrite
CG	(calcite) grey/brown limestone
CS	(calcite) sparite
CH	skeletal remains (such as fossilised shells)
CX	(calcite) single-crystals
DS	dolospar
DX	single dolomite-crystal
QX	quartz single-crystal
MX	Mud (argillaceous) with crystals

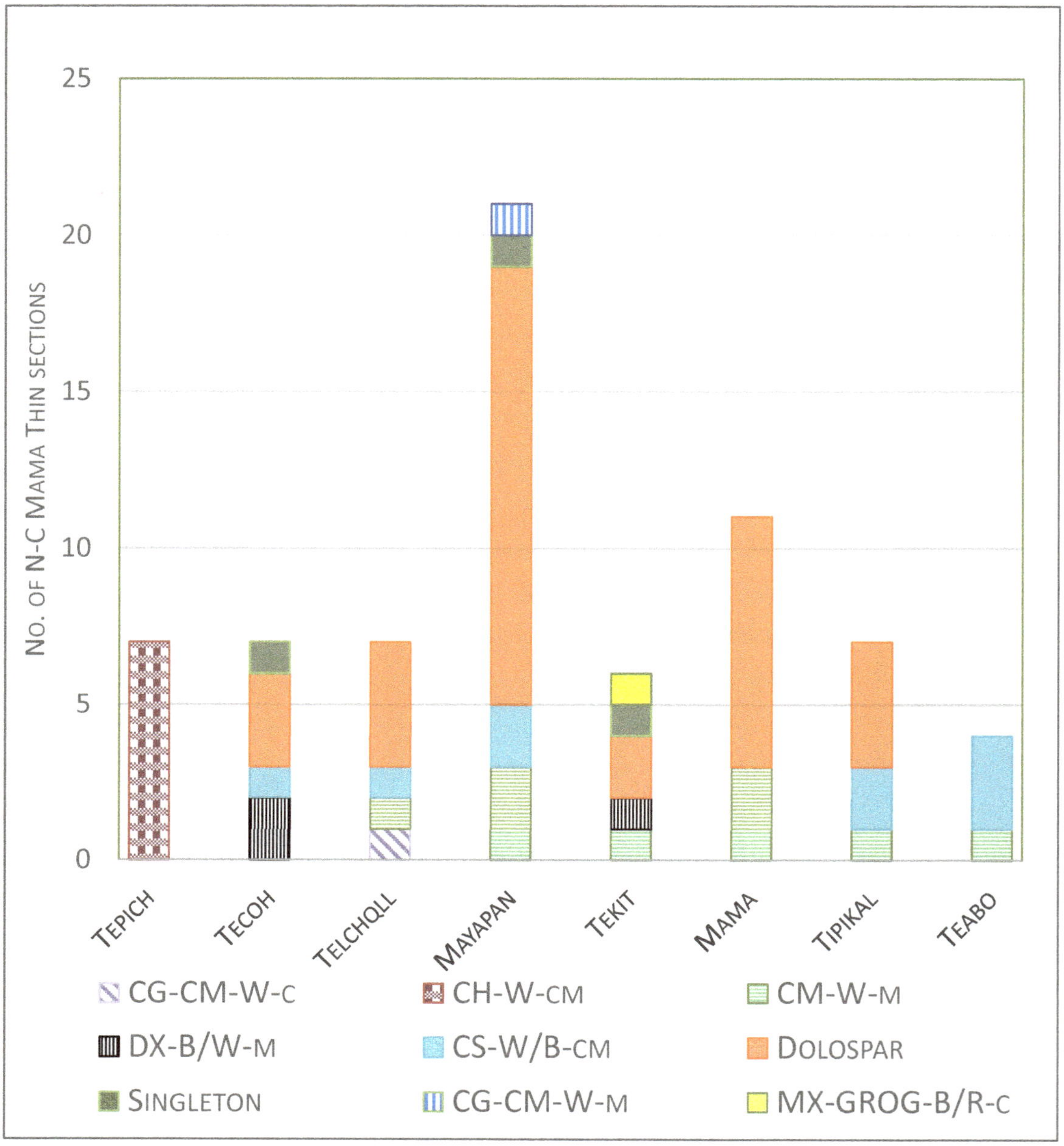

Figure 9-12. Frequency of red-slipped Mama fabric classes within N-C sites.
Key to Fabrics:
CG-CM-W-c = DarkMicrite-Micrite-Coarse; CM-W-m =Micrite-W-Medium;
CH-W-cm = Bioclast-W-Coarse/Medium; Dolospar = List of fabrics in Table 9-2
CS-W/B-cm = Sparite-W/B-Coarse/Medium; DX-B/W-m = Dolomite-B/X-medium;
CG-CM-W-m = DarkMicrite-Micrite-Medium; MX-GROG-B/R-c = MudXtal-Grog-Coarse
Key to inclusions: CalcXtal = single calcite crystals, DarkMicr=dark micrite, W= single dolomite crystal

Table 9-2. Presence(X)/absence in Mama Samples of the Various Dolospar Fabrics

Fabric Class	Texture	Fabric Code	Presence/Absence
Dolospar-W-Medium	Medium	DS-W-m	**X**
Dolospar-W-Hard	Medium to coarse	DS-W-mc-h	**X**
Dolospar-W-Coarse	Coarse	DS-W-c	x
Dolospar-B-Coarse	Coarse	DS-B-c	-

Another group of samples in Figure 9-12 found at multiple north-central sites predominantly contains finely crystalline sparite, or fabric Sparite-W/B-Coarse/Medium (CS-W/B-cm). This group presents high variability in its petrographic attributes (section 8.4) and may represent multiple sources with a common basic recipe using medium-to-coarse particles of sparitic limestone.

The remaining key group of samples also found at multiple sites is fabric Micrite-W-Medium (CM-W-m) in which micrite dominates (section 8.5). Similar to the sparite fabric, this micrite fabric catches samples that are not homogeneous in their petrographic characteristics but share what appears to be a basic recipe of medium-grained micrite. The above mentioned fabrics, their association with sites, ceramic types, and to other fabrics can be better observed using Table 8-19.

In sum, multiple fabrics were found in the Mama samples, but the bulk comprises fabrics dominated by medium-grained particles composed of finely crystalline dolospar, micrite, and sparite. With the unaided eye or very low magnification (10X), they are difficult to set apart due to their very fine to finely crystalline crystal size (see section 8.2.1 with scales and measurements).

9.1.5 Red-slipped Mama Potters' Groups

This section deals with the number of potters' groups that produced the Mama vessels sampled for this study. One fabric class represents a common technology, or way of doing things, for the construction of the pottery clay and a potential potters' group, although the possibility exists that from different ways of doing things, the same fabric class might result. One or many groups of potters could have been sharing a technology. For instance, one fabric class with samples within two chemical groups may indicate one technology but two raw materials sources.

Table 9-3 shows the chemical composition groups associated with each fabric class and their distribution through the sites. This table is breaking down further the information in Figure 9-12 to show potential differences in the selection of raw materials (or processing) by including the chemical divisions within the fabric classes. In regard to the samples that were not assigned to any chemical group (unassigned), the chemical composition of one of these samples is not necessarily similar to the composition of another unassigned sample, and in Table 9-3, they are shown as potentially having different chemical compositions. For instance, in the case of the Micrite-W-coarse fabric, unassigned samples having this fabric were found at four sites. It is possible that four or more potters groups existed. However, it is also possible that they all represent only one potters' group. The minimum number of potters' groups working with micrite and unassigned material is only one.

The red-slipped Mama samples found at north-central sites have a chemical composition (Table 9-3) within Group 1, Group 2, or are unassigned, with the exception of one sample within Group 5. Samples within Group 1 and Group 2 were found at many sites and have a widespread distribution; therefore, these groups by themselves are not useful in identifying zones of procurement. However, inspection of the combinations of fabric classes (second column), chemical groups (third column), and contextual information can inform us about the number of potters' groups, or even tell us something about the location of the raw material sources of a group of samples.

In this study, one potters' group is inferred by the combination a technological class (fabric class) and a potential zone of procurement (chemical group), although the differences in chemical compositions may also be due to the chemical heterogeneity of one source. Contextual information, such as patterns in the types of vessels or ceramic varieties is also considered when inferring potters' groups (the associations between fabric classes and ceramic varieties have been presented in chapter eight for individual type of fabrics at the end of the corresponding section and for all the fabrics in Table 8-19).

The last column in Table 9-3 is used to indicate that, for a given combination of petrographic and chemical characteristics, at least one distinct potters' group (PG) can be inferred from the data. For instance, the micritic fabric Micrite-W-Medium (CM-W-m) was found at Tekit and Mayapán with a chemical composition falling within chemical Group 1. Therefore, at least one potters' group (PG) was producing red-slipped pots with these characteristics; granting that the number of potters' groups could have been higher. A code was given to the potters' group (PG#3, for instance) to expedite when referring to the combination fabric and chemical group. PG#3 only represents the minimum number of potters' groups because other interpretations are also plausible. It is possible that several groups in a chemically homogeneous Group 1 area could have been producing these vessels. Because of the lack of other contextual information, such as different types of vessels at Mayapán and Tekit that would help to divide PG#3 into further potters' groups, only one potters' group was assigned.

It is important to clarify that the actual location where these vessels were produced was not necessarily at Mayapán or Tekit, which are locations of deposition. The production location is not known because neither the petrographic characteristics nor the chemical group is distinctive in regard to location. The table shows that the pottery clay for a group of samples was constructed with a distinctive technology and within a given chemical fingerprint. A larger sample, however, might show wider variations.

Table 9-3. Chemical Groups within Fabric Classes and Sites, showing Suggested Minimum No. of Potters' Groups for North-central Mama (red-slipped vessels)

Type of Fabric	Fabric Class Code and Name	Ch. Grp	Tepich	Tecoh	Telchaquillo	Mayapán	Tekit	Mama	Tipikal	Teabo	Minimum No. of Potters' Groups (PG)
Bioclast	CH-W-cm (Bioclast-W-Coarse/Medium)	1	X								PG#1
Dark Micrite	CG-CM-W-c (DarkMicrite-Micrite-Coarse)	1			X						PG#2
Micrite	CM-W-m (Micrite-W-coarse)	1				X	X				PG#3
		uk				X		X			
		u/a			X						PG#4
		u/a						X			
		u/a							X		
		u/a								X	
Dolospar	DS-W-m (Dolospar-W-Medium), DS-W-mc-h (Dolospar-W-Hard)	1		X	X	X	X	X	X		PG#5a PG#5b
Sparite	CS-W/B-cm (Sparite-W/B-Coarse/med)	1			X				X		PG#6
		2							X	X	PG#7
		uk				X					
		u/a		X							PG#8
		u/a								X	
Fine Dolomite Crystals	DX-B/W-m (Dolomite-B/W-Medium)	2		X			X				PG#9
Argillac./ Crystals and Grog	(*) MX-GROG-B/R-c (MudXtal-Grog-Coarse)	5					X				PG#10

Key: u/a = chemically unassigned; uk=unknown (no chemically analysed)

From Table 9-3, it can be said that the production of red-slipped Mama vessels at the north-central sites is associated with several potters' groups. Going down one by one in the rows in the table, the potters' groups that might have built Mama jars and cajetes are reflected in the following patterns:

(a) At least one potters' group (PG#1) is associated with the Bioclast-W-Coarse/Medium (CH-W-cm, section 8.8.1) fabric within Group 1 found at Tepich, although many potters' groups working with homogeneous raw materials could have created the same pattern.

(b) One potters' group (PG#2) at Telchaquillo is associated with two Mama samples within fabric DarkMicrite-Micrite-Coarse, or GM-CM-W-c (section 8.9.2) and Group 1.

(c) As discussed earlier, at least two potters' groups are associated with a group of Mama samples having a composition within fabric Micrite-W-Medium (CM-W-m, section 8.5.1). At Mayapán and Tekit, this fabric was found within chemical composition Group 1, pointing to, at least, one potters' group (PG#3). However, chemical homogeneity over a wide area or an import from another area could produce the same pattern. This fabric was also found at Telchaquillo, Mama, Tipikal, and Teabo (Table 9-3) but with unassigned chemical composition. Likely, another potters' group existed (PG#4).

Fabric Micrite-W-Medium is not petrographically homogeneous, presenting a wide variation in the frequency of micrite and the presence/absence of secondary inclusions. Contextual information about this fabric in regard to the types of vessels manufactured with it shows that the bulk of samples come from red-slipped Mama and Navula (but not Yacman) vessels (section 8.5.2). Based on the petrographic and chemical heterogeneity and the association with two ceramic types, it may represent a common technology shared by several potters' groups across sites. This common technology is composed of a basic recipe that involves medium-grained particles of a white, opaque, relatively friable carbonate rock in a very light-coloured paste. However, the data do not rule out other explanations, including the possibility that one group of potters supplied other centres from a location with local chemical variations in raw materials, or that the processing of the clay (such as levigation or clay mixing) affected the chemical composition of what was originally a chemically uniform raw material. What we do know is that all of these vessels with Micrite-W-Medium micrite fabric were made using a common basic recipe based, most likely, as argued in the previous section, in the processing of sascab.

(d) Two potters' group (PG#5a and PG#5b) may be associated with red-slipped Mama samples with medium-grained dolospar fabrics: Dolospar-W-Medium (DS-W-m, section 8.3.2) and Dolospar-W-Hard (DS-W-cm-h, section 8.3.3). These two fabrics are always associated with chemical Group 1 at all sites and have high petrographic homogeneity. As discussed in sections 8.33 and 8.3.5, they may be one compositional fabric, and their differences in organic matter and hardness may be related to differences during firing. One or two potters' group at one location or one production zone producing Mama is a likely scenario (section 8.3.3). However, the existence of many groups working from several locations in an area presenting a homogeneous chemical composition cannot be ruled out.

(e) At several sites, some of the Mama samples were found within a finely crystalline sparitic fabric (Sparite-W/B-Coarse/Medium or CS-W/B-cm, section 8.4.1) that falls within chemical Group 1 (PG#6) at Tecoh and Tipikal, chemical Group 2 (PG#7) at Tipikal and Teabo, or unassigned (PG#8) at Telchaquillo and Tipikal. Similar to micritic fabrics, sparitic fabrics are not petrographically homogeneous (based on variations of frequency of sparite, presence/absence of secondary inclusions, porosity, silt frequency, and other characteristics). The heterogeneity of petrographic and chemical compositions indicates that many potters' groups producing at different locations is a likely scenario.

(f) In some of the Mama samples from Tecoh and Tekit, a fine-grained fabric with single dolomite crystals (Dolomite-B/W-Medium or DX-B/W-m, section 8.7.1) was found. Its chemical composition falls within Group 2 (PC#9), which is a combination particular to these two sites. There are only four samples in this fabric. The characteristics of this fabric containing mostly fine dolomite crystals, which are not common, suggests that their presence at two sites and in the same chemical group makes production by one group of potters a likely scenario.

9.1.5.1 Summary

Summarising the findings of this section, the combination of fabric classes, chemical compositions, and contextual information is the principal method of identifying potters' groups in this study. It can be said from the list of different fabric classes within the various chemical compositions that the red-slipped Mama vessels sampled in this study were produced by many groups of potters. At Tepich, Tekit, Tecoh, and Telchaquillo, fabrics were found that are likely local with limited distribution; acknowledging that a more extensive sampling may change this interpretation. Nevertheless, the bulk of Mama production comprises vessels having dolospar, sparitic, and micritic fabrics.

Dolospar, micrite, and sparite fabrics have similarities in their external appearance, widespread distribution across north-central sites, and association with Mama (Table 8-19). They have in common the presence of medium-grained particles comprising a white, relatively friable, micro to finely crystalline carbonate rock as the main inclusions in a light-coloured paste. The undifferentiated optical characteristics of carbonate minerals—small crystal size, and general light paste colour—impart to the fabrics a visual similarity. Given that the three widespread fabrics share these characteristics and are associated with red-slipped vessels, it indicates that a basic recipe was shared by at least these three potters' groups for the production of Mama, and probably also Navula.

The compositional heterogeneity of micritic and sparitic fabrics and their presence at several sites can be interpreted as many potters' groups at different locations. The data, however, do not rule out other explanations (given under micritic and sparitic fabrics).

The group of red-slipped Mama with medium-grained dolospar fabrics differs from the micritic or sparitic samples in its internal, and across sites, petrographic and chemical (Group 1) homogeneity. Therefore, this may indicate one group of potters working from one location and supplying other centres without completely ruling out a widespread dolospar tradition under chemical homogeneity.

9.1.6 Forming and Finishing Techniques of Red-slipped Mama

To contribute to the composite picture of ceramic production, which is one of the main objectives of this study, in this section, techniques that might have been used in forming and finishing Late Postclassic vessels from north-central Yucatán are investigated. In addition to the pottery samples in this study, five almost complete vessels were examined.

The analytical method applied to infer ancient techniques is non-destructive and based on the inspection, visual and tactile of surface features. This method is based on work (such as that in Rye 1981, p.58) showing that some surface features in sherds or vessels correlate to specific forming or finishing techniques. However, a given feature can result from different techniques, and operations can erase the traces of operations applied before them (Roux and Courty 1995, p.18; Rye 1981, p.58).

Research on Maya ceramics forming techniques is sparse, to say the least. There is, however, few but important research related mostly to the southern Classic period pottery. Reents-Budet (1994:210, 231) examined complete and almost complete Classic polychrome vessels of the southern lowlands. These vessels are primarily cylindrical in form, but plates are also found. She described the vessels as handmade evidencing the use of different techniques depending on the vessel form. The polychrome cylindrical vessels show evidence of a base started using a pancake of clay in the characteristic crack where the first coil was attached to the bottom (Reents-Budet 1994:210, 231). The interior of the cylindrical vessels shows discernible horizontal ridges that were not totally obliterated by scrapping indicating to Reents-Budet that coiling was the technique to form the cylinders' walls. On the other hand, Reents-Budet observed that press moulds might have been used to form plates due to the uniformity of the plates walls.

As regard to jars, large red-slipped Terminal Classic period jars of the Petexbatum (southern lowlands) area show a breakage pattern characteristic of coiling, a breakage along the joints between coils in a kind of spiraling pattern (Rice 1985:128 cited in Foias 2004:140). Foias (2004) sees evidence that these jars were made in sections: a concave base probably made first in a simple mould; the vessel body was made by coiling and attached to the base; and the jar neck was made separately and attached to the vessel. Classic polychrome plates seem to have been formed by pressing into a mould, based on the uniformity of the wall thickness (Reents-Budet 1994:231).

There is no evidence in current research of the use of the potter's wheel or any other device that would have worked as a wheel by ancient Maya potters. As of the current research, Maya vessels were hand built. For instance, studies of Classic period lowland pottery seeking clues of forming techniques have not found indications of the use of a wheel (Reents-Budet 1994).

There is, however, a device called *kabal*. It is a wooden cylinder or a pottery-saucer, rotated by the potter's feet, that is used in modern Yucatecan times (Thompson 1958, p.76-80, 88). It may not have been used in antiquity because the word *kabal* did not appear in the early Maya-Spanish dictionaries until the 17th century (Thompson 1958, p.76). In addition, there is no Maya word for the *tabla* or board that provides a flat area in which to turn the *kabal* (Thompson 1958, p.76).

Maya pottery is unglazed. Clay slips were used instead. At Mayapán, the unslipped pottery was almost always smoothed but never burnished (Smith 1971, p.23), and the slipped pottery was burnished to attain the lustrous finish that can still be seen.

The rest of this section presents the surface analyses to examine the forming process. The objective of the surface analysis is to identify features and configurations of features (Livingstone and Viseyrias 2010) in pottery that will give indications of the techniques used throughout the different stages of pottery making; granting that the lack of diagnostic features does not discard the possibility that a given technique was applied. The techniques are usually grouped by the main process in manufacturing such as forming or firing.

Ethnographic work has shown that the study of forming techniques is useful in identifying social divisions. In central Yucatán, Thompson (1958) found that the forming methods used to start large jars is a community specialisation. In the highlands of Guatemala, Reina and Hill (1978) found that the way large vessels are started divides along linguistic groups. In Senegal, Gelber (2005) found two types of traditions related to the techniques used to start the vessels.

The link between technological choices and potters' identities may be more related to socio-economic interaction than to identity. In a study in which many cultural aspects, from learning networks to population density, were compared to clay preparation technology, Livingstone-Smith (2000) concluded that of the contemplated factors, settlement patterns and distribution zones (that matched the identity of potters only partially) were the factors that better fit the technological areas. This indicated to him that the technological areas closely matched areas of intense socio-economic interaction bounded by unpopulated areas.

Different approaches can be taken for the reconstruction of forming techniques including X-ray and visual inspection of surface features. In this study, the latter approach was taken. Based on ethnographic and experimental work, certain visual features or combinations of features may point to a specific technique (Rye 1981, Livingstone and Viseyrias 2010). Some of the surface features related to different forming methods mentioned in this section are described below.

9.1.6.1 Use of a Concave Mould or Support

A concave support or mould can be itself made as a ceramic vessel manufactured with the desired shape and size. Gelbert (2005) identified four main macrotraces diagnostic of this technique.

1. Usually, the part of the pot that rests on the support is thicker than the rest. Given that the base of the vessel rests on the support, the potter has difficulties in obtaining a uniform thickness.
2. The support leaves a circular mark around the vessel. This happens if the diameter of the support is small in relation to the rest of the vessel.
3. The vessel's base is slightly concave. A small depression is formed on the exterior wall of the vessel's bottom. This occurs because the pressure exercised by the hand against the clay is insufficient. This trace can be erased by adding clay later.
4. Depressions on the interior wall that are produced by the fingers while opening the clay or pressing it against the mould or support.

9.1.6.2 Use of a Convex Mould

The mould itself can be a ceramic vessel manufactured for this purpose. It could be a plate or a bowl (Sheppard 1958, p.63). Gelbert (2005) identified four macrotraces that are very diagnostic of this technique.

1. Imprints or furrows at the bottom of the interior wall. They are caused by using a tissue to cover the mould, or by the uneven surface of the mould.
2. Uniform curving of the bottom. According to Gelbert, this technique allows even the less capable potters to obtain a perfectly rounded vessel.
3. Very compacted paste. Pressing against the mould while thinning the pot leads to compaction of the paste.
4. Thin and uniform vessel. The uniformity of thickness is obtained while pressing the clay with a tool against the mould to thin the vessel.

9.1.6.3 Coiling

The term refers to building vessels by superimposing rolls of clay (Shepard 1958, p.57). Indications of coiling (Rye 1981, p.67) include the following:

1. Variations in wall thickness.
2. Irregular sherd edges with meandering contour.
3. Step-like fracture or cubic facets.
4. Separations along coil lines.

The description of the features is presented in combination with a drawing of the outline of the vessel following Livingstone and Viseyrias (2010, p.134-139). Table 9-4 lists the surface features observed. A description is included for terms that maybe un-familiar.

9.1.6.4 Surface Analysis of Bowl I (# 9150, Table 8-18)

This is a small to medium size (height, 12cm; diameter mouth, 15 cm) Mama variety bowl (Figure 9-13) with *inturned wall* and mouth, *sub-hemispherical body, round bottom with concave area, red-slipped surface* on the outside, *unslipped surface* on the inside, *smooth interior surface.*

The surface features observed in this vessel are listed in Table 9-5. As expected by the above presentation, this vessel does not show the signs of being thrown in a potters' wheel, such as spiral marks on the inside bottom caused by the hand or thumb pushing to open the vessel (Rye 1981, p.75).

.

Table 9-4. Features used in Surface Analysis

Abrupt walls: abrupt walls rising from the vessel bottom making a composite form
Attached neck, neck do not come from body
Burnished
Conical feet: feet are attached to some vessel with various forms, conical-shaped feet
Composite form: usually applies to deep dishes (cajetes) with round bottom (sub-hemispherical body) and relatively planar walls
Circular groove: groove going around exterior wall of vessels or interior of cajete
Diagonally burnished surface
Direct rim: a rim that does not turn inward or outward
Even thickness
Evenness
Flat bottom
Horizontally burnished surface: burnishing marks are horizontal around the vessel.
Horizontal bump: this is a thicker area going horizontally around the vessel.
Horizontal depression: this is a horizontal depression encircling the parts or whole vessel.
Horizontal un-even area: un-even horizontal area going around the vessels.
In-turned wall: opening of sub-hemispherical vessel is constricted by inturn just below the rim, producing small shoulder.
Nearly cylindrical neck: Jar neck approximates a cylinder.
Ring base with concave area
Round bottom
Round bottom, bottom with concave area: this is a vessel that would have a rounded bottom if were not because there is a rounded concave depression on the outside where the bottom rests.
Sub-hemispherical body: most non-jar vessels approximate a hemisphere.
Subspherical body: jar bodies approximate a sphere.
Smooth interior surface: even, compact, smooth feeling surface.
Smooth exterior surface: even, compact, smooth feeling surface.
Smoothing marks
Red-slipped surface: red slip was applied.
Un-evenness on thickness
Unslipped surface

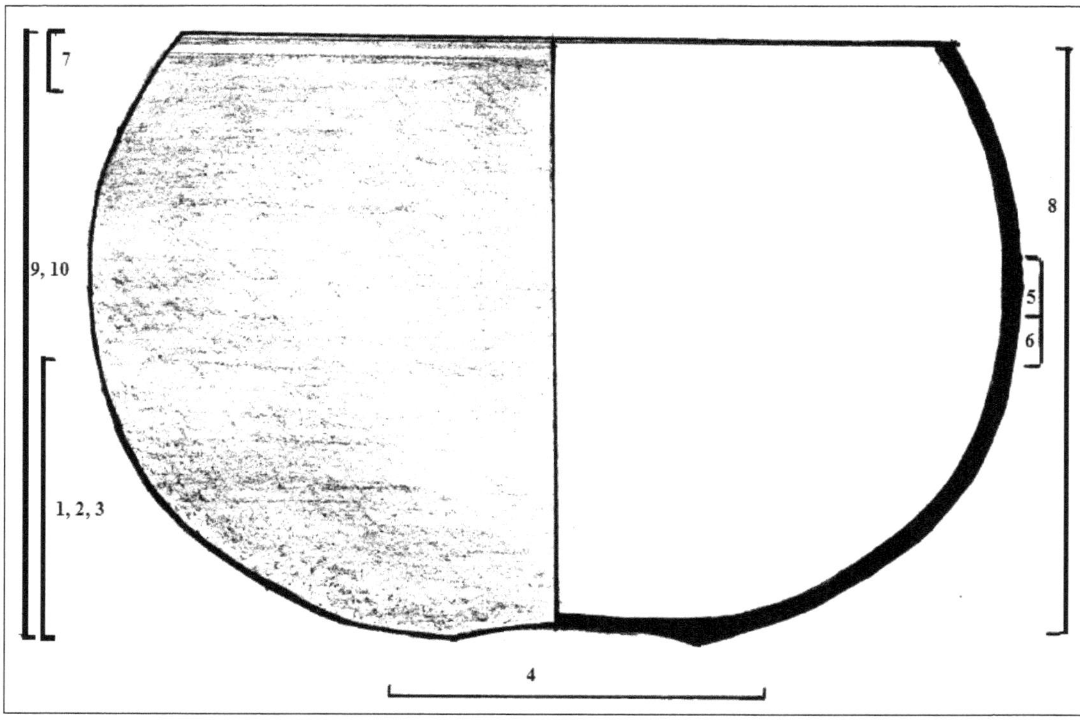

Figure 9-13. Outline of Bowl I showing location of surface features

Table 9-5. Surface Features of Bowl I

(1)Smooth interior surface in lower part of the body, smooth exterior surface in lower part of the body
(2)Evenness of lower part of body
(3) Rounded bottom
(4) Ring base with concave area
(5) Horizontal un-even area, at seven cm from bottom. Diagonal *smoothing marks* can be felt in some parts of this horizontal area.
(6)Horizontal depression on inside wall just under *horizontal un-even area*
(7)Horizontal bump, just below the rim.
(8)Smoothed interior surface
(9)*Red-slipped surface* on the outside
(10)Horizontally burnished surface on the outside.

The lower half of the vessel shows smooth interior and exterior walls (1) and evenness (2) without bumps or depressions. The vessel probably was not started on a flat area. The vessel has a rounded bottom (3). To achieve a rounded bottom, the vessel could have been started in a flat area, and the excess trimmed to obtain a rounded bottom. This technique was documented in Yucatán (Thompson 1958). However, when this technique is used, the curvature of the bottom varies significantly (Thompson 1958, p.51). This is not the case with Bowl I, which has a uniform curvature without discontinuities. Because of this, it is probable that it was started in a rounded surface.

It is probable that the vessel was started in a concave mould. According to Gelbert (2005), a very diagnostic feature of the use of a concave mould is a slightly concave area observed on the outside bottom of the vessel (section 9.1.6.1). Bowl I presents a concave rounded area at the bottom (4), which is visible from the outside. This feature is otherwise difficult to explain. This feature and the very rounded bottom of the vessel make it likely that it was started on a rounded, concave mould.

Comparing this technique to ethnographic work, Thompson mentions that, in addition to the trimming technique, round bottom vessels were also started on a concave surface (Thompson 1958, p.51). Bowl I shows evidence of being formed in two sections. When the lower part of this pot was built up to 7 cm, it was left to partially dry. This would explain the difference between the even lower part and the uneven band area (5) going around the vessel, in addition to the smoothing marks in this area that were not totally obliterated. Other than this band, there are no surface traces on the interior or exterior of this pot up to the seven cm high mark.

The depression (6) and the horizontal uneven area (5) at seven cm indicate that the potter could not totally erase his actions, likely because the bottom was already semi-dry (Thompson 1958, p.52). This uneven horizontal band joins the lower part of the vessel to the next addition. This type of feature related to a section of the pot left to dry has been observed by Thompson (1958) in Yucatán vessels during the manufacture of pots.

The top half of the body is smooth and even without traces. Right below the rim, there is a horizontal bump (7) going around the vessel. It has the appearance of the superimposition of a coil (Gelber 2005, Figure 8a), but could also be the result of pressing down with the fingers while forming the lip.

The surface was smoothed (8), slipped on the exterior only (9), and burnished (10), presenting horizontal burnishing marks, which indicates that it may have been turned as it was burnished.

9.1.6.5 Surface Analysis of Cajete I (# 9151, Table 8-18)

This is a whole Mama tripod cajete (Figure 9-14) with *abrupt walls, round bottom,* three *conical feet, a red-slipped surface* on the whole interior and exterior wall. The exterior of the bottom *has* an *unslipped surface*. The surface features (1 to 9) are summarised in Table 9-6.

This vessel presents a very rounded bottom without indications that the roundness was formed by trimming. The rounded bottom has very uniform curving and even thickness throughout the bottom (1). As discussed before, the rounded bottom and the uniform curvature (without indication that the round shape results from trimming) may indicate that the vessel was started on a rounded concave or convex surface (section 9.1.6.1). The use of a concave surface, however, conforms to ethnographic data (Thompson 1958). This surface also supports the vessel while it is built.

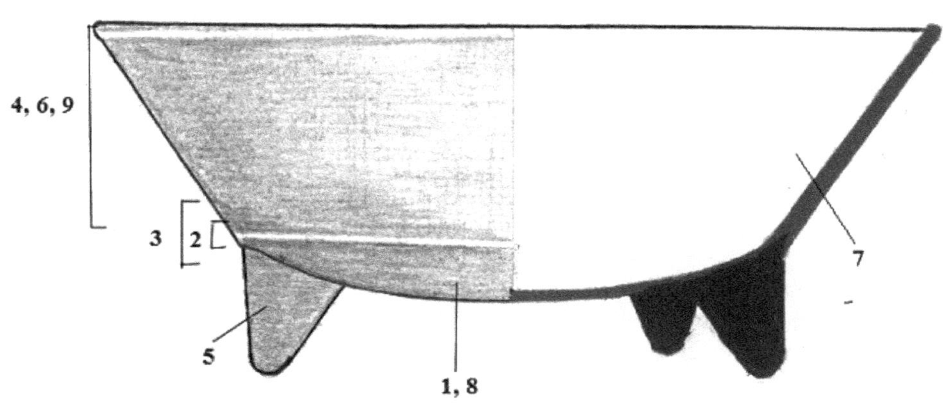

Figure 9-14. Outline of Cajete I showing location of surface features

Table 9-6. Surface Attributes of Cajete I

(1) Even thickness, very even throughout the *rounded bottom*, with uniform curving.
(2) Horizontal bump, where rounded bottom joins the start of the wall
(3) *Un-evenness* on thickness between bottom and walls, with walls somewhat thicker than the bottom.
(4) Even thickness, very even, throughout the wall.
(5) Conical feet, three.
(6) *Red-slipped surface outside*
(7) Red-slipped surface inside
(8) Unslipped outside bottom
(9) Diagonally burnished surface. The burnishing strokes are diagonal.

This pot shows evidence of having been made in two sections. Extra clay, probably a coil was added to help join the bottom to the walls, explaining the horizontal bump (2) going around the vessel, the visible disruption, and the thicker area protruding from the wall where the bottom and walls join. These indications of a join in addition to the walls that rise abruptly from the bottom and the un-evenness and difference in thickness (3) between the rounded bottom and the walls indicate that the bottom and walls are two sections.

It is not clear how the walls were built. A very even thickness throughout the walls is observed (4). The walls may be the result of using a long slab of clay or of coiling, which was well smoothed. The conical feet were added while the bottom was already leather-hard, judging by the lack of deformation on the rounded bottom. The cajete was smoothed and slipped outside (6) and inside (7), while the outside bottom was left unslipped (8). The vessel was then burnished (9) inside and out. Given the diagonal marks, the cajete was not rotated while it was burnished.

9.1.6.6 Surface Analysis of Cajete II (# 9152, Table 8-18)

This is a partial tripod Mama cajete with missing feet (Figure 9-15), *composite form*, lightly *round bottom, abrupt walls, circular groove* delimiting interior separation of bottom and walls, *red-slipped surface* on interior and exterior wall. The surface features (1-7) for this cajete can be found in Table 9-7.

This vessel presents a very rounded bottom without indications that the roundness was formed by trimming. The rounded bottom has uniform curving, and even thickness throughout the bottom (1). As discussed before, the rounded bottom and the uniform curvature (without indication that the round shape is the result of trimming) may indicate that the vessel was started on a rounded concave or convex surface (section 9.1.6.1).

In contrast to Cajete I, there are no bottom-to-walls joining marks. Any joins that may have occurred were erased. This cajete could have been made entirely in a mould. This would explain the even thickness throughout the walls (2) and the lack of evidence for a joining of the bottom and the walls. The walls are slightly thicker than the bottom (3), which indicates that, if a mould was used, it was probably a convex rather than concave mould (section 9.1.6.1). Alternatively, the walls may have been coiled, smoothing the joins to the bottom and the coil marks.

A circular groove (4) in low relief is observed on the interior of the cajete where the bottom turns into a wall. This groove could be a mark left by a mould. In addition, the circularity of this groove could have been accomplished in a variety of ways including pressing an object against the vessel while the cajete was rotating or by lightly pressing the lip of an upside-down vessel against the interior of the cajete. The cajete was slipped on the outside (5), inside (6), and burnished (7).

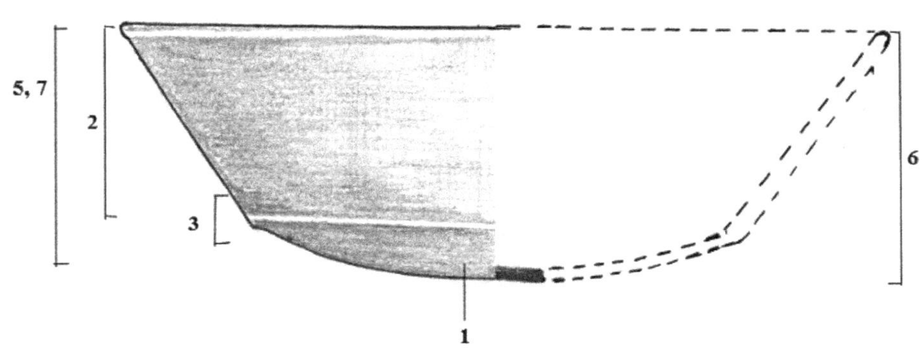

Figure 9-15. Profile of Cajete II showing location of surface features

Table 9-7. Surface Attributes of Cajete I I

(1) Rounded bottom, with even thickness and uniform curving throughout the bottom.
(2) Even thickness throughout the walls.
(3) Un-evenness on thickness between bottom and walls, but it is hardly noticeable.
(4) Circular groove in the interior (not shown). It is perfectly circular.
(5) *Red-slipped surface outside*
(6) Red-slipped surface inside
(7) Burnished surface. The burnishing strokes are diagonal.

9.1.6.7 Surface Analysis of Large Slipped Jar I (# 9153, Table 8-18)

This is a medium to large (height of body 20 cm, neck 7 cm) Mama variety jar (Figure 9-16), missing the whole bottom and almost the whole left side of the pot, *subspherical body, pear-shape body, flat bottom, horizontal handles, narrow mouth, abrupt neck, medium-high neck, slightly outcurving neck, flaring neck, red-slipped surface* on outside, *unslipped surface* on the inside. This type of jar is described by Smith (1971, p.80) as one of the main Late Postclassic forms at Mayapán. It was found near the small site of Chan Pixyax (four km west of Mayapán). Table 9-8 includes the surface features (1 to 10) observed in this jar. The interior lower half presents vertical smoothing marks (1). The lower part is thin (5 mm) and of even thickness (2) up to 9 to 10 cms from the bottom. There are no visible indications of coiling. The thin walls and even thickness may indicate the use of a mould. The interior marks are not consistent with the use of a convex mould. The vessel was started on a flat surface, and could have been started by a variety of methods in which coiling and a concave mould concur with ethnographic data.

Given that the bottom of the vessel is flat and missing, a join existed that attached the bottom to the vessel wall. The bottom and the walls were built as two separate sections. At nine to 10 cm from the bottom, a horizontal uneven area (3) encircles the pot. This indicates that the vessel was built up to that height, and then, it was left to partially dry. This would explain the very uneven two to three cm wide horizontal band where the thick joining coil can still be detected. These types of surface features have been attributed to the dryness of the bottom half, which makes the coil marks difficult to obliterate (Thompson 1958, p.52).

The second half was added, but there is no evidence of the technique used, given that the top half presents even thickness (4) of around 7 mm, evenness throughout (5) and no smoothing marks (6).

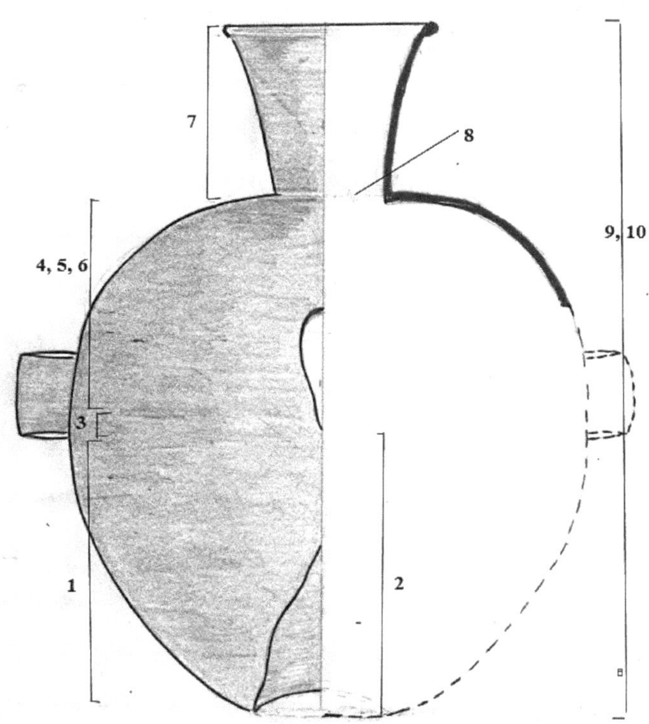

Figure 9-16. Profile of Large Slipped Jar I showing location of surface features

Table 9-8. Surface Attributes of a Large Red-slipped Jar I

(1) *Smoothing marks,* vertical, in interior lower half, bellow the horizontal bump
(2) *Even thickness*, around 5 mm, up to to 9-10 cm high.
(3)Horizontal un-even area, three cm wide around vessel starting at around 9-10 cm (around half body height) from bottom.
(4) Even thickness, around 7 mm, of pots top half
(5) *Evenness* of whole upper half
(6) No *smoothing marks* on upper half
(7) *Even thickness* of neck,
(8) *Attached neck*
(9) *Red-slipped surface outside*
(10) Burnished outside surface.

The neck has even thickness (7) and is attached (8) to the body. It does not come from the vessel. It appears to have been formed by adding a tall, flat slab to the interior of the opening. Given that one section of the body is broken, it is possible to look from the body up to the neck and observe that it is cut clean.

Without considering the horizontal handles, the jar appears to have been built in at least four sections: a lower part up to 9 – 10 cm, most likely leaving an opening in the bottom, the bottom of the pot, the top half of the body, and the neck. After the first 10 cm were built, it was left to dry. The neck was added because it was not built from the body.

9.1.6.8 Surface Analysis of Large Slipped Jar II (# 9154, Table 8-18)

This is large (height 28 cm, diameter 30 cm, neck 8 cm) Mama variety jar, found near the small site of Chan Pixyax (four km west of Mayapán), *subspherical body, round bottom, bottom with slightly concave area, horizontal handles, narrow mouth, abrupt neck, medium-high neck, slightly outcurving neck, flaring neck, red-slipped surface* on outside, *unslipped surface* on the inside. Table 9-9 includes the attributes (1 to 12) observed in this jar.

Table 9-9. Surface Attributes of a Large Red-slipped Jar II

(1) *Rounded bottom, bottom with slightly concave area*
(2) *Smooth interior surface up to 6 cm, no bumps, no smoothing marks*
(3) Slight *un-evenness on the exterior,* vertical, up to 8 cm, likely smoothing marks
(4) *Horizontal un-even area*, furrows, with *horizontal bump* at around six cm encircling the pot, 2-3 cm wide
(5) *Horizontal un-even area with,* at around 15 cm encircling the pot, furrows, with *horizontal bump*
(6) *Horizontal un-even area*, furrows, with *horizontal bump* at around eight cm from the neck
(7) *Even thickness* of neck
(8) no *horizontal bumps* on neck
(9) Attached neck
(10) *Handles attached*
(11) *Red-slipped surface outside*
(12) *Burnished outside surface.*

The vessel has a rounded bottom with a slight concave depression (1). The interior is smooth (2) without smoothing marks. There are smoothing marks on the lower exterior half of the vessel (3). The lower part is thin (5 mm) and of even thickness up to nine to 10 cm from the bottom. It is unclear how the vessel was started. The slightly concave bottom, a difficult-to-explain feature, could be explained with the use of a mould (Gelbert 2005). A mould would also explain the rounded bottom.

The bottom was built up to around six cm and left to dry, judging by the horizontal two cm uneven area (4) with light furrows going around the vessel. These types of surface features have been attributed to a dryness of the bottom half, which makes coil marks difficult to obliterate (Thompson 1958, p.52).

The vessel was built up to 15 cm and left to dry somewhat. This is indicated by the second slightly uneven area (5) with horizontal bumps and ripples at 15 cm going around the vessel.

The vessel was built up to the shoulder (eight cm from neck) and left to dry, as shown by a slightly uneven area, with horizontal bumps encircling the pot.

The neck presents an even thickness (7). It was made as a separate section, because it can be observed that it is a separate section attached (9) to the interior wall of the rim body, and it does not grow from the body. It appears to have been formed by adding a tall, flat slab to the interior of the opening. The handles were attached (10). The vessel was smoothed and slipped (11) and burnished (12).

Jar II was built in sections: the first six cm may have been manufactured in a mould or support. There were three drying episodes. The neck was made separately and attached to the interior of the opening. The handles were attached.

Summarising these findings, the analysed vessels were all hand built. They were constructed using coils, slabs, and moulds. Different techniques were used for the different types of vessels, probably using combinations of mould and coils in some of the vessels.

One of the techniques encountered is very distinctive of modern Yucatecan potters: the open base technique with which the large Jar I appears to have been started. Thompson (1958), in his ethnographic work at nine towns of northern Yucatán, remarks that this is the technique used for most vessels (although modelling is used for some small vessels). The potter starts the vessels by coiling around an open base on the edge of the *kabal* leaving a hole at the bottom. The coil thickness varies depending on the size of the vessel but the diameter is usually between 4 and 6 cm. To fill the hole at the bottom of the forming vessel the potter presses a lump of clay into the hole and on top of the kabal, spreading the paste toward the interior edge of the vessel.

Foster (1967) did not find this technique in his extensive ethnographic work in central México. The open base technique is not used in Guatemala either, with two exceptions (Reina and Hill 1978). One is the Chorti-speaking area in southeast Guatemala. The other is the town San José Petén, located in Lake Petén Itzá, Guatemalan lowlands. Its Maya inhabitants are descendants of Terminal Classic/Early Postclassic migrations from northern Yucatán (Roys 1962).

9.1.7 *Summary*

A prospection made searching for deposits of clay around Mayapán was not successful. However, marls are common, and samples were taken with the objective of testing their suitability to function as a viable pottery clay. The location and description of these raw materials can be found in Table 5-4. Of the marls collected, firing tests (section 9-4) demonstrated that two of the samples were suitable for pottery making: sample #1 from the road Mama-Chumayel and sample #11 from road by Mayapán, showing that raw materials suitable for pottery production are found in the vicinity of Mayapán, and probably near other north-central sites.

Most red-slipped pottery samples contain medium-grained particles composed of micritic to finely crystalline sparry calcite or dolomite. Divisions of grain size between red-slipped Mama and unslipped Yacman jars, and frequency of inclusions between these two ceramic varieties, indicate that the coarser particles were added by potters.

The source of the particles in red-slipped vessels is most likely sascab, based on comparisons of pottery fabrics (such as Dolospar-W-Medium, Micrite-W-Medium, and Sparite-W/B-Coarse/Medium), raw materials (micritic and sparry carbonate inclusions), and ethnographic work (Thompson 1958; Rendón 1947, 1948a; Morales Valderrama 2005). Further processing of the raw materials collected from the area would have been necessary to construct a pottery clay that is similar in grain size and frequency of inclusions to the samples in this study, probably by crushing and sifting the very coarse to coarse particles of micritic and sparry carbonate rocks found in sascab.

Surface analysis of five red-slipped vessels showed that they were hand built in sections likely using coils for the jars and moulds for the rounded bottom of cajetes. The lower parts of jars up to seven to 9 nine cm were built and left to dry before continuing with the next section.

Many potters' groups were involved in the production of the red-slipped Mama vessels sampled from the north-central sites. Production of Mama most likely occurred at multiple locations. Based on the petrographic and chemical variability of micritic and sparitic sascab fabrics, they may represent several potters' groups at various locations. However, it cannot be ruled out that production of Mama with micritic fabrics, for instance, occurred at one location that presents a varied composition.

Mama with dolospar fabrics, contrastingly, presents chemical and petrographic homogeneity across sites. They may represent centralised production, but alternative explanations exist including local production at the different sites, that results in a homogeneous fabric by using raw materials with chemical compositions within Group 1, along with fragments of dolospar. The location of production of vessels with dolospar fabrics will be further discussed in the next chapter.

9.2 Reconstructing Production: North-central, Plain Navula

The plain, unslipped Navula jars and cajetes sampled for this study are part of one of the most abundant Tases ware, the Unslipped Navula Ware. Together with Mayapán Red Ware, they make up most of the Late Postclassic assemblage from north-central sites (Smith 1971a). Most researchers have divided these unslipped ceramics, based on the probable intended function of the vessel, into utilitarian and ceremonial (Smith 1971a; Ochoa 2007). Some of the utilitarian forms include open-mouth jars, deep dishes or *cajetes*, flat vase dishes, and restricted orifice bowls (Smith 1971). Of the many varieties and forms within the unslipped Navula Ware, this section deals with jars of the Navula variety. Some Navula *cajetes* (Figure 3-7) were included for comparison given that the intended function of jars and *cajetes* had to differ greatly.

Three aspects of production for plain Navula samples differ markedly from what was discussed about the red-slipped samples in the previous section, and they are the focus of this section: (1) raw materials selection in regard to the types of coarse inclusions found; (2) the fabric classes identified in the plain Navula samples; and (3) the number and location of the potters. In the previous section about red-slipped Mama samples, the clays within the area and the techniques related to raw materials processing were addressed, and given that they, apparently, also apply to plain, unslipped Navula samples, and that no additional information has been found pertinent to Navula, they are not discussed again in this section.

9.2.1 *Raw Materials Selection for Navula: Coarser Inclusions*

The objectives of this section are to examine the inclusions present within the fabric classes of Navula samples and to compare them to the local raw materials in order to draw inferences as to their origin. Sixty thin sections of unslipped plain (Navula) jars and cajetes were analysed, comprising multiple types of fabrics. The distinctive inclusions or types of fabrics are shown in Figure 9-17. They are mainly dolospar (finely crystalline sparry dolomite), bioclasts (skeletal remains), dark micrite, single crystals, and sparite (finely crystalline sparry calcite). Figure 9-17 contains the percentage of the total of Navula samplescharacterized within each type of fabric. That more unslipped, plain samples were found to have dolospar types of fabrics may be due, at least in part, to more samples being analysed from Mayapán than from any other site (61 of 188 thin sections), and that at this site, dolospar fabrics comprise most (64 per cent including Payil, or foreign samples) of the samples.

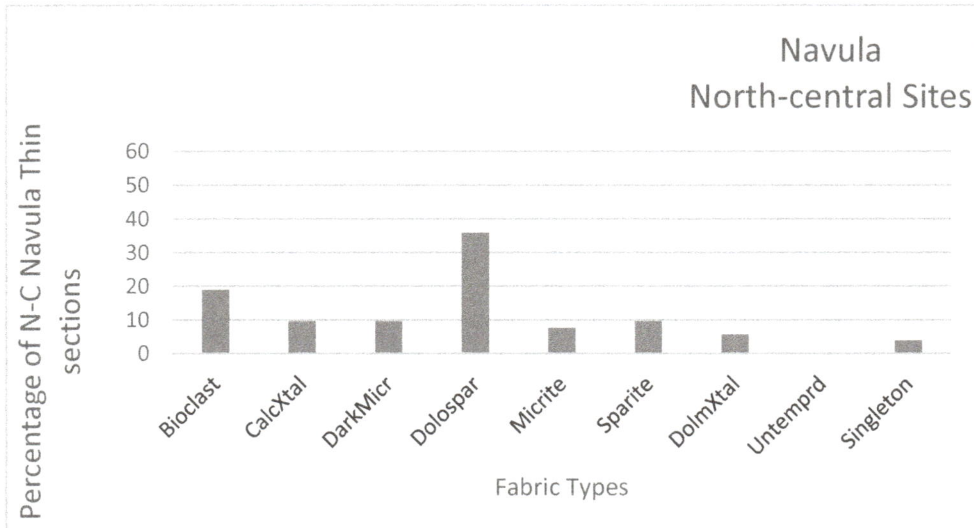

Figure 9-17. Percentage that fabric types represent of total Navula thin sections (n=53)
Key to inclusion types: CalcXtal = single calcite crystals, DarkMicr=dark micrite, DolmXtal= single dolomite crystals

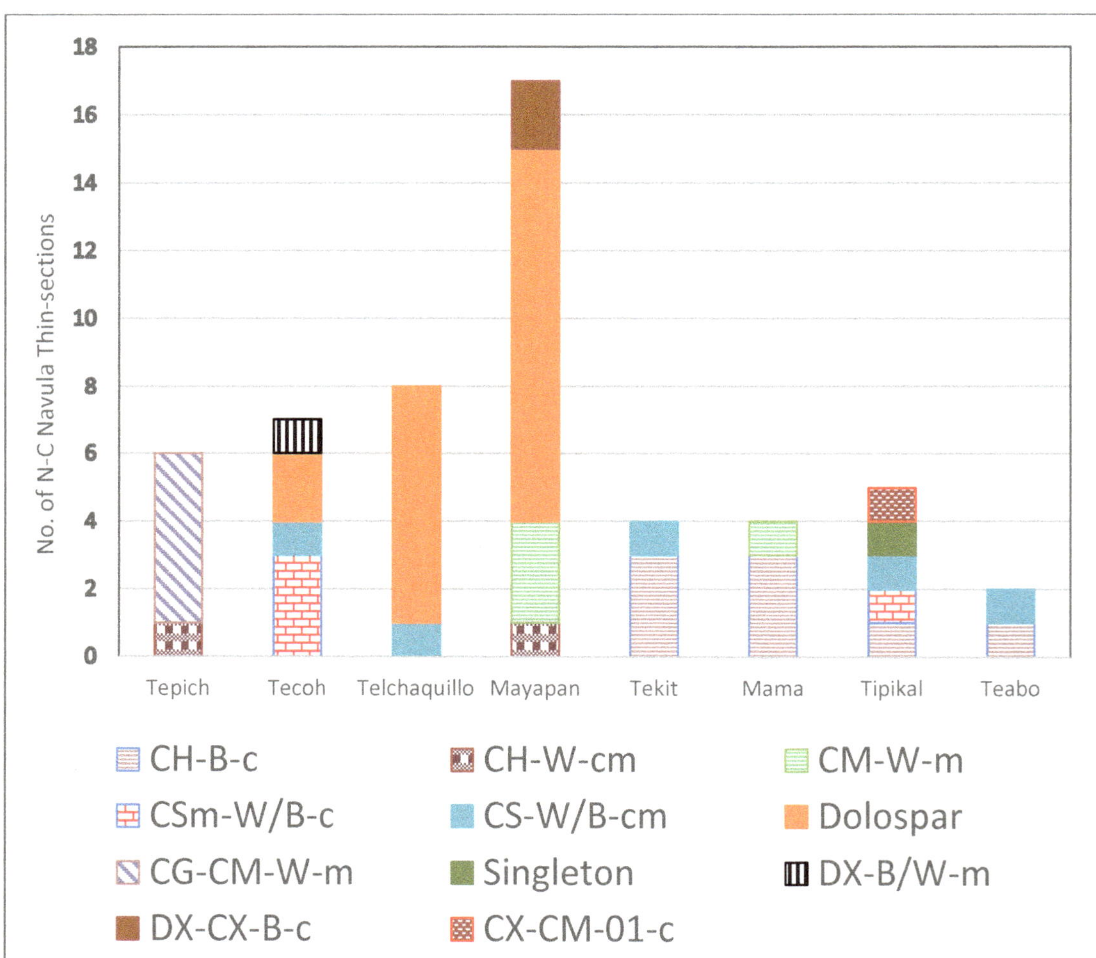

Figure 9-18. Frequency of fabric classes of unslipped, plain Navula within sites.

Key to fabric classes (Dolospar= fabrics in Table 9-10):
CG-CM-W-m=DarkMicrite-Micrite-Medium CH-B-c = Bioclast-B-Coarse
CH-W-cm= Bioclast-W-Coarse/Medium CM-W-m=Micrite-W-Medium
CS-W/B-cm= Sparite-W/B-Coarse/med CSm-W/B-c = Sparite-m-W/B-Coarse
CX-R/B-c= CalcXtal-R/B-Coarse CX-CM-01-c = Calcite-Micrite-01-Coarse
DX-CX-B-c = Dolomite-Calcite-B-Coarse; DX-B/W-m=Dolomite-B/X-medium

From the previous section, it can be recalled that the fabrics of red-slipped (Mama) samples are mostly composed of dolospar, micrite, or sparite, and it was argued that they are associated with sascab. When compared to the types of fabrics found in red-slipped (Mama) samples (Figure 9-7) of the north-central sites, Navula samples were found in greater variety of the types of fabrics, including fabrics with bioclasts, single crystals (calcite and dolomite), dark micrite, or singleton fabrics (un-grouped), indicating a greater variability of types of inclusions and sources of raw materials.

9.2.2 The Fabric Classes of Plain, Unslipped Navula Jars

The fabric classes and sites at which plain, unslipped Navula samples were found are presented in this section. One fabric class represents a common technology for the construction of the pottery clay, acknowledging that it is possible that from different ways of doing things the same fabric class will result. When one fabric class is found in multiple sites, it can be argued that they share a common technology, or way of doing things.

Figure 9-18 breaks down the information in Figure 9-17 to show on a site-by-site basis the frequencies of the fabric classes found in plain, unslipped Navula samples. Figure 9-18 shows that a variety of fabric classes was found within the unslipped plain Navula samples from the north-central sites.

One way of looking at the variety of Navula fabrics is to divide them into those containing mainly inclusions of micro to finely crystalline calcite or dolomite and those with mainly larger crystals. Among the finely crystalline crystal category are those containing:

1. micritised fossil shells (usually calcite homogeneous, section 8.1.3) or Bioclast-W-Coarse/Medium (CH-W-cm),
2. microcrystalline calcite Micrite-W-Medium (CM-W-m),
3. finely crystalline dolospar fabrics, or Dolospar-B-Coarse, Dolospar-W-Coarse, and Dolospar-W-Medium, and
4. finely crystalline sparry calcite or Sparite-W/B-Coarse/Medium (CS-W/B-cm).

In hand specimens, except for some dolospar tempered samples, it was not possible to set them apart. They all look, for the most part, like white micritic limestone. One example of a sparitic fabric in a hand specimen is illustrated in Figure 6-10. It required using the binocular or polarising microscope to recognise the different crystal textures. It is highly likely that potters were not able to set them apart and may have used them indiscriminately. However, each of these fabrics was found at different sites, as can be observed in Figure 9-18 and Table 8-19. At Tepich, plain, unslipped samples were also found within a fabric containing medium-dark micrite particles as a main inclusion, or DarkMicrite-Micrite-Medium (CG-CM-W-m).

Other groups of Navula samples contain distinctive particles that in a hand specimen looked like translucent or semi-translucent rocks. These samples comprise fabric classes in which the main or diagnostic particles are coarse-to-very-coarse crystalline calcite or dolomite, as well as sparry particles in which each crystal is at least medium-crystalline (one example in a hand specimen is illustrated in Figure 6-15). Among these fabrics are those containing as main inclusions:

1. medium-crystalline sparry calcite, or Sparite-m-W/B-Coarse (CSm-W/B-c),
2. coarsely grained single crystals of dolomite, or Dolomite-Calcite-B-Coarse (DX-CX-B-c), and
3. coarse-to-very-coarse single calcite crystals or Calcite-Micrite-01-Coarse (CX-CM-01-c).

Two fabrics, apparently, do not fit within these two sets. One is a coarse shell fabric that was found south of Mayapán, or Bioclast-B-Coarse (CH-B-c). Although the polarizing microscope showed skeletal remains, in a hand specimen, they appear angular, whitish, smooth, and opaque to semi-translucent. They may have looked like a crystalline rock to potters. The second fabric appears in a hand specimen as fine-grained with fine dark inclusions that the microscope revealed as iron-altered, rhombohedral, fine, single crystals of dolomite, or fabric Dolomite-B/W-Medium (DX-B/W-m).

Table 9-10. Presence/absence of Dolospar Fabric Classes Found within Navula

Texture	Fabric Class	Fabric Class	Presence
Coarse	Dolospar-B-Coarse	DS-B-c	X
Coarse	Dolospar-W-Coarse	DS-W-c	X
Medium	Dolospar-W-Medium	DS-W-m	X

At Mayapán, most Navula samples (Figure 9-18) were found with the dolospar fabrics. Unslipped and plain Navula jars or cajetes with dolospar fabrics were not found at Tepich or south of Mayapán (Figure 1-4). Table 9-10 shows that in contrast to the red-slipped samples discussed earlier in which the samples taken are dominated by one grain size (medium), Navula samples were found to have medium and coarse textures, acknowledging that the number of samples is small and that more extensive sampling can yield different interpretations.

At Mapapán, a fabric with coarse, clear, single crystals of dolomite and rare calcite, or Dolomite-Calcite-B-Coarse (DX-CX-B-c), a fabric with fossilised homogeneous micrite (bioclasts) that re-fired to white, or Bioclast-W-Coarse/Medium (CH-W-cm), and a micrite Micrite-W-Medium (CM-W-m) fabric were also found in Navula samples.

9.2.3 Plain, Unslipped Navula Potters' Groups

The chemical composition potentially gives information about zones of raw material procurement, and in this section, chemical composition is analysed in combination with the fabric classes, seeking patterns to infer the number and even location of the potters' groups. Table 9-11 reiterates the information in Figure 9-18 and, in addition, incorporates the chemical variability of the unslipped, plain Navula samples.

Samples that could not be assigned to any of the chemical groups, or unassigned, are also included in the Table 9-11. The composition of one unassigned sample is not necessarily similar to other unassigned samples, and in Table 9-11 they are shown as having different chemical compositions from other unassigned samples. As expressed in the previous section, the ceramic production technologies implied by the fabric classes, when examined in relation to the chemical variations, may reveal different potter's groups and probable zones of procurement of the raw materials.

Table 9-11. Fabrics and Chemical Compositions of Unslipped, Plain Navula Jars from the North-central Sites

	Fabric	**Ch Gr (*)**	**Tepich**	**Tecoh**	**Telchaquillo**	**Mayapán**	**Tekit**	**Mama**	**Tipikal**	**Teabo**	**Minimum Number Potters' Groups (PG)**
Micrite	Micrite-W-Medium (CM-W-m)	1				X					PG#1
		u/a						X			PG#2
		u/a				X					
Bioclasts	Bioclast-W-Coarse/Medium (CH-W-cm)	1	X			X					PG#3
Dolospar	Dolospar-W-Coarse(DS-W-c), Dolospar-B-Coarse(DS-B-c), Dolospar-W-Medium (DS-W-m)	1		X	X	X					PG#4
Finely-crystalline Sparite	Sparite-W/B-Coarse/Medium (CS-W/B-cm)	2		X			X		X	X	PG#5
		u/a			X						PG#6
Bioclasts	Bioclast-B-Coarse (CH-B-c)	2					X		X		PG#7
		u/a						X			PG#8
		u/a					X				
Medium-crystalline or Larger	Sparite-m-W/B-Coarse (CSm-W/B-c)	2		X					X		PG#9
		5		X							PG#10
	Dolomite-Calcite-B-Coarse (DX-CX-B-c)	uk				X					PG#11
Dark Micrite	DarkMicrite-Micrite-Medium (CG-CM-W-m)	uk	X								PG#12

KEY:u/a = chemically unassigned; uk=unknown (no chemically analysed)

The sites in Table 9-11 are arranged from north to south, showing a division in the sub-regional distribution of Navula samples based on their chemical groups: samples within Group 1 were not found south of Mayapán (with one exception), while samples within Group 2 were found south of Mayapán and at Tecoh. This geographical and chemical division indicates that different potters' groups likely produced the Group 1 and Group 2 pottery, and that the pottery of these groups may have been distributed differently.

Further examination of the combination of fabric classes and chemical groups shown in Table 9-11 indicates that the plain, unslipped samples likely represent vessels made by multiple potters' groups.

(a) At Mayapán, Navula specimens share the medium-grained fabric Micrite-W-Medium (CM-W-m, section 8.5.1) with red-slipped Mama. One sample from the site of Mama also shares this fabric. Although micritic Navula from Mayapán and the site of Mama may represent one group of potters, it is probable that two or more groups were involved (PG#1, PG#2). This proposition is based on that, at Mayapán, Mama and Navula samples have both composition within Group 1, while, at the site of Mama, Navula and Mama are not assigned to any chemical group (Tables 9-3 and 9-11). Given that CM-W-m was found in Mama and Navula vessels almost exclusively, it may represent a common basic recipe for producing these vessels at different locations.

(b) At least one group of potters (PG#3) is inferred at Tepich and associated with Navula samples with the bioclast fabric Bioclast-W-Coarse/Medium (CH-W-cm) and composition within Group 1. This is a fabric that at Tepich is shared with Mama samples, and it is found at Tepich only, with one exception from Mayapán (which can be seen more clearly in Table 8-19). Several groups of potters in a homogeneous raw material environment producing vessels with Bioclast-W-Coarse/Medium characteristics is a possibility.

(c) At least on potters' group or one production zone is associated with the Navula samples having a dolospar fabric, and found at Mayapán, Telchaquillo, and Tecoh. Navula vessels with dolospar (always chemical Group 1) were produced without a recognizable pattern regarding grain size. For this reason, these potters were not divided according to the dolospar fabrics (which are divided by texture, mainly), and the minimum number of potters' group that could have been producing Navula with dolospar is one, PG#4.

(d)At least one or two groups of potters are associated with a sparitic fabric Sparite-W/B-Coarse/Medium. This fabric was found at Tecoh, Tekit, Tipikal, and Teabo within chemical Group 2 (one potters' group or PG#5). It was also found at Telchaquillo in which the samples are chemically unassigned (a second potters' group or PG#6). This fabric is not homogeneous across the group in regard to porosity, silt frequency, colour of the micromass, or the presence/absence of other particles such as micrite, skeletal remains, red concretions, or discrete dolomite crystals. Navula and Mama share this fabric. At Tipikal and Teabo, the ceramic varieties Mama and Navula share chemical compositions (Group 2), while at other sites, no match was found.

The heterogeneity of Navula petrographic characteristics with this fabric may indicate one or two potters' groups working within a very heterogeneous raw materials environment. Alternatively, the heterogeneity of petrographic attributes within, mostly, one chemical group, (Group 2) may indicate little chemical variability across several raw materials locations. In this last scenario, that Yacman is not found in this fabric indicates the preferential use of this type of finely crystalline limestone for the construction of Navula (and Mama), and that a widespread tradition may have existed.

(e) A bioclasts or skeletal remains coarse fabric, Bioclast-B-coarse or CH-B-c, within Group 2 was found at Tekit and Tipikal pointing to one potters 'group PG#7. A second potters' group (PG#8) is inferred from this fabric being found at Tekit and Mama with unassigned chemical composition. An alternative explanation is that Group 2 raw materials with the characteristics of this fabric are found at Tekit and Tipikal and that several potters' groups at different locations may have existed. In the samples taken, Bioclast-B-coarse fabric was found in unslipped vessels only (Table 8-19), which, in the last proposed scenario, indicates the preferential use of this fabric across locations for the construction of unslipped vessels.

(f) One or two potters' groups at a minimum, (PG#9 and PG#10), may have been associated with a crystalline fabric Sparite-m-W/B-Coarse (CSm-W/B-c) with medium-crystalline sparry calcite. This fabric was found in samples from Tecoh and Tipikal having a Group 2 composition, and in samples from Tecoh having a Group 5 composition. More potters' groups may be involved if raw materials with similar characteristics and chemical compositions are found at the different locations. These potters were preferentially selecting this material for unslipped Navula (and Yacman, see next section) vessels (Table 9-11, Table 8-19).

(g) At least one potter's group (PG#11) is associated with a coarse dolomite fabric, Dolomite-Calcite-B-Coarse (DX-CX-B-c) that was found at Mayapán and at Tekit in samples from Yacman and Navula unslipped jars.

(h) One potters' group (PG#12), at least, may be responsible for a fabric found at Tepich only (with one exception) containing mainly dark micrite, Dark Micrite-Micrite-Medium (CG-CM-W-m). This is a dark, microcrystalline fabric that Navula shares mostly with the red-slipped Mama.

9.2.4 *Summary*

When compared to Mama (red-slipped), Navula samples were found in a greater variety of fabric types including fabrics with bioclasts, dolospar, sparite, micrite, single crystals (CalcXtal and DolmXtal), dark micrite (DarkMicr), or singleton fabrics, indicating a greater variability of types of inclusions and potential sources of raw materials. These fabrics may represent different potters' groups. The above mentioned fabrics, their association with sites, ceramic types, and to other fabrics can be better observed using Table 8-19.

Navula samples also have fabrics with crystalline coarse particles, such as the fabrics of many Yacman samples. Adding to the variety, two fabrics are found in Navula almost exclusively, Bioclast-B-Coarse (CH-B-c) and DarkMicrite-Micrite-Medium (CG-CM-W-m), granting that the number of samples is small and a more extensive sampling can change this interpretation.

At Mayapán, Mama and Navula (and Yacman) samples share fabrics having dolospar inclusions. Forming techniques are not discussed for Navula because vessels of this type were not available for analysis.

9.3 Reconstructing Production: North-central, Unslipped, Striated Yacman

The striated, unslipped Yacman jars are a variety of the Unslipped Navula Ware (Figure 9-19). Yacman and Navula are the two varieties of this ware included in this study, differentiated by the light striations that Yacman vessels present. The Yacman form included in this study is also the most common—an open-mouth jar thought to be for cooking.

Three aspects of unslipped Yacman jar production differ markedly from what was discussed about the red-slipped samples in section 9.1, and they are the focus of this section: (1) raw materials selection concerning the types of coarse inclusions found; (2) the fabric classes identified in the plain Navula samples; and (3) the number and location of the potters. In a previous section about red-slipped Mama vessels, the clays within the area were discussed, and, given that this discussion also applies to Yacman samples and that no additional information has been found pertinent to Yacman, they are not discussed again.

Figure 9-19. Reconstructed Yacman vessel from cave Yok Dzono (from Brown 2017)

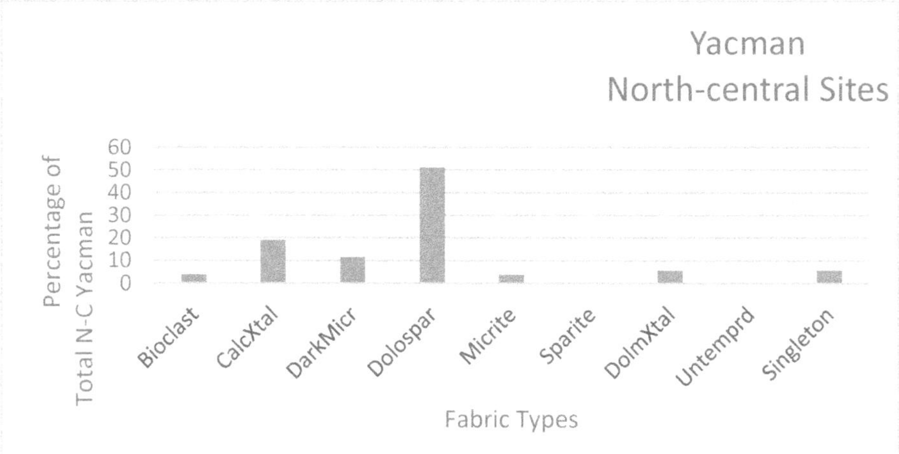

Figure 9-20. Percentage that the types of fabrics represent of total Yacman samples (n=53)

9.3.1 Raw Materials Selection for Yacman: Coarser Inclusions

The types of fabrics found in striated, unslipped open-mouth *ollas* can be broadly divided by the main or diagnostic inclusions. This broad categorisation is shown in Figure 9-20, which shows the different types of fabrics in the Yacman samples. Figure 9-20 shows that:

1. Samples having fabrics containing coarse clusters of dolospar comprise a significant portion of the total fabrics found in striated jars. Given that most dolospar found is finely crystalline (or each crystal size around 0.02 to 0.06 mm), the appearance in a hand specimen of the crystals comprising the grains varies from white and dull (like micrite) to clear, depending on grain-size and degree of alteration.
2. The contribution of micrite and sparite fabrics is insignificant.
3. Fabrics with larger, from medium to very-coarse, single crystals of calcite or dolomite as the main or diagnostic inclusions make a significant contribution to the total variety of fabrics found in striated, unslipped Yacman samples. Due to their larger size, the individual grains can be detected in a hand specimen as translucent or semi-translucent, smoothed, planar or blocky, angular inclusions.

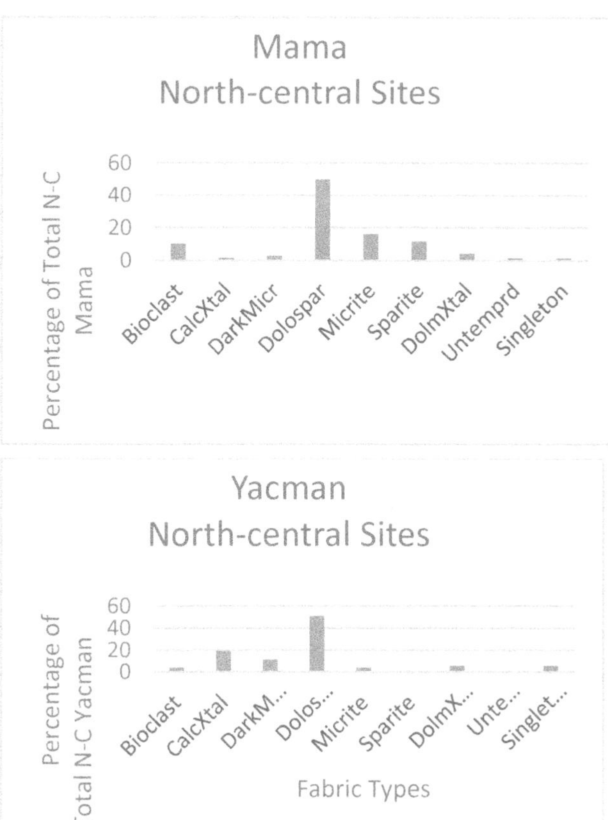

Figure 9-21. Comparison of Mama and Yacman fabrics

In Figure 9-21 (that combines Figures 9-7 and 9-17), a pattern for the north-central sites is shown in which single calcite crystal fabrics comprise a negligible percentage of the Mama samples for this area, whereas they comprise a significant proportion of unslipped striated samples throughout the north-central sites. If fabrics with dolomite crystals are added (DolmXtal), the proportion of fabrics with translucent rocks is even higher.

It is apparent that the fabric classes found in the striated, unslipped Yacman samples, when compared with red-slipped Mama, overlap only on the dolospar fabrics and have little in common with all other fabrics.

9.3.2 Raw Materials Selection and Processing for Yacman: Hi'

Ethnographic studies in Northern Yucatecan show that potters use *hi'* to refer to the translucent rock added to the pottery clay that is essential for the construction of cooking pots (Thompson 1958, Rendón 1948b). *Hi'* is composed of calcite crystals (Arnold 2008, p.192; Thompson 1958, p.70). Potters obtain calcite nodules from caves and veins in the rocks (Thompson 1958, p.70). The translucent rocks found in pottery samples are therefore most likely the same substance.

Rendón (1947, p.116) described the processing applied to hi'. The rock is cut from the *canteras* in transportable portions and taken to a place just outside the potter's house. In some towns, to obtain *hi'*, limestone is burned, as described by Thompson (1958:70) at Lerma, Maxcanú, and Ticul.

Figure 9-22. Grinding stone (about 1m in diameter) from Yucatán (from Thompson 1958, Figure 16f)

Thompson suggested that burning is performed in order to break down the calcareous cement that binds the crystals to form a pile of granules. While he didn't know the exact temperature of this firing, he thought it was not high enough to decompose the crystals (calcinar), which gave the pile of granular rock a grey colour instead of the original white colour of ground hi' (1958:70).

Rendón (1947, p.116) observed that the decision to heat the crystalline rock before grinding is a community specialisation. At Maxcanú and Becal, *hi'* is not heated but ground as it was taken from the *canteras*.

The grinding is done using a hard, smooth-surface limestone boulder (Thompson 1958, p.68) of about 1m in diameter called a mux (Figure 9-22) that is moved back and forth. Thompson (1958, p.68) and Rendón (1947) described that the mux was used to roll over the temper to grind it. The mux cannot be lifted due to its size and weight. After grinding the *hi'*, the potter sieves it. In the north-central region, *hi'* is not as available as sascab. Towns such as Becal and Tepakam had to import *hi'* from Maxcanú and Celcetok (Thompson 1958, p.66).

9.3.3 *The Fabric Classes of Striated, Unslipped Yacman Samples*

This section presents the fabric classes and sites in which the striated, unslipped Yacman vessels sampled for this study were found. Each fabric class represents a common technology for the construction of the pottery clay that can be found at multiple sites, sharing a common technology, or way of doing things. Figure 9-23 breaks down the information in Figure 9-20 to show on a site-by-site basis the frequency of fabric classes.

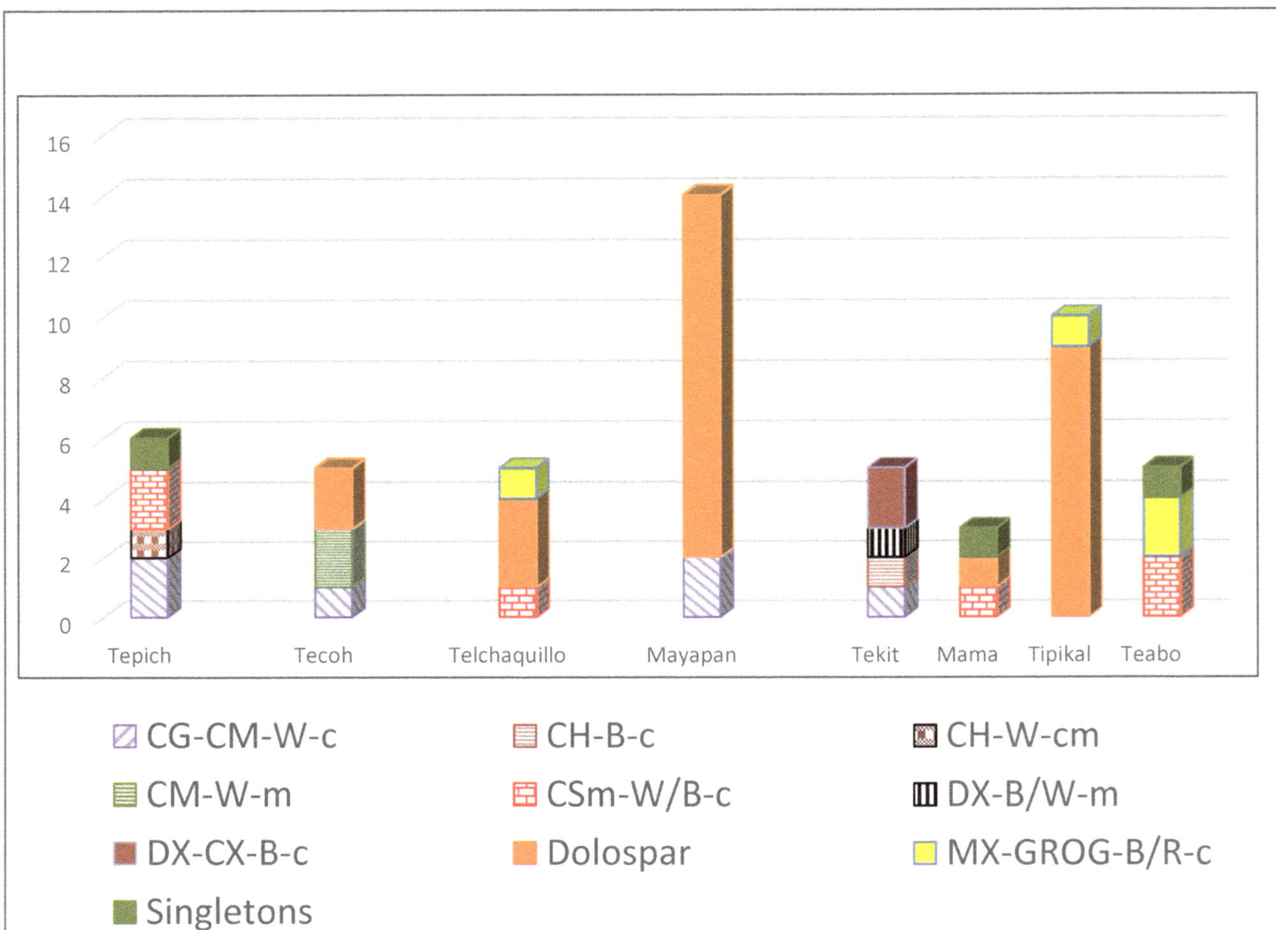

Figure 9-23. Frequency of Yacman samples within the different fabric classes (n=59).
Key to inclusions: CalcXtal = single calcite crystals, DarkMicr=dark micrite, DolmXtal – single dolomite crystal
Key to Fabrics:
CG-CM-W-c = DarkMicrite-Micrite-Coarse
CH-B-c = Bioclast-B-Coarse
CH-W-cm = Bioclast-W-Coarse/Medium
CM-W-m = Micrite-W-Medium
CSm-W/B-c = Sparite-m-W/B-Coarse
DX-CX-B-c = Dolomite-Calcite-B-Coarse
DX-B/W-m = Dolomite-B/W-Medium
MX-GROG-MDX-B/R-c = Calcite-Grog-B/R-Coarse

Yacman fabrics can be divided into those with main inclusions of micro to finely crystalline calcite (or dolomite) and those with larger crystals. The larger crystals range from medium to very coarsely crystalline, which usually look translucent to semi-translucent in a hand specimen. The fabrics dominated by micritic to finely crystalline inclusions are the following:

1. Bioclast-W-Coarse (CH-W-c), with micritised fossil shells
2. Micrite-W-Medium (CM-W-m) fabric, or micritic limestone
3. Dolospar-W-Coarse (DS-W-c) and Dolospar-B-Coarse (DS-B-c), similar fabrics differing in their refired colour, containing finely-crystalline sparry dolomite (dolospar)
4. DarkMicrite-Micrite-Coarse (CG-CM-W-c), with dark micrite fabric

Of these fabrics, dolospar fabrics stand out because they were consistently found in the thin sections from multiple sites. They are in the majority of striated samples from Mayapán, Telchaquillo, and Tipikal. The sites at which these fabrics were found coincide with the sites at which the red-slipped (Mama) samples with dolospar fabrics were found: these were only north-central sites with the exception of Tepich and Teabo. They were not identified at the eastern sites, Cobá, Chac Mool or Culubá.

Figure 9-24. Yacman sample from Mayapán

Almost 25 per cent of Yacman thin sections from the north-central area were found to have larger calcite or dolomite crystals. As has been mentioned, this type of proportion does not represent the site or assemblage. They represent the variability of the samples. The different crystalline fabric classes found in Yacman are many:

1. Sparite-m-W/B-Coarse (CSm-W/B-c), containing coarse particles composed of medium-crystalline calcite
2. Calcite-Micrite-02-Coarse (CX-CM-02), particles composed of coarse single calcite crystals
3. Dolomite-Calcite-B-Coarse (DX-CX-B-c), with coarse-to-very-coarse discrete particles of dolomite
4. Dolomite-B/W-Medium (DX-B/W-m), a fabric with individual fine-to-medium-crystalline single dolomite crystals
5. Calcite-Grog-B/R-Coarse (MX-GROG-MDX-B/R-c), a fabric with coarse dark argillaceous inclusions with embedded finely crystalline calcite and, in some, grog (crushed sherds).

9.3.4 Striated, Unslipped Yacman Potters' Groups

In this section, the chemical compositions are analysed in combination with the fabric classes and contextual information to seek patterns in order to infer the number, and even the location, of the potters' groups. Table 9-12 reiterates the fabric class and site information in Figure 9-23 and, in addition, incorporates the chemical variability of the Yacman samples. Samples that could not be assigned to any of the chemical groups, or unassigned, are also included in the table.

The last column in Table 9-12 shows the minimum number of potters' groups that can be inferred from the data. For instance, Yacman samples were found to have coarse, dark micrite particles as main inclusions, or fabric DarkMicrite-Micrite-Coarse (CG-CM-W-c) at Tecoh and Tepich within Group 1 chemical composition, and at Mayapán and Tekit within chemical composition Group 2. The potters' group producing vessels with fabric DarkMicrite-Micrite-Coarse and composition within Group 2 most likely was not the same as that building vessels within Group 1. At least one potters' group is likely associated with this fabric within chemical Group 1 (PG#1). PG#1 represents the minimum number of potters' groups because other interpretations are also plausible. For instance, it is possible that several groups in a homogeneous Group 1 chemical area could have been producing vessels within CG-CM-W-c fabric. Due to the lack of other criteria, including contextual information such as the existence of different types of vessels at Tepich compared to Tecoh that help divide the data into further potters' groups, only one potters' group was assigned to the combination of fabric CG-CM-W-c and Group 1.

The number of Yacman samples in this study that are not from Mayapán is small, but it is still possible to observe from the data in Figure 9-23 and Table 9-10 that several potters' groups for Yacman jars may have existed. The production of Yacman probably occurred at many sites, given the number of fabrics and the different sites in which they were found. However, some of this variation may be due to potters at one location working with compositionally heterogeneous raw materials.

Table 9-12. Fabrics and Chemical Compositions of Striates, Unslipped Yacman Jars from the North-central Sites

Type of Fabric	**Fabric Class**	**Ch Gr (*)**	**Tepich**	**Tecoh**	**Telchaquillo**	**Mayapán**	**Tekit**	**Mama**	**Tipikal**	**Teabo**	**Minimum No. Potters' Groups (PG)**
Dark Micrite	DarkMicrite-Micrite-Coarse (CG-CM-W-c)	1	X	X							PG#1
		2				X	X				PG#2
Bioclasts	Bioclast-W-Coarse/Medium (CH-W-c)	1	X								PG#3
Dolospar	Dolospar-W-Coarse(DS-W-c), Dolospar-B-Coarse(DS-B-c)	1		X	X	X		X	X		PG#4a PG#4b
Mediun-crystalline sparite	Sparite-m-W/B-Coarse (CSm-W/B-c)	1	X		X			X			PG#5
		2								X	PG#6
Coarse single dolomite	Dolomite-Calcite-B-Coarse (DX-CX-B-c)	5					X				PG#7
Argillac./ calcite & grog	Calcite-Grog-B/R-Coarse (MX-CROG-B/R-c)	5			X				X	X	PG#8

Key : u/a = chemically unassigned; uk= unknown (it was not chemically analysed)

There are, however, petrographic and chemically homogeneous fabrics found at multiple sites: dolospar fabrics DS-W-c and DS-B-c, differing in fired and refired colour, and the argillaceous fabric Calcite-Grog-B/R-Coarse. Based on the homogeneity observed in these fabrics, they may have been produced at one location and moved to other north-central sites. An alternative explanation would be the existence of a widespread technological tradition over a chemically homogeneous region in which the raw materials are sourced at each site for the construction of the pottery clay for unslipped, striated pots.

9.3.5 Summary

The translucent rocks found in pottery samples are most likely the substance called *hi'* by local modern potters. Due to the large size of the crystals, the individual grains can be detected in hand specimen as translucent or semi-translucent, smoothed, planar or blocky, angular inclusions.

The samples show that multiple potters' groups for Yacman jars existed. The bulk of Yacman thin sections have composition within dolospar and coarse crystalline fabrics. Differing from the unslipped Navula jars, fabrics in which micrite (white in hand specimen) or finely crystalline sparite dominate were largely absent in the Yacman samples. However, Yacman was found to have coarse particles of the finely crystalline dolospar fabrics. Examination of Table 8-19 indicates that Mama and Yacman fabric classes do not contain the same type of inclusions (this is to say, they do not share the same type of fabrics), with one exception: fabrics with dolospar inclusions are common to Mama and Yacman.

The variety of crystalline fabrics in the Yacman samples from throughout the north-central sites may represent a common basic recipe based on coarse particles of translucent to semi-translucent rocks for the production of these jars. The production may have occurred at many sites because, with the exception of dolospar and the Calcite-Grog-B/R-Coarse fabric (MX-CROG-B/R-c), Yacman fabrics were found to have limited distribution at one or two sites. An alternative scenario to consider is that the different macro crystalline fabrics are the product of one or a few potters' groups, in which variations in the types of crystals represent local variations in the availability of crystalline calcite or dolomite. However, we know from ethnographic work that these crystals are not available at all north-central localities (Thompson 1958) and are imported into some pottery towns, which would add to the variety of crystalline fabrics.

9.4 Firing of Samples from North-central Sites

The aim of this section is to investigate and present the processes that successfully take raw material from the north-central area and change it into durable ceramics. To achieve this goal, in this section, the behaviour of local material when exposed to temperature is investigated through experimentation from two points of view: one, effects of temperature on carbonate inclusions; and second, effects on the plastic or clay portion.

To start this section, background information on the effects of temperature on calcite, dolomite, and clays is presented. Next, experiments on the effects of temperature on briquettes containing various grain sizes of carbonate inclusion are described and the results presented. This is followed by a description of experiments and the results of a firing schedule aimed at determining whether samples of marl sascab, red soils, and grey clay might become ceramic, the characteristics that they might take, and the conditions under which these may occur.

The Maya potters who manufactured the ceramics analysed in this research were dealing with carbonate rocks. Potters need to raise the temperature during firing to a point at which pots lose their plasticity and become ceramic. Up to this state of research (section 3.1), Maya pots appear to have been fired in open fires. No direct evidence for the production of utilitarian ceramics has been found in the study region.

In his ethnographic work on pottery towns of Yucatán, Thompson (1958) describes in most towns the use of a traditional kiln that is built as a half dome (Figure 9-25). Given that his potter informants do not have a Maya word for it, calling it *horno,* Thompson believes that this type of dome is not pre-Hispanic. In addition, in several instances, potter informants remembered the time when pottery was open fired, usually in a square against an *albarrada*, a dry-laid stone or rubble wall (Thompson 1958, p.97). Thompson (1958, p.97) remarks that (but no reference is given) Eduardo Toro Quiñones found one of those open kilns at Dzununcán in which two more walls were added to the *albarrada* to build an 'unroofed and open-ended square enclosure. The vessels are stacked inside and covered with firewood'.

Figure 9-25. Ethnographic half dome kiln, Yucatán (Thompson 1958, Fig 26g)

9.4.1 Effects of Firing on Calcite Inclusions

Calcite and high temperatures are a concern to potters requiring a careful control of the firing process because this mineral, through a reaction known as calcination, changes into lime and water. Lime absorbs atmospheric humidity (water vapour), rehydrating the lime into the higher volume calcium hydroxide $Ca(OH)2$ (Rye 1981, p.33), which expands and potentially, depending on the amount of lime and the degree of calcination, destroys the pot and crumbles it.

Various temperatures have been proposed for the start of calcite calcination, with some considering that calcite starts to decompose at 750°C (Rye 1981; Bronitsky and Hamer 1986; Feathers 1989), while some have suggested that the decomposition starts at lower temperatures (Rice 1987a, p.103–105)

Table 9-13. Different States of Conversion to Lime when Different Calcite Varieties are Fired for Six Hours (data from Shoval and Beck 2005)

Calcite Crystal-size Varieties	**600°C**	**650°C**	**700°C**	**750°C**
Micrite (chalk)	Almost complete	Totally complete		
Sparry limestone	Starts	Almost complete	Totally complete	
Coarse Single calcite Crystal	Starts			Almost complete, crystal shape remains

Differences in the calcination starting point suggested by different researchers may be explained by the experiments on firing conducted by Shoval and Beck (2005). They showed that conversion to lime is greatly affected by the crystal size of the calcite. More specifically, their experiments demonstrated an inverse relationship between the exposed crystal surface areas of individual calcite grains and their temperature required for conversion to lime. The calcite particles with the least exposed surface area are coarse single crystals (also called mono-crystalline), followed by the multi-crystalline sparry variety, and finally the microcrystalline (micrite) aggregates of calcite. Shoval and Beck (2005) found that, after six hours of heating, these three calcite varieties start to de-carbonate at 600°C. Micrite totally decomposed at 600ºC, sparite at 650ºC, and mono-crystalline calcite at 750ºC (Table 9-13), showing that calcite in the form of individual single crystals is more stable at higher temperatures.

Six hours of heating is longer than the duration of most open fires, acknowledging that the actual firing conditions of ancient Maya pottery are still unknown. Reina and Hill (1978, p.114) report a firing time of 20 minutes for one *tinaja* (water vessel) at San Miguel Acatán, while Thompson (1958, p.97) reports firing times of 2 to 3 hours in a half dome kiln in Yucatán. Differences in firing times are the result of potters' working knowledge of what works and what does not work with the materials used. A good portion of the samples in this study contain dolomite in addition to calcite, and as is discussed later in this section, dolomite also decomposes following a different reaction than calcite and at different temperatures.

9.4.2 Effects of Firing on Dolomite Inclusions

The changes that calcite experiences with temperature have received considerable attention in archaeology. Less information is found in the literature about the effects of rising temperatures on dolomite in relation to ancient pottery. Dolomite (Ca (Mg, Fe) (CO3)2) in comparison to calcite (CaCO3), is slightly harder (Gautam 2001, p.153), 3.5 to 4 in Mohs scale. Dolomite and calcite are similar in that they both decompose with enough temperature, with dolomite dissociating into calcite and magnesium oxide with temperatures above 650°C (Britton et al. 1952a, 1952b).

There is no agreement in regard to the details of the mechanisms for the dissociation reaction of dolomite (McIntosh et al. 1990). The mechanisms are complex with many factors affecting the reaction including the atmosphere, temperature, the rate of heating, and the particle size (McIntosh et al. 1990). In general, McIntosh et al. observe that dolomite calcination proceeds in two stages. In the first stage, dolomite dissociates into calcite and magnesium oxide and in the second, if the temperature is brought up to a point at which carbon dioxide is driven off, the calcite will calcinate, producing calcium oxide with a considerable reduction in volume due to the loss of carbon dioxide.

CaMg(CO,), -> CaCO, + MgO + CO (Stage 1)
CaCO, -> CaO + CO (Stage 2)

McIntosh et al. (1990) showed that the temperature at which dolomite decomposes could vary widely depending on the atmosphere surrounding the particles: if the carbon dioxide produced cannot escape or the atmosphere is rich in carbon dioxide, the temperature of calcination (second inflection) will be raised. In turn, the atmosphere is affected by the heating rate and the mass of the sample. In an example illustrated by McIntosh et al. (1990, Figure 1) for the decomposition of dolomite into calcite under different atmospheres, decomposition starts at about 600°C. From 600°C up to 750°C, for all atmospheres, 40-50 per cent of dolomite decomposes rather quickly. Beyond around 750°C and under oxidation, decomposition speeds up, and it is complete at 800°C. On the other hand, at 750°C and with a carbon-rich atmosphere, decomposition reaches a plateau of 40-50 per cent until the temperature is raised over 900°C. The temperature of complete decomposition of dolomite into calcite varies with the amount of CO and reaches as high as 900°C. According to Xie et al. (2016, p.45) the complete decomposition of dolomite generally occurs at temperatures higher than 800°C. After 800°C, the remaining calcite turns into lime (Trindade 2009, p.349) and dolomite and calcite disappear.

9.4.3 A Firing Experiment with Raw Materials from the North-central Area

A limited testing was performed on raw materials collected from the north-central area with the objective of investigating technological processes that may have been used to produce a successful firing given the materials on hand. Two of the main divisions observed in the fabrics of pottery samples from the north-central area (discussed in this and in previous chapters) are between the types of carbonate rock varieties found in the pottery (micrite and sparry in red-slipped vessels and single crystals in a good part of the unslipped, in particular Yacman) and between medium and coarse fabrics.

The results of Shoval and Beck (2005) discussed above showed an inverse relationship between the crystal size of the carbonate crystals (micrite, sparite, coarse single crystals) making up a particle and the efficiency of calcination. Maya potters may, therefore, have been aware of differences in the firing between sascab (micrite, spar) tempered vessels and coarse single-crystal tempered vessels.

While experimenting with the raw materials to fire a successful briquette, it became apparent that the particle size has an important role in the efficiency of calcination and on the subsequent destruction of the briquette. To investigate the grain sizes and temperatures that produce a successful outcome, briquettes were prepared with the plastic raw materials collected in the field (Table 5-4). Three types of briquettes were prepared with samples of marl sascab and subjected to a firing schedule. The firing conditions are summarized in Table 9-14.

Case A. One briquette was prepared with sample #11 (Table 5-4), which is the plastic sascab found close to Mayapán. Petrographic analysis (Table 8-32) showed that the original raw material had hardly any silt, fine, or medium particles. The inclusions were composed of coarse-to-very-coarse dolospar and some sparite.

For the construction of the briquette for case A, the raw material was lightly ground and sifted through a 0.5 mm mesh. The resulting briquette had a fine-to-medium texture (Case A, Table 9-14).

Case B. A second group of briquettes was made of the marl sascab raw material samples (#1, #2, #4, #11, #16, #18) including sample #11. Petrographic analysis (Table 8-32) had shown that plastic sascab samples contain very coarse clusters of very-finely crystalline sparry calcite, except clay #11, which contains coarse to very coarse particles of dolospar. For the briquette construction, pebbles and other extremely coarse particles were removed manually, just enough to permit building the briquettes.

Case C. One last group of briquettes was made with red clayey soils (Table 8-32), also containing coarse-to-very-coarse carbonate particles.

The samples were fired at 650°C, 700°C and 750°C in an oxidising atmosphere for 45 min (Paragon Quickfire 6 test kiln). A starting target temperature of 650 °C and a time of 45 minutes were selected based on Table 9-15 containing Rye's (1980) decomposition temperatures and times for different types of clays. The results of the experiments are outlined in Table 9-14 and summarized next by firing temperature.

Temperature of 650°C. The marl sascab briquettes (Case A and B) calcined to an extent at 650°C; the degree of which could not be measured, but calcination may have started from a lower temperature. This calcination did not change the shape of the briquettes. That calcination occurred was demonstrated by submerging a piece of the briquette in water: bubbles are released from the reaction of the water with the quicklime formed during calcination. The test for excessive and destructive hydration of quicklime was done by letting the briquettes stay in the open air for a few days. All the briquettes maintained their form.

Temperature of 700°C. All of the briquettes kept their form when exposed to the air.

Temperature of 750°C. Upon firing, two different outcomes were witnessed: the fine to medium-grained briquette kept its form, while the coarse-to-very-coarse briquettes, or Case B and C, disintegrated within two days.

Based on this experiment, the damage that calcination can inflict on pottery is dependent on grain size, among other factors such as crystal area. Fine-to-medium-grained briquettes did not crumble at 750°C (Figure 9-26a) while very-coarse briquettes were rapidly reduced to a pile of fragments (Figure 9-26b).

Table 9-14. Changes to Plastic-sascab Briquettes depending on Temperature and Size of Sascab Inclusions

Particle Size of Sascab Inclusions	**Type of Sample**	**650°C.**	**700°C.**	**750°C.**
Case A: Particle size = fine-to-medium (sascab #11)	(a) Plastic portion = marl sascab (b) Coarse portion = fine-to-medium sascab inclusions	- calcination of sascab's carbonate mud portion - briquette keeps form	- briquette keeps form	- briquette keeps form
Case B: Particle size = coarse to very coarse (sascab #1, #2, #4, #11, #16,#18)	(a) Plastic portion = marl sascab (b) Coarse portion = coarse-to-very-coarse sascab inclusions	- calcination of sascab's carbonate mud portion - briquette keeps form	- briquette keeps form	- calcination of coarse and very coarse inclusions - briquette crumbles
Case C: Particle size = coarse to very coarse (red clay #6, #8)	(a) Plastic portion = plastic red soil (b) Coarse portion = coarse-to-very-coarse carbonate inclusions	- briquette keeps form	- briquette keeps form	- calcination of coarse and very coarse inclusions - briquette crumbles

Table 9-15. Clay Decomposition Temperature (data from Rye 1981, p.11; U.S.G.S.)

Clay	**Decomposition Temp (C)**	**Firing time required**
Kaolinite	415	weeks
Kaolinite	550	30 mins
Kaolinite	585	very rapid
Halloysite	558	very rapid
Montmorillonite	678	very rapid
Palygorskite and sepiolite	400	30 min

Figure 9-26. (a: left): briquettes keeping their form at 700°C; (b: right): obliterated coarse-to-very-coarse briquette at 750°C (#2)

This experiment illustrates the major difficulties and challenges that potters encountered when carbonate inclusions were part of the firing equation. Shoval and Beck (2005) showed that coarse mono-crystals calcinate at 750°C, but also, that they will keep their form. This experiment showed that the calcination does not necessarily cause the destruction of the briquette, rather, when fired at a given temperature, destruction is related to grain size.

The divisions in grain size and the truncation of most grain sizes larger than 1mm indicate that the coarser inclusions were most probably sifted (section 9-1) and potters controlled the size of the particles. The grain size (mode) of Yacman cooking pots, coarse, differs from red-slipped Mama, or medium. This control may have been related to the potters' working knowledge of the firing conditions. This implies that Yacman firing conditions may have been different from red-slipped Mama conditions.

But grain size is only one of several factors in the samples that appear related to firing performance, such as the presence of macro crystals in some of the Yacman pots. As seen in section 9.4.1, larger single crystals are more stable, and calcination starts at a higher temperature. Even after calcination, they appear to expand less than smaller crystals. On the other hand, as shown in the experiment in this section, a smaller grain size protects the vessels from destruction when the calcination temperature is reached because the temperature can be increased to accelerate the decomposition of the clay and reduce the firing time.

9.4.4 *Effects of Firing on Clays*

Pottery making requires that pots be fired at a temperature at which the clay decomposes and becomes ceramic, losing its plasticity. During firing, a clay undergoes a series of modifications, including dehydration, dihydroxylation, decomposition and new phase formation (Trindate et al. 2009, p.345), that change its organisation and eventually its chemical composition. Unless the clay is decomposed completely, the clay may become plastic when exposed to water, or rehydroxylation.

Potters need to raise the temperature during firing to reach the point at which the clay starts to sinter, and either maintain the temperature for a relatively long time until the clay has transformed or, if possible, raise the temperature quickly for a more rapid transformation. The presence of calcite and dolomite fragments in the pottery clay puts limitations on the temperatures that can be reached without damaging the pottery.

There are three main variables that are crucial in firing: maximum temperature, atmosphere, and the rate at which the temperature rises (Rye 1981, p.110). The temperatures at which fired clay starts to decompose and the time required for the decomposition of some of the different types of clays are listed in Table 9-15.

XRD of a marl sascab near Mayapán (Shannon 2017, included in Appendix A for easy reference), #11, found that the clay portion is mostly composed of montmorillonite, and kaolinite to a lesser degree. The temperature of decomposition of this clay is rather high when compared to other clays that have been documented in the area (section 4.3). However, this marl presents a mixture of montmorillonite and kaolinite, and the properties of this mixture likely are not the same as the properties of the individual clays.

Table 9-16. Changes to Palygorskite and Kaolinite with Temperature (compiled from Xie et al. 2016; Trindade et al. 2009)

	350 °C		**500 °C**	**550 °C**	**800 °C**	**900 °C**
Palygorskite (Xie et al. 2016)	- Paly partial dehydration	- paly anhydride appears		- paly & - paly anhydride disappear, - keeps form	paly morphological characteristics disappear	
Kaolinite (Trindade et al. 2009)	Natural kaolinite		- Kaolinite disappears - amorphous metakaolinite apears			
			Mica-like (illite/muscovite) phase appears		Mica almost disappears	Mica disappears

Table 9-17. Transformation of Sample Raw Materials into Ceramics

Test Case (#sample)	**Type of Raw Materials**	**Changed to Ceramics at 700 °C, for 45 min?**	**Changed to Ceramics at 600 650°C, for five hrs?**
Case A (#11) marl sascab, fine-to-medium texture	(a)Plastic portion = clayey carbonate mud, or marl sascab (b) Inclusions • Particle size = fine-to-medium; • Crystal size = finely-crystalline sparry	No	Yes
Case B (#1, #2, #4, #11, #16,#18) marl sascab, very coarse rock fragments	(a)Plastic portion = clayey carbonate mud, or marl sascab (b) Inclusions • Particle size = coarse-to very coarse; • Crystal size = finely-crystalline sparry	No	**#1 = Yes** #2 = No #4 = No **#11 = Yes** #16 = No #18 = No
Case C (#6, #8) red clayey soil, coarsely grained	(a)Plastic portion = clayey red soil (b) Inclusions • Particle size = coarse-to very coarse; • Crystal size = finely-crystalline sparry	No	Not tested
Case D (#3, #7) red clayey soil, fine grained	red clayey soil, no coarse inclusion, likely no carbonates	**Yes**	
Case E (#19a) Light grey clay (Chapab)	Fine clay probably with carbonate mud (some change into lime ocurred)	No	low resistance to break

As can be recalled, two of the common clays in Yucatán are kaolinite and palygorskite (section 4.3, section 9.1.1). At 500-550°C for 30 min, kaolinite largely disappears as a crystalline phase (but remnants may be present depending on the atmosphere) and an amorphous meta-kaolinite and a mica phase appear (Trindade et al. 2009). This phase is stable up to approximately 800°C. A minor mica phase was detected in a Mayapán pottery sample (Shannon 2016).

Palygorskite dehydration stars at the very low temperature of 350°C (Frini-Srasra and Srasra 2008), and palygorskite anhydrate appears. The remaining palygorskite and this new phase transform into an amorphous substance at around 550-600°C (Xie et al. 2016, p.44), but retain the fibrous morphological characteristics of palygorskite until about 800°C. These changes are shown in Table 9-16.

An experiment on firing was conducted to determine whether the raw materials from the area (marl sascab samples, red-soils, and one grey clay) might become ceramic, the characteristics that they might take, and the conditions under which this occurs. Briquettes, with characteristics summarised in Table 9-17, were fired at 700°C for 45 minutes and at 600-650°C for five hours.

Based on the previous discussion on the effects of temperature on micritic and sparry calcite and dolomite, as summarised in Table 9-14, the temperature reached by the pots should be equal to or higher than the temperature needed to decompose the clay, but less than the temperature indicated in Table 9-14 for a given particle size. For instance, from Table 9-14, a vessel built with medium-grained sascab inclusions can be fired to 750°C, whereas a vessel containing coarse sascab can be fired only up to 700°C. It is important to notice that the clay matrix is a marl and that, even if coarse particles are removed, a carbonate mud is present. Experiments on the effects of an oxidising versus a reducing atmosphere were not performed.

9.4.4.1 Firing at 700°C for 45 Minutes

At 700°C for 45 minutes, none of the marl sascab briquettes lost their plasticity, even though they were firm to the touch. When they were submerged in water for one day to test whether they had become ceramic, the briquettes recovered their plasticity, as demonstrated by their clay portion dissolving into a very fine mud. The reasons for this could be many. One likely reason is that, given that they are marls, the amount of clay is not sufficient to bond the material when sintered, preventing it from becoming ceramic. Another reason is that the firing time has to be longer than indicated in Table 9-15 because they are marls and not pure clay.

On the other hand, when fired at 700°C for 45 min, two red clays (Case D: #3, #7) without coarse carbonate inclusions became ceramic. Carbon was deposited on these red clayey briquettes due to the amount of organic matter in the red soils from the area (Shang and Tiessen 2003). These two briquettes were re-fired at 750°C to examine their appearance after burning off the organic material. Clayey red soils are common in the north-central area and represent an option available to potters.

Ancient potters were aware that red soils could be used for pottery. An odd-looking tile-like device thought to be a candle shield has been found at Mayapán/Telchaquillo. Potters, however, did not select these clays for the construction of their vessels despite being fast-firing and harder than typical Mayapán ceramics. Additional experimental testing of these clays is needed to show whether large vessels can be successfully built with them.

9.4.4.2 Firing at 600-650°C for Five Hours

To test whether allowing for more firing time resulted in a transformation of the raw materials into ceramics, the marl sascab briquettes (including sample #19a, Table 5-4) were fired at 600-650°C for five hrs. Table 9-17 summarises the results of this experiment.

Sample #1 (marl from Mama–Chumayel road, Table 5-4) and sample #11 (marl from a Mayapán sascabera matching the chemical compositions of Mayapán pottery) became ceramic with good resistance to breaking. The raw materials in marl samples #2, #4, #16, #18 probably do not have enough clay to bond the material because these did not change into ceramics.

The briquette made with sample #19a (Table 5-4) from a mine near Chapab (Figure 5-4) also became ceramic. This mine supplies Ticul potters with sascab for temper (Arnold 2005a) and contains a critical ingredient for their pottery clay, *sacalum* (Arnold 1971). This ingredient was shown to be palygorskite (Arnold 2005a). Although Sánchez del Río et al. (2009) took a sample from this mine that was 100 per cent palygorskite, the sample taken for this research, although very fine, is palygorskite with a carbonate mud component demonstrated by the slight fizzling when submerged in water after firing.

Sample #19a kept its form throughout the tests mentioned earlier. However, when the briquette was bent with the hands, its resistance to breakage was low, and it broke easily. This may be related to the briquette becoming very light in weight after firing, most likely due to the loss of water. The loss of weight was not measured, but a comparison between sample #19a with #11 after firing showed that #19a had close to half the weight of #11.

9.4.5 Discussion

The research presented in this section showed a varied temperature and rate of transformation of carbonate rocks into lime depending on several factors that required the attention from part of the potters. Lime hydrates to calcium hydroxide, taking up more volume than calcite. The expansion can disintegrate or break a vessel. Dolomite decomposes first into calcite, which, with further heating, follows the calcite decomposition path. At a given moment, dolomite, calcite, and lime may all coexist.

Current research also shows that the efficiency of decomposition is inverse to the area exposed by the crystals (Shoval and Beck 2005). Coarse, single crystals have lower surface area and begin to decompose at higher temperatures (lower decomposition rate), while micrite or carbonate mud have higher surface area and start to decompose at lower temperatures.

Experiments performed in this study showed that particle size is relevant to the expansion caused by the crystallisation of quicklime that may damage or even break a pot: the larger the particle, the greater the damage, with the frequency of large particles most likely playing a role in this process. This study also showed that the temperature reached by the pots should be equal to or higher than the temperature needed to decompose the clay, but not higher than 700°C, unless the fabric is finely grained or the particles present are coarse discrete crystals, such as *hi'*.

Table 9-18. Fabrics found at Cobá, Chac Mool, and Culubá within Ceramic Varieties

Fabric Class	Chemical Group	Tepich	Tecoh	Telchaquillo	Mayapán	Tekit	Mama	Tipikal	T eabo	Cobá (*)	Chac Mool	Culubá (**)
CALCITE S-XTALS												
CX-QZ-B/R-f	4, u/a				**P P P P**						**P P P P P**	
CX-QZ-B-c											**N**	
QZ-R-m											**N**	
CH-QZ-B-f											**P**	
CH-R-f	3											**P**
DX-CX-m-h	3											***M M***
CX-CM-01-c	1,2,3,5, u/a				**X**			**N**		***M M M*** **N**	**XXXX**	***M M M M M M M***
CS-W/B-cm	1,2,5, u/a	**X**	***M*** **N**	***M*** **N** **X**	***M M***	**N**		***M M*** **N**	***MM M*** **N**	***M M M*** **N**		
CX-CM-02-c	3,5										**N** **Y Y Y**	***M*** **Y Y** **V**
CSm-W/B-c	1,2,5, u/a	**Y Y**	**N N N**	**Y**			**Y**	**N**	**Y Y**	**N**		

KEY: *M* = Mama and Cancun variety ((*)=Cancun; (**)=Mama and Cancun), N = Navula (plain), **P** = Payil, **Y** = Yacman (striated), X = Xcanchakan, ; u/a = unassigned

Experiments with local sascab showed that some of the marl sascab samples collected from the area are suitable for the production of ceramics. These marls are of light colour, from off-white to light yellow brown. A short firing of 45 minutes transformed the red soils into ceramics, but this was not so for marl sascab and the light grey clay sample from Chapab. However, a slow firing at 600-650 °C for five hours transformed two of the sascab samples into ceramics: sample #1 (Table 5-4) from the road Mama-Chumayel and #11 from a sascabera by Mayapán.

Even when calcite was not present, as was the case in most Late Classic ash-tempered pottery, and to maintain the glossiness of burnished slips, the pots might have been fired at low temperatures (Rice 2009b), probably not exceeding 700°C (Taylor 1982 cited in Rice 2009b). This suggests that for slipped vessels, the limiting temperature factor was probably maintaining the lustre of the slip (Rice 2009b), while the limiting factor for unslipped vessels was the disintegration of the pots due to calcination. It is very clear from the results of the experiments performed that the raw materials from the north-central area of Yucatán presented great challenges to potters. Nevertheless, that they were successful in controlling temperature and preserving the vessels' lustre can be appreciated in the result of their work.

9.5 Fabric Classes of the Eastern Sites in this Study

A reconstruction of ceramic techniques used in samples from the eastern sites (Coba, Chac Mool, and Culubá) was not possible because no raw materials were collected from this area and the number of samples is very small. Instead, this section focuses on examining the fabric classes found at the eastern sites, seeking technological patterns, and on the comparison of these fabrics with fabrics of samples from the north-central sites. Table 9-18 illustrates the fabric classes and ceramic varieties found in samples from the eastern sites of Cobá, Chac Mool, and Culubá, as well as the fabric classes shared between eastern and north-central sites.

Based on this limited number of eastern Cobá, Chac Mool, and Culubá samples, the following general patterns in the data can be inferred from Table 9-18:

1. At these sites, all samples classified as Mama or Cancun (a Tases variety used for pink, harder Mama-like samples), Navula (plain, unslipped), Yacman (striated, unslipped), or Xcanchakan (cream-slipped that started during the Terminal Classic, included for comparison) are characterised by their coarse texture.
2. The division by grain size of red-slipped versus striated, unslipped samples observed at north-

central sites was not observed in the samples from Chac Mool and Culubá (no Yacman samples from Cobá).

3. The most common inclusions found in samples classified as Mama (or Cancun), Yacman and Navula, and Xcanchakan from Cobá, Culubá, and Chac Mool are coarse calcite single crystals. In contrast, in the north-central sites, macro crystalline calcite crystals were found in unslipped samples.
4. Payil samples from Chac Mool contain fine-grained calcite and quartz. One from Culubá does not contain quartz (no samples from Cobá).

The samples from Mama and Cancún ceramic varieties from Cobá and Culubá, as well as the cream-slipped Xcanchakan from Chac Mool shared a basic recipe. This recipe consists primarily of coarse discrete calcite crystals and micrite lumps, or fabric Calcite-Micrite-01-Coarse (CX-CM-01-c).

Fabric Calcite-Micrite-01-Coarse has a varied chemical composition, pointing to a common technology with multiple sources, as shown in Table 9-19 by the chemical groups of this fabric at the different sites. This fabric represents a common manufacturing technique or basic recipe for constructing pottery clay for some of the coarse slipped samples found at eastern sites including red-slipped Mama (or Cancún) and the cream-slipped Xcanchakan.

A variant of the Calcite-Micrite-01-Coarse fabric is also coarse. It differs from this fabric in that it is very porous. This fabric has been named Calcite-Micrite-02-Coarse (or CX-CM-02-c). Eight out of ten samples in this fabric are unslipped plain (Navula) or striated (Yacman) samples. One more sample in this fabric is the only Vista Alegre sample (# 181, Culubá) included in this research for comparison. Vista Alegre is an unslipped ware that started prior to the Postclassic and that is particular to the eastern sites.

Table 9-19. Chemical Variability of Fabric Calcite-Micrite-01-Coarse (CX-CM-01-c)

Fabric Class	Site Name	Chemical Group	Variety	Sample Code
CX-CM-01-c	Tipikal	2	Navula	281
CX-CM-01-c	Mayapan	unassigned	Xcanchakan	328
CX-CM-01-c	Culuba		Cancun	215
CX-CM-01-c	Culuba	2	Mama	209
CX-CM-01-c	Culuba	2	Mama	208
CX-CM-01-c	Culuba	2	Mama	204
CX-CM-01-c	Culuba	2	Mama	203
CX-CM-01-c	Culuba	3	Mama	210
CX-CM-01-c	Culuba	unassigned	Mama	207
CX-CM-01-c	Coba	1	Navula	365
CX-CM-01-c	Coba	1	Cancun	385
CX-CM-01-c	Coba	2	Cancun	354
CX-CM-01-c	Coba	5	Cancun	388
CX-CM-01-c	Chac Mool		Xcanchakan	373
CX-CM-01-c	Chac Mool		Xcanchakan	372
CX-CM-01-c	Chac Mool	5	Xcanchakan	371
CX-CM-01-c	Chac Mool	5	Xcanchakan	370
CX-CM-01-f	Mayapan	unassigned	Payil	135
CX-CM-02-c	Chac Mool		Navula	378
CX-CM-02-c	Chac Mool		Yacman	387
CX-CM-02-c	Chac Mool	5	Yacman	366
CX-CM-02-c	Chac Mool	unassigned	Yacman	367
CX-CM-02-c	Culuba		Mama	206
CX-CM-02-c	Culuba	3	Yacman	191
CX-CM-02-c	Culuba	3	Yacman	200
CX-CM-02-c	Culuba	unassigned	Vista Alegre	181

Regarding Payil, with the exception of two unassigned samples, Mayapán and Chac Mool Payil samples share the same distinctive petrographic characteristics and chemical composition (Group 4). These characteristics are found only in Payil and Palmul (incised Payil) sherds. As discussed in previous chapters, these samples did not originate at these sites. They may have originated in northern Belize around Laguna de On.

However, Payil samples in this study may have been produced by multiple groups of potters. Two Payil fine grained samples, one from Culubá (#349) with chemical composition within Group 3 and another from Mayapán (# 135), have chemical and petrographic characteristics that differ from the rest.

Xcanchakan is one of the ceramic varieties of Peto Cream Ware, a buff-slipped type of ceramics that started during the Terminal Classic (Smith 1971a) and continued into the Late Postclassic. Smith (1971) considered that the coarse carbonate rock temper in Peto presaged the composition to come during the Late Postclassic. Eight samples were selected for petrographic analysis, and six of them for chemical analysis.

The results of the analysis of this small sample indicate that there is no one Xcanchakan fabric but many. Xcanchakan seems to reflect local traditions and the availability of raw materials. Xcanchakan sherds found at north-central sites have fabrics indistinguishable from other north-central slipped samples, such as Dolospar-W-medium (DS-W-m) within Group 1 and Sparite-W/B-Coarse/Medium (CS-W/B-cm) within Group 2. Similarly, Xcanchakan sherds found at Chac Mool (no Xcanchakan samples from Cobá and Culubá were analysed) have a fabric that is indistinguishable from other slipped Mama or Cancún samples found at eastern sites, containing single calcite crystals as main inclusions, or Calcite-Micrite-01-Coarse (CX-CM-01-c). Xcanchakan found at Chac Mool may represent a common technology for the construction of the buff-slipped Xcanchakan and the red-slipped Mama (and Cancún) pottery at eastern sites. Its chemical variety points to several sources and producers using a common technology.

In sum, the findings of this section indicate that the fabric classes of the samples from utilitarian varieties (Mama, Yacman, Navula, and Xcanchakan), when found at the north-central sites, are largely different from the fabric classes of varieties found at Coba, Culubá, and Chac Mool. At the eastern sites, the samples analysed are coarsely grained, and the division by grain size among utilitarian red-slipped and striated Yacman jars present at north-central sites was not observed in the samples from the three eastern sites. A division between coarsely grained utilitarian wares (that include Mama, Yacman, Navula, and Xcanchakan varieties) and a fine ware (Payil) is observed. In regard to Xcanchakan (cream-slipped), the results of the analysis of this small sample indicate that there is no one Xcanchakan fabric but many. Xcanchakan sherds share fabric classes with other utilitarian ceramic types from the sites in which they were found.

9.6 Summary

In this chapter, it was shown that the main three utilitarian types of vessels in this study, Mama, Yacman, and Navulá, were produced at many centres by many potters' groups. The fabrics of these vessels greatly differ when compared between the north-central and the three eastern sites. At the north-central sites, there is a correlation between these types of vessels and the types of tempers found in the samples. This correlation may be related to the function of the vessels. At Chac Mool and Culubá, the same types of materials were used for all utilitarian vessels sampled, whereas at Cobá, fabrics were typical of the east and north-central region, apparently sharing traits with both major regions.

Concerning the manufacture of red-slipped Mama vessels, most contain medium-grained particles composed of micrite, finely crystalline sparry calcite (sparite), or sparry dolomite (dolospar). Differences in divisions of grain size and frequency of inclusions between ceramic varieties indicate that these particles were added by potters. The source of all these types of particles is sascab.

A general recipe was used for constructing the clay body of Mama vessels. It consists of grinding (or whacking with a club) the carbonate rock (sascab, most likely) to a medium grain size and sifting the results. The pottery was hand built. The surface analysis of one large Mama jar found at a small site four km from Mayapán shows that it was made in sections, apparently using the open base technique that has also been described ethnographically for large jars (Thompson 1958). Moulds could have been used to build the rounded bottom of bowls and cajetes. The examined large jar was built up to 7-10 cm in height and left to dry before continuing with the upper section of the vessel. The long neck was added to the body.

Regarding Navula, when Navula is made with sascab (micrite, sparry calcite, dolospar), it displays fabrics similar to most red-slipped vessels. However, a good portion of the Navula samples comprises a variety of fabrics not found in the Mama samples, including fabrics in which bioclasts, calcite or dolomite single crystals, or dark micrite dominate. Navula fabrics display a considerable variability in the petrographic and chemical characteristics, which points to many producers probably in as many locations. Furthermore, while a relative uniformity in grain size was observed for Mama (grain size has mode of medium) and Yacman (grain size has mode of coarse), no patterns were observed for Navula.

The fabrics of Mama, Navula, and Yacman ceramic varieties intercept in the dolospar fabrics. The raw materials and the chemical Group 1 are the same for the dolospar fabrics within these three types of samples. They differ in the technology applied. Mama samples with dolospar have a fabric that is medium-grained (mode), many with organic material, whereas the Yacman samples are coarsely grained and no organic material is observed. There is another distinction. Mama and Navula with dolospar fabrics refired to a white colour (with a few exceptions), but about half of the Yacman samples containing dolospar fabrics refired to a light reddish brown. Differences in refired colour can be related to differences in the clay sources.

Concerning Yacman vessels, in contrast to Mama and Navula samples, micrite and finely crystalline sparite fabrics were not found in the Yacman samples (with two exceptions). Instead, fabrics with large medium to very coarse single crystals of calcite or dolomite make a significant contribution to the total variety of fabrics found in the striated, unslipped Yacman samples. Due to the larger size of the crystals, many times the individual grains can be detected in a hand specimen as translucent or semi-translucent particles. The translucent rocks found in pottery samples are most likely the substance called *hi'* by local modern potters, which is collected in nodules and requires grinding. Yacman was also found to have a grey micrite fabric. In contrast to the dolospar fabrics, the crystalline and grey micrite fabrics have varied composition indicating several potters' groups.

Some Mama and Yacman vessels were made with similar dolospar fabrics, which makes it possible to compare aspects of the firing technology of these two types of vessels. In some of the Mama samples, the matrix shows impregnation and concentrations of brown material near to only one of the sherd's walls, while close to the other wall, the colours are clear, indicating that the brown material is most likely organic. The colours are clear in Yacman samples. It is then likely that Mama was fired at a lower temperature or experienced a shorter firing than Yacman. Most likely, the shorter firing of the Mama vessels was done to protect the lustre of the slip.

That the ancient potters who produced the vessels in this research were successful in controlling temperature and preserving the vessels' lustre can be appreciated in the result of their work. Slipped samples maintain their lustre even today and show little firing defects, such as spalling or popping of carbonate particles. Most are soft and friable, though, and this may be the result of using marls.

It is clear that the raw materials from the area present significant challenges to potters. In this section, through experimentation, a firing schedule that works for the marls from Mayapán (#11) and Mama-Chumayel road (#1) was found. An XRD analysis of sample #11 found montmorillonite and kaolinite in this marl. It is also clear that the potters had developed an empirical understanding of what worked. Given the material constraints of working with carbonate materials and that clays, marls, or sascab found at different pottery towns varied, most likely what worked in one location may not have worked in another.

In the next chapter, the techniques inferred in this chapter are further examined in combination with other contextual information, such as sites, to advance new hypotheses about the interrelations and organisation in which the potters were embedded, including the existence of technological traditions, the organisation of production, and ceramic exchange.

Chapter 10

PATTERNS OF LATE POSTCLASSIC CERAMIC PRODUCTION

The previous chapter focused on seeking patterns indicative of ceramic production techniques applied in the construction of the vessels sampled in this research. A reconstruction of the techniques during ceramic manufacture was proposed for the jars and cajetes of the three main varieties in this study: the red-slipped Mama and the unslipped Navula and Yacman. However, it is not known whether the findings of this study apply to all the ceramic varieties in the Mama or Navula group. The number of potters' groups that might have been involved was also examined. This chapter examines the techniques applied in the context of the distribution or geographical location of the pottery, the types of vessels (e.g., slipped versus unslipped) covered by the different variety names, and the assumed function (cooking versus non-cooking) of the vessels, seeking patterns that reflect the existence and scope of technological traditions, the organisation of production, and the movement of pots. Having gathered and analysed the data of this research, the chapter ends with the evaluation of the hypotheses that guided this study.

10.1 Technological Traditions

In this section, archaeological and ethnographic data are compared to show that Late Postclassic technological traditions can be identified, and that they extend into the present. It will be argued that ceramic production, and the choices made in regard to the various aspects of production, were guided by these traditions. These traditions had a sub-regional scope that does not match the spatial area of the ceramic sphere to which they are related, i.e., traditions at the north-central sites differ from those at Cobá and Culubá, which are also in the Tases ceramic sphere.

Based on the results, within the north-central sites, at least two technological traditions can be proposed. One was related to the construction of striated open-mouth jars and the second to the construction of red-slipped jars and cajetes (the forms included in this study). Table 8-19 combines the tables presented in chapter eight for most of the fabrics. It shows the fabric and site associations, as well as their associations with ceramic technology traditions. This table is used in the following discussion on the existence of these two traditions. In Table 8-19, enclosed in red are fabrics comprising the red-slipped (Mama) tradition, while in green are the fabrics involved with the striated, open-mouth (Yacman) jar tradition.

10.1.1 Sascab Red-slipped Tradition

One characteristic of Mama samples from the north-central sites is the homogeneity in their external characteristics of composition including the general appearance of the inclusions, grain size, and even colour of the paste. In a hand specimen, most red-slipped jars and cajetes are characterised by frequent to common white and chalky or finely crystalline carbonate inclusions, with a medium (mode) grain size, and light fired colours that are usually light reddish brown, light red, grey, or pinkish grey.

Researchers have observed that, during the northern Late Postclassic period, the form and attributes of decorative styles were very similar across sites. This similarity extends over vast areas of the Yucatán Peninsula into current-day Belize, and the states of Yucatán, and Quintana Roo (Pendergast 1985, p.240; West 2002, p.178; Rathje 1975; Sabloff and Freidel 1975). These similarities have been present for a long period that extends into present times. This is illustrated by comparing modern ethnographic water-carrying and storage vessels to Late Postclassic ones. For instance, modern vessels from (a) San José Petén (Reyna and Hills 1978, Figures 41f, 41d); (b) Belize (Thompson 1930, Plates XVIIa and XVIIb); and (c) northern Yucatán (Thompson 1958, Figures 40a, 38h) show similarities among themselves and with Late Postclassic vessels from northern Yucatán (Smith 1971, Figure 38).

Sascab in the forms of dolospar (50 per cent), micrite (16 per cent), and sparite (12 per cent) represents the bulk of the constituents of the fabrics found in red-slipped Mama samples from the north-central area. This may account for the perceived homogeneity in composition. With the naked eye or very low magnification (10X), it is difficult to set these fabrics apart.

The sascab fabric classes involved were presented in chapter eight and discussed in relation to Mama in section 9.1. Briefly, one of the sascab fabrics in Mama samples is Micrite-W-Medium (CM-W-m, section 8.5.1) found within Group 1 at Mayapán, Tecoh, and Tekit. It is also found in chemically unassigned samples at other sites (Mama, Telchaquillo, and Tipikal). Another Mama sascab fabric is the finely crystalline sparitic Sparite-W/B-Coarse/Medium (CS-W/B-c, section 8.4.1) fabric within Group 2 found at Tipikal and Teabo. Chemically unassigned samples were found at Telchaquillo and Tipikal. Lastly, the sascab fabric Dolospar-W-Medium (section 8.3.2) and its hard variation, Dolospar-W-Hard (section 8.3.3), both within chemical Group 1, are found at most north-central sites.

The production of the red-slipped Mama samples at the minor north-central sites comprised multiple potters' groups and, most likely, several different procurement zones of raw materials. Nevertheless, across sites, Mama samples present a commonality of surface characteristics, including similar macroscopic compositions that make it difficult to set them apart in a hand specimen examination. Throughout the north-central area, potters conformed to a mental schema of how red-slipped vessels should look.

Concerning to the temper added to the pottery clay, white opaque inclusions (sascab) were preferred. It was shown in this study that choices existed as to the colour of the clay and that red clayey soils could have been a viable alternative to the light coloured clays found in the pottery, although more experimental data are needed to examine the viability of the use of red clay in large vessels. Nevertheless, potters apparently selected clay materials exhibiting a light colour as part of the tradition. Although this colour may have been the result of the low iron in the composition of the raw materials of this north-central area (section 7.3.1 and section 7.3.2), other choices exist, shown by a Mayapán ceramic type (one non-vessel form) made using red clay. Also part of the tradition is an overall medium-grain size for the inclusions. Mama potters may have aimed to achieve a medium grain particle.

The materials were selected, and techniques were applied from an array of available choices according to an overarching technological tradition. The tradition included, primarily, the use of sascab as temper, the process of aiming for a medium texture, and the selection of light coloured raw material for the construction of the paste.

In addition, other stages of ceramic manufacture were also part of the tradition. Of the ceramic production stages, forming techniques identified by Thompson (1958) and also found in this study are important components that help with the identification of the traditions: for instance, coiling in sections with intermediate drying stages or the very Yucatecan characteristic of starting large vessels with the open base technique. This study also found that the round-bottom cajetes analysed in this study (section 9.1.6) were apparently started using moulds. This indicates that the principal sascab tradition was sub-divided by type of vessels and forming techniques, suggesting that potters may have divided their work accordingly.

The principal technological tradition may have had different branches depending on the forming techniques used for different type of vessels. However, the types of vessels analysed in this study are not from the same site. Surface feature analysis of two cajetes from Mayapán showed that the rounded bottom cajetes might have been started using moulds (section 9.1.6). Analysis of two large jars from a minor site four km from Mayapán suggests that one of the jars was started using the open base technique (it is not clear how the second jar was started). Future research comparing different types of vessels within the same site can clarify whether there was a division of potters' work by forming techniques.

The open base technique detected in the analysed jar has been documented in modern northern Yucatán (Thompson 1958) and modern San José Petén, Guatemala by Reina and Hill (1978, p.141-145). San José Petén is a town culturally linked to Postclassic northern Yucatán that remained in relative isolation from European influence, and it was not conquered by the Spanish for close to three hundred years (Reina and Hill 1978, p.141-145). The shapes of their vessels are similar to vessels produced during the Late Postclassic in northern Yucatán (Reina and Hill 1978, p.141-145).

Within the sascab red-slipped technological tradition, it is likely that local variations existed in the way things were done to accommodate variations in local materials. For instance, at Mayapán, a marl could have been used as pottery clay. The experiments performed in this study with marls of the area showed that firing for 45 min would not transform them into ceramics. A longer time of several hours is required. However, 45 min could have been sufficient at sites at which deposits of pure clay, kaolinite for instance, exist. In addition, the use of a marl as clay body at some localities would have required a more labourious and complex processing, such as grinding and sifting, than when a clay(s) is used.

10.1.1.1 Ethnographic Data

The ethnographic data indicates that modern Yucatecan potters preferred sascab for the construction of the pottery clay of non-cooking pots (Thompson 1958, Rendón 1947), in particular when the town specialised in the production of water containers, such as at Tepekán, Uayma, and Valladolid (Thompson 1958). Table 10-1 compiles ethnographic observations about the use of sascab and *hi'* at northern Yucatán towns that made pottery (Thompson 1958).

Table 10-1. Tempers used during the 1950s in Yucatán by Town and Vessel Type (compiled from Thompson 1958)

Town	Types of vessels that are red-slipped	Town specialisation	Temper for Non-cooking pots'	Temper for Cooking pots
Lerma, Campeche	All vessels are unslippped		hi'	hi'
Tepakán, Campeche	Only large water-storage jars	Water container	sascab	hi' (when *ollas* made)
Becal, Campeche	Only stamp-decorated water containers	Cooking pots	sascab (cuut or very fine, if non-cooking made)	hi' (imported from Maxcanú)
Maxcanú, Yucatán		Cooking pots	hi'(when non-cooking made)	hi'
Ticul, Yucatán	Non-cooking	Town-potters' separation by temper	sascab	hi'
Mama, Yucatán	Only water storage. For all other slipped, a creamy slip		sascab and hi' mixture	sascab and hi' mixture
Izamal, Yucatán	Most non- cooking pots.		sascab and hi' mixture	sascab and hi' mixture
Uayma		Non-cooking	sascab and hi' mixture	hi'
Valladolid	All non-cooking pots	Non-cooking	sascab	sascab

Many of these modern Yucatán vessels, in particular if they were water vessels, had red-slip applied, whereas cooking pots were never slipped. At Ticul in particular, Arnold (2008) observed that the temper added was associated with the vessel function: a white earth to water vessels (storage and carrying), *hi'* (ground crystalline calcite) to cooking pots, and a recipe that varied for cajetes, figurines, and other non-water vessels.

10.1.2 Translucent Rock Cooking Pot Tradition

Construction of some of the open-mouth Yacman (unslipped, striated) jars, followed a technology associated with the use of translucent rocks as temper within a coarse and open (porous) paste. Medium to very coarsely grained translucent calcite or dolomite crystals were found in close to 25 per cent of the Yacman thin sections from the north-central area (section 9.3). The proportion is higher (57 per cent) if samples with dolospar fabrics are excluded (for it will be argued in the next sections that they are imports to the minor centres). Several fabrics were found to have in common inclusions that would have looked translucent to the naked eye. The fabrics and their samples within sites are encircled by a green line in Table 8-36. Some of the fabrics are not petrographically homogeneous, presenting differences in the minerals involved (calcite versus dolomite) and inclusion textures, including crystal size, overall particle colour, and translucency.

The use of translucent crystals for the construction of unslipped jars has been documented (Smith 1971b, p.171-172) from the time prior to the Late Postclassic, the Terminal Classic, and summarised in Table 10-2. This is an ancient tradition in northern Yucatán as shown by Anna Shepard (Smith 1971b, p.169-172) in the petrographic analysis of sherds from Uxmal, Kabah, and Chichén Itzá. These sites had their major occupancy earlier than the apogee of Mayapán.

Table 10-2 shows a clear division of slipped versus unslipped wares and forms by the textural characteristics of the tempers used; divisions that Smith (1971a, p.2) thought were probably related to function. Unslipped jars at Puuc sites and Chichén Itza contained clear crystalline or grey particles, while the slipped vessels primarily contained cryptocrystalline (micrite), saccharoidal (finely crystalline sparite), or volcanic ash.

Table 10-2. Types of Inclusions and Vessel Forms at Chichén Itzá, Uxmal, and Kabah during the Terminal Classic (data compiled from Smith 1971b, p.171-172)

	Chichén Itzá	Uxmal and Kabah	Chichén Itzá	Uxmal and Kabah
Inclusions Types/ Finish	**slipped**	**slipped**	**un-slipped jars**	**un-slipped jars**
Inclusions Types				
Clear Crystalline			X	X
Grey			X	x
Lamellar			X	
Cryptocrystalline (or micrite)		X		
Saccharoidal (or sparry)		X		
Volcanic ash	X	X		
Others				
Potsherd		x		
Clay lump		x		

10.1.2.1 Ethnographic Data

The practice of adding a translucent rock to the clay was documented in early colonial times in the Pío Pérez dictionary (Pío Pérez 1877). This is a Yucatec-Spanish/Spanish-Yucatec dictionary compiled in 1898 but based on various earlier dictionaries including the Ticul dictionary, a Spanish-Maya dictionary completed in 1690. The dictionary describes the substance that in Maya language is called *hi'* as a 'kind of transparent rock that potters grind to mix with clay and build pots for the kitchen.'

In modern times, the use of translucent rocks as temper for the construction of northern Yucatán cooking jars or *ollas* has been documented in Yucatán ethnographic sources repeatedly (Thompson 1958, Rendón 1947, p.116, Arnold 2008). Table 9-23 compiles ethnographic observations (Thompson 1958) about the use of sascab and *hi'* at northern Yucatán pottery towns. It shows that *hi'* was used consistently for the production of cooking pots. At Ticul, this rock has been documented as ground crystalline calcite (Arnold 2008). Reina and Hill (1978, p.141-145) documented the use in modern San José Petén, Guatemala of a substance added to the clay for ollas. San José Petén is a town culturally linked to northern Yucatán (Reina and Hill 1978, p.141-145). They collect quartz and feldspar and call it *hi'*. This *hi'* is heated (as it is in some northern Yucatán towns) and ground before adding it to the clay.

Ultimately, the selection of temper may be related to the intended function of a vessel. Given that Yacman jars are thought to be for cooking (Brown 1999a) and that ethnographic work tells us that Yucatecan slipped vessels are not used on fire (Thompson 1958), the division between the sascab and translucent-rock technological traditions reflected in the north-central samples may be related in part to cooking versus non-cooking function of the pots.

Not only in Yucatán but also in other parts of the world coarse calcite has been preferred for the construction of cooking pots. For one, limestone as temper is an excellent solution to the problem of thermal expansion when heating. For instance, in other areas of North America, limestone temper was preferred for cooking pots (Bronitsky and Hamer 1986; Hoard et al. 1995; James Feathers 1989). Experiments have shown (Rye 1981) that the inclusion of particles with thermal expansion similar to the clay represents a desirable thermal solution, and calcite is among those inclusions (Rice 1987a, p.229). That the thermal coefficients are similar between calcite and clay means that the differential expansion due to the heating and cooling of the vessels is minimal (Rye 1981:33) and that under heating and cooling conditions, calcite and clay expand and contract together.

Potters prefer not only limestone but coarse crystalline limestone. Coarse crystalline calcite performs better than fine limestone and other tempers in regard to the thermal expansion (Hoard et al. 1995:830). And, in addition, as mentioned in the previous chapter, experiments (Shoval 1993) have demonstrated the inverse correspondence between the crystal surface areas exposed and the temperature required for total conversion to lime. Coarse crystals (as opposed to micrite and finely crystalline spar) have less exposed area and the conversion to lime occurrs at a higher temperature. In other parts of the world (Shoval 1993, p.269) and northern Yucatán, potters preferred coarse mono-crystalline (coarser single-crystal) calcite for the construction of cooking pots, going to great lengths to obtain it.

In sum, this study has shown that even in an environment dominated by limestone, choices of different materials exist. Texture provides most of the variation. Ancient and modern potters make choices from a wide array of variables. For instance, potters selected sascab for the construction of slipped vessels. However, grey/dark micrite is available, as found in some unslipped vessels. The selection of sascab, from other possible technological choices that could have functioned equally well (Sackett 1990), is part of the north-central Yucatecan technological style.

Two technological traditions were recognised. Mama and Yacman produced at the local level were guided by sub-regional Yucatecan traditions, the red-slipped sascab and the translucent rock (*hi'*) cooking pot traditions that, in the case of the sascab tradition for red-slipped vessels, resulted in a textural uniformity of the pottery fabric to the naked eye. Red-slipped jars and cajetes were produced in the various north-central communities. Some of the red-slipped samples found at the different north-central centres are similar in petrographic and chemical composition to those found at Mayapán. In the next section, it is argued that they were not locally made but imported to the lesser centres from Mayapán.

10.2 Ceramic Production Organisation at North-central Sites

The organisation and the location of production of utilitarian ceramics at the north-central sites are the topics of this section. In the previous chapter, manufacturing techniques for the production of red-slipped Mama, and the unslipped Navula and Yacman were identified. In this section, the fabric classes and the techniques applied, as discussed in the previous chapter, are examined in the context of their distribution and the types of vessels involved, aiming to identify patterns from which the organisation of ceramic production for utilitarian vessels at north-central sites can be inferred.

10.2.1 Production Organisation

Many groups of potters and zones of procurement were involved in the production of utilitarian ceramics in the north-central region. Data supporting this proposition were presented in sections 9.1 to 9.3, consisting of the several fabric classes and chemical groups. For instance, the red-slipped Mama producers include potters using different fabrics at many sites with varied chemical compositions and most likely representing several potters' groups: a bioclast (or skeletal remains) fabric found at Tepich, a fine-to-medium dolomite fabric at Tekit and Tecoh, and micritic fabric (likely local) at several locations. Mama was also found with dolomite fabrics throughout most north-central sites.

Among the fabrics found at the north-central sites, the homogeneity in petrographic and chemical attributes of the dolospar fabrics stands out. In Table 9-19, and sections section 9.1 and 9.3 dealing with the reconstructions of red-slipped Mama and unslipped Yacman production respectively, it was shown that for these two types of vessels, the dolospar fabric classes appear at many sites with homogeneous petrographic and chemical characteristics. Mama and Yacman with dolospar fabrics are characterised into two different fabrics mainly differing in their texture, while Navula with dolospar does not appear to follow a grain size pattern.

Table 10-3. Distribution of Non-dolospar Types of Fabrics within Ceramic Type

	Micrite (sascab)	**Sparite (sascab)**	**Bioclasts (only Tepich)**	**Bioclasts**	**Discrete Crystals**	**Grey Micrite**
Mama	X	X	X			
Yacman					X	X
Navula	x	x		X	x	X

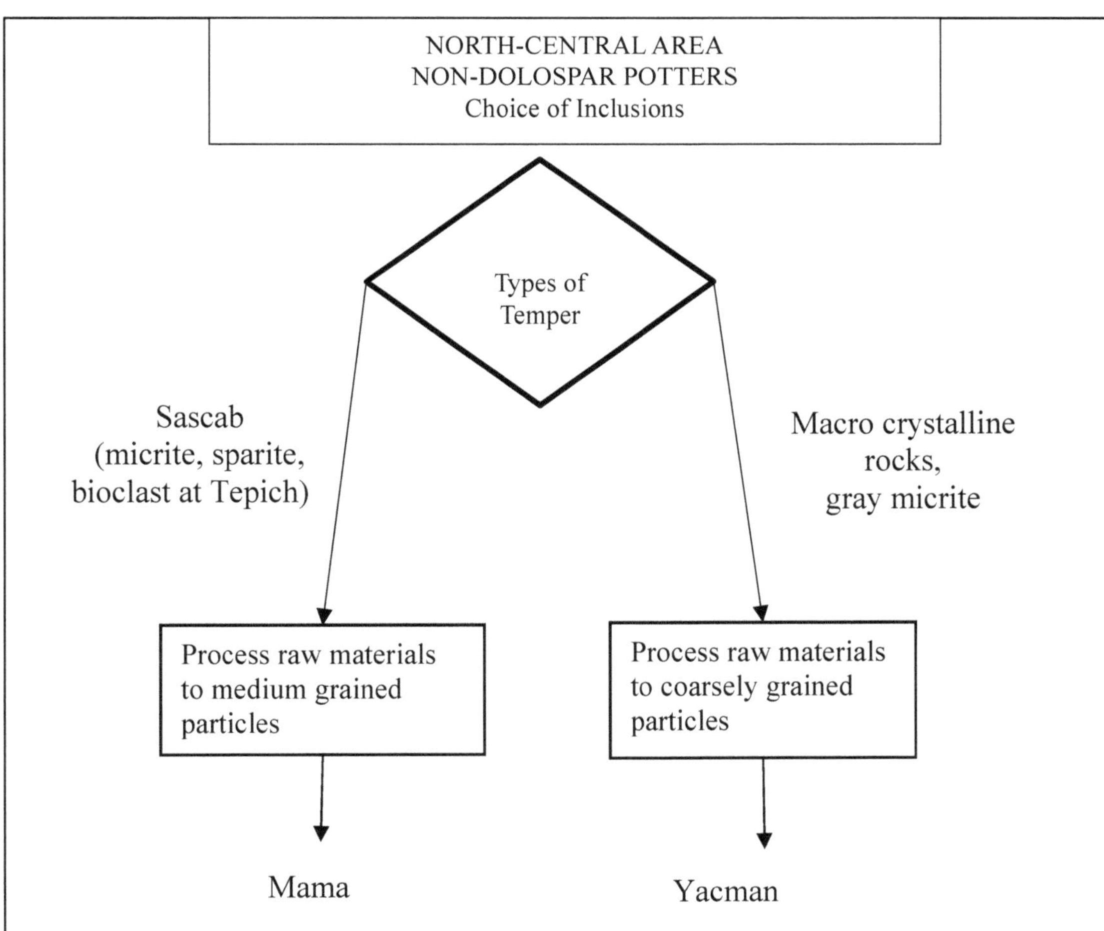

Figure 10-1. Chart with choices made by potters working with non-dolospar inclusions

Non-dolospar Mama and Yacman fabrics are varied and found at all locations, as presented in sections 9.1 and 9.3. They are either not petrographically homogeneous or not chemically homogeneous, indicating that many potters' groups or many sources may have been involved. Table 10-3 illustrates a very broad categorisation based on chapter nine results of the main non-dolospar groups of fabrics for Mama, Yacman, and Navula samples grouped by the primary type of inclusion present. The table shows, for instance, non-dolospar fabric Mama having mainly sascab and, at Tepich, having a bioclast (which may very well be a type of sascab) fabric.

In the vessels sampled having non-dolospar fabrics, there is a correlation between some of the types of vessels (red-slipped or striated *olla*) and the types of temper. If different kinds of materials and techniques imply that different potters' groups were involved, then Mama and Yacman with non-dolospar fabrics were produced by distinct sets of potters.The patterns or divisions summarised in Table 10-3 are illustrated in Figure 10-1. Potters working with non-dolospar materials divided their work based on the raw materials, grain size, and type of vessels. The potters can be broadly divided into two categories: those working with sascab and those working with either macro-crystals or grey micrite. Based on the samples containing micritic or sparitic sascab, it appears that sascab potters processed the clay and temper to obtain a medium-size (most of the time) overall texture to produce red-slipped vessels.

The table 9-3 (previous chapter) gives an idea of the minimum number of potters' groups involved in the production of Mama jars and cajetes at the north-central sites. Other groups of potters worked with grey micrite or macro crystals to construct a coarse fabric used for the production of striated open-mouth Yacman jars. The table 9-12 shows the potential groups of potters involved in the production of the Yacman vessels sampled. Given the variability of Navula fabrics, Navula may have been produced by the same or different sets of potters. Table 9-11 suggests what the minimum number of groups involved in the production of Navula jars and cajetes might have been.

If samples with dolospar fabrics are included in Table 10-3, the result looks like Table 10-4. This table shows that dolospar potters differed from the rest in many ways. They used the same kind of temper for all types of vessels. Dolospar potters divided their production into red-slipped Mama and unslipped Yacman *ollas* based not on the types of materials added to the clay (as non-dolospar potters did), but only on the grain-size. Red-slipped Mama samples are mainly medium-grained while Yacman jars are coarse.

It is doubtful that potters could have differentiated one type of sascab from the others. In this study, it took a stereomicroscope and a polarising microscope to do so. Therefore, potters did not select dolospar sascab as opposed to micrite or sparite sascab to organise their work in a given manner.

The differences between dolospar and non-dolospar potters are likely due to the availability of the types of sascab at some locations but not at the others. For instance, given that the three types of sascab cannot be differentiated with the unaided eye, if the three of them were found at one location, it would not be possible to have the distribution illustrated in Table 10-4 in which Yacman is only found having dolospar sascab.

Figure 10-2 illustrates the divisions observed in the production of Mama and Yacman with dolospar fabrics. Navula within dolospar fabrics appears to have been made without any grain-size discrimination, but the number of Navula samples is small, and further sampling and research may uncover new associations that were missed in this study.

In summarising, two orientations to the organisation of the north-central Mama and Yacman potters are inferred from the data.

At the north-central communities, one arrangement reflects potters divided by groups producing red-slipped Mama vessels or striated Yacman ollas. As regard to Mama and Yacman, there is a correlation between the type of temper, grain size, and vessel type. The group or groups of potters producing Mama preferentially used sascab as temper. Potters producing Yacman used macro crystals or grey micrite. Navula did not follow a pattern, and it may have been produced by different potters' groups. In addition to these two arrangements, a few groups producing Mama or Navula did not adhere to these patterns just described. This correlation of raw materials and vessel types implies that a division of the groups of potters by raw materials and type of vessels needs to be considered.

A division of potters by raw materials and vessel types is very similar to that found by Thompson (1958) at Yucatán pottery towns. What is more, he found that not only the potters but also most towns were divided. Thompson (1958) found that Yucatecan potters of a given pottery-making town tended to produce non-cooking vessels, using primarily, sascab as temper while other pottery-making towns preferentially produced cooking pots, usually, using *hi'* as temper. What is more, at Yucatán towns, Thompson (1958) observed that many aspects of ceramic production were divided by community or town, in what he called a community specialisation. The community specialisations observed by Thompson included the types of vessels made, mainly divided into cooking and non-cooking pottery-making towns, the forming techniques, in particular, the technique used to start the vessels, and the type of temper used for cooking and non-cooking pots.

Table 10-4. Distribution of All Types of Fabrics within Ceramic Types

	<-----Non-Dolospar Fabrics-----							<-Dolospar>
	Micrite (sascab)	**Sparite (sascab)**	**Bioclasts (Tepich)**	**Bioclasts**	**Discrete Crystals**	**Grey Micrite**		**Dolospar (sascab)**
Mama	X	X	X					X
Yacman					X	X		X
Navula	x	x		X	x	X		X

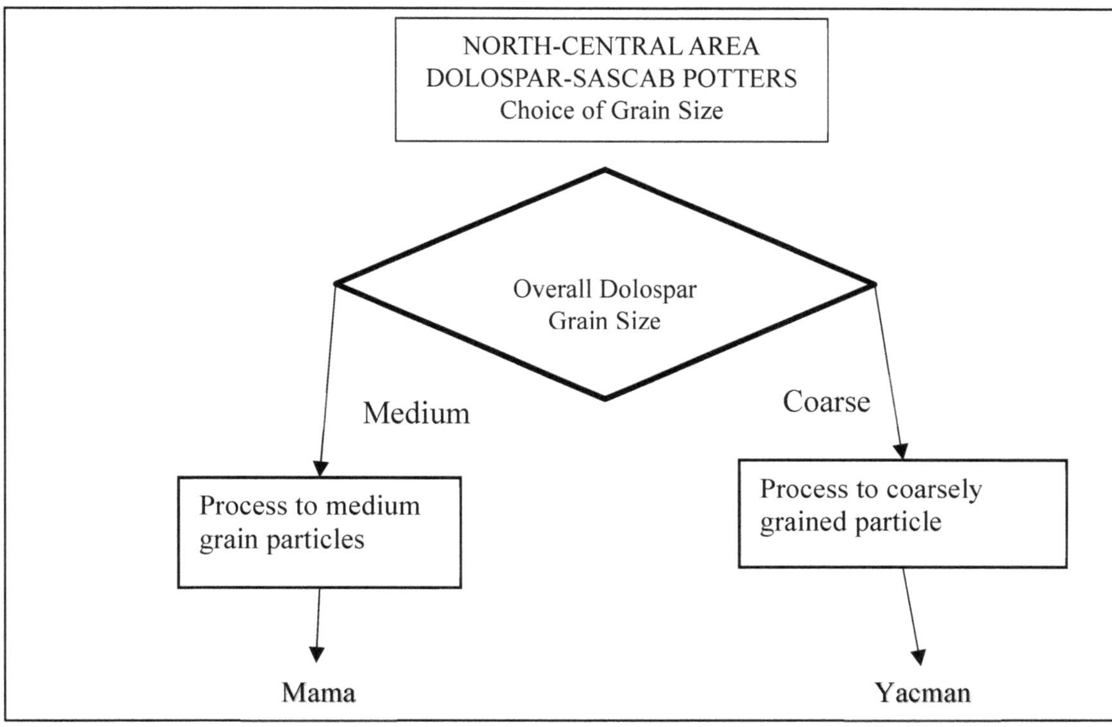

Figure 10-2. Choices made by potters working with dolospar sascab

Even the gender of who made the pottery was a community specialisation. At Lerma, Becal, Maxcanú, Izamal, and Valladolid men made the pottery, while, at Tepekán and Mama, women made the pottery. There was a tendency for men to be the potters if the town specialised in cooking pots. At Palenque, Rands and Bishop (1980, p.42) found indications of a division of ceramic production, apparently, at the community or the zonal level. In the different communities or zones, certain vessels dominated production.

Given the petrographic and chemical homogeneity of dolospar fabrics across sites, the next section examines the data to draw inferences about the location and number of producers that may explain this homogeneity.

10.2.2 Dolospar Fabrics: Production Location

Another point that needs to be explained is the production of potentially significant quantities of vessels with dolospar fabrics characterised by their homogeneity. Out of 56 thin sections from utilitarian vessels from Mayapán, 39 have dolospar fabrics (70 per cent). Although proportions of samples such as this are not representative of the proportions at the site level due to sampling restrictions and strategy, they are indicative of the variability, or lack of it, of the fabrics from Mayapán. Mayapán vessels, as represented in the samples, are fairly homogeneous in regard to the type of materials found in most of them, sascab. Most of this sascab is dolospar sascab. Micritic sascab is present to a lesser degree. In addition, samples from Mayapán are homogeneous in regard to their chemical composition, with two exceptions.

One scenario that explains the significant amount of samples with similar chemical compositions of Mama (or Yacman) with dolospar fabrics across sites is one in which the raw materials to produce vessels with dolospar fabrics within chemical Group 1 are found at most north-central sites. Under this scenario, a technological tradition existed at Tecoh, Telchaquillo, Mayapán, Tekit, Mama, and Tipikal in which some potters (for some Mama with non-sascab fabrics were found) worked with local materials and preferred sascab (dolospar, micrite, or sparite) for the construction of red-slipped Mama vessels. As mentioned before, at the different localities, it is doubtful that potters could have distinguished one type of sascab from others (for instance, dolospar sascab from sparite sascab).

The fabrics found in Yacman vessels present a problem for this scenario. At the different sites, Yacman vessels were found to have fabrics that include dolospar sascab, semi-translucent or smooth angular particles, or grey/dark micrite and a few instances with bioclasts. Of the 62 Yacman thin sections (62) only two (#332, #333 from Tecoh) have sascab different from dolospar, micrite in this case, as main inclusions. If Yacman was locally sourced and manufactured and sascab was used, potters at these minor centres apparently preferred dolospar to micrite and sparite sascab. However, the three varieties of sascab (dolospar, sparite, micrite) are impossible to differentiate from each other using the unaided eye. The plain Navula jars and cajetes present another problem to the local sourcing of dolospar scenario, because at Teabo, Tipikal, Mama, and Tekit, these vessels were found to have micrite or sparite sascab, among other fabrics, but never dolospar sascab. It is not possible, if dolospar sascab is locally available at these sites, that potters could have selected dolospar sascab for Yacman and avoid it for Navula. The Navula samples having dolospar sascab were found at Mayapán and vicinity, Telchaquillo and (two samples at) Tecoh.

Another scenario that could explain the homogeneity in the chemical composition of Mama (or Yacman) with dolospar fabrics across sites is one in which red-slipped Mama, and unslipped Yacman and Navula with these characteristics were sourced and produced at a mineralogically (dolomite) and chemically (Group 1) homogeneous zone or zones. At the dolospar production zone, the pottery fabrics of all types of vessels were prepared using the same temper, dolospar sascab.

This scenario accounts for the patterns in the distribution of Yacman and Navula with dolospar fabrics observed at minor centres. Mama and Yacman from the dolospar production zone were taken to the minor north-central sites, while at this centres Yacman was not produced using sascab. This scenario explains the presence of Yacman with dolospar sascab at the minor centres, while Yacman with other types of sascab is not found. This scenario also explains how Mama, which was also produced at the minor centres using local sascab (micrite, sparite), is found with all types of sascab at the minor centres. In this scenario, the plain Navula from the dolospar production zone did not travel far. In this scenario, the dolospar centre produces Mama, Yacman, and Navula. Therefore, the producing centre presents these three ceramic varieties, and it is expected, that they are the most at this centre. This points to Mayapán or Telchaquillo as the producing centres given that they are the only sites in which the three studied ceramic varieties were found with dolospar fabrics and, at the same time, these ceramics are the bulk of the samples.

The propositions that the utilitarian vessels having dolospar fabrics were produced at one location, the sources for the raw materials are chemically and mineralogically homogeneous, and that this location was Mayapán (or a place in its vicinity) has support from a combination of evidence that, although no one is conclusive, altogether weigh in favour of this view. The evidence includes:

(a) A local marl (#11, Table 5-4) collected near Mayapán matches the chemical composition of most vessels sampled from Mayapán, which is overwhelmingly within Group 1.

(b) The chemical composition of raw materials collected near Tekit, Mama, Teabo, Tipikal, and other localities does not match the composition of pottery produced with dolospar within Group 1 (granting that more clay prospection could find Group 1 clay at other localities).

(c) Thin-section petrographic analysis of the marl sascab collected near Mayapán (#11) identified sparry clusters of dolomite (dolospar). Dolospar was not found in the rest of the thin-sections made from the collected raw materials.

(d) When having dolospar fabrics, samples taken from Navula jars or cajetes are found only at Mayapán and its vicinity (Telchaquillo and two samples at Tecoh). Therefore, Navula jars and cajetes with dolospar fabrics are types of vessels associated with the Mayapán area and no other area (acknowledging that this interpretation could be the result of the limitations of the sampling).

(e) An Euclidian distance (Pierce and Glascock 2015) comparing each specimen in this study to every one of the 2,071 northern Yucatán specimens in the MURR database, of which only 51 (2.5 per cent) are from Mayapán, found that samples in this study from Mayapán are more similar to the 51 MURR samples from Mayapán than to any other sample.

(f) In the samples from Mayapán, dolospar is the most common fabric.

(g) Lastly, the greater variety of ceramic types at Mayapán when compared to the secondary sites reinforces the proposition that Mayapán was or was near the main production centre of red-slipped and unslipped ceramics (Masson 2015, p.54).

In sum, production at one centre or one production zone of vessels with dolospar fabrics, as discussed before, explains better the particular distributional patterns encountered.

10.3 Movement of Pots

In this section, the variability of fabrics and other aspects of technology are examined, seeking patterns that reflect whether pots have been moved over long or relatively short distances. Movement over a short distance is studied through the examination of the variability of pottery and raw materials within the north-central sites, while long-distance movement is studied through patterns indicating that non-local production may be involved.

10.3.1 Short-distance Movement of Pots

Given the lack of direct evidence for production, such as kilns or wasters, the movement of vessels is investigated by examining the chemical and mineralogical fingerprints. The criteria that must be met to show local production at a given centre and the movement of the pots to another were discussed in section 5.5 in relation to the proposed Hypothesis B (red-slipped pots were centrally produced). Although that discussion refers to one ceramic type, it is valid as a guide in a more general discussion. In sections 9.1, 9.2, and 9.3, potential potters' groups, based mainly on shared chemical and petrographic compositions, were discussed. Of these groups, those found at more than one site are examined further in this section, for they may be indicative of the movement of vessels.

To investigate the movement of vessels, the first step is to establish whether the chemical or petrographic compositions are different at the different sites. The chemical analysis found relatively low variability in the samples from north-central sites, assigning the samples to only three chemical groups (Group 1, Group 2, and Group 5) that geographically largely overlap. However, when the chemical groups were examined in combination with the petrographic classes, greater variability was found (section 9.1, 9.2, 9.3). Only for those samples found at more than one site with composition falling within one single chemical and petrographic group can the argument of movement between sites be made.

Perhaps the most convincing cases are shown by a few instances in which a fabric class is found with the same chemical group at two sites. The following instances share the same chemical group and petrographic fabric across sites:

Case a: the micritic fabric Micrite-W-coarse within Group 1 found at Mayapán and Tekit.

Case b: the sparitic fabric Sparite-W/B-Coarse/Medium within Group 2 found at Teabo and Tipikal.

Case c: the dolomite crystals fabric Dolomite-B/W-Medium within Group 2 found at Tecoh and Tekit is another example.

Case d: Mama and Yacman samples with dolospar fabrics, and all within Group 1, found at many of the north-central sites.

Shared chemical and petrographic characteristics between samples at different sites could be the result of movement of vessels. It also could result from local production at the different sites under a common tradition and chemical homogeneity. Importation from somewhere else can also produce this pattern. The matter is complicated by the fact that, when sites are located relatively close together, ceramics could have been locally manufactured while potters could have shared one source of raw material. Further evidence needs to be present to establish local production or to eliminate some of these alternatives. For example, inferences can be made about the sharing of resources based on the distance between sites.

Arnold (2005b) has provided, based on ethnographic work at 117 sites, empirical metrics for the maximum distances that a potter might have travelled to obtain raw materials. The maximum distance that approximately 90 per cent of potters would have travelled on foot while carrying raw materials is seven kilometres. Therefore, localities 14 or fewer kilometres apart may have shared resources.

Table 10-5 shows the distances between north-central sites. Based on Arnold's empirical metric, in the case of vessels with fabric Micrite-W-coarse with a chemical composition in Group 1 found at Mayapán and Tekit (case a), most likely, raw materials were not shared (distances > 14 km). It is possible that vessels with these characteristics were transported from one of these sitesto the other, but local production at each site is a possibility.

In case c, most likely, raw materials were not shared (distances > 14 km). It is possible that, for instance, vessels with fabric Dolomite-B/W-Medium (within Group 2) from Tekit were transported to Tecoh. However, it is also possible that they were independently produced at these two sites with similar petrographic and chemical characteristics. In the case (b) of the sparitic fabric Sparite-W/B-Coarse/Medium within Group 2 from Teabo and Tipikal, there is the possibility of resource sharing with local production, in addition to the scenarios mentioned for cases (a) and (c).

Lastly, in the case of Mama and Yacman samples with dolospar fabrics found at many of the north-central sites all within chemical Group 1, or case d, the data presented allows us to make a strong argument (section 10.2.2) for the production of these types of vessels in the vicinity of Mayapán.

Table 10-5. Aerial Distance between North-central Sites

Site	Neighbouring Site	Km
Tepich	Tecoh	14.5
Tecoh	Telchaquillo	10.7
Telchaquillo	Mayapán	1.8
Mayapán	Tekit	16.4
Tekit	Mama	8
Mama	Teabo	9.7
Mama	Tipikal	6.2
Tipikal	Teabo	6.2

The implication of this argument is that these vessels, when found at sites other than Mayapán, are the result of the movement of these vessels from this centre to the periphery. The mechanisms of distribution by which they might have been transported are discussed in the next chapter.

From the data of this study, Mayapán may have received few vessels from other north-central sites. This is inferred not only from the limited petrographic variety of the samples from Mayapán (70 per cent dolospar) as discussed in the previous section but also from the variety of chemical compositions at the minor sites that are not observed in Mayapán samples. Most utilitarian samples (Mama, Yacman, Navula) found at Mayapán have compositions within Group 1. Out of 32 utilitarian samples (Payil samples not included) from Mayapán that were chemically analysed, 26 (81 per cent) fall within Group 1, four are unassigned, and only two have a composition within Group 2. In contrast, at other sites, such as Tecoh or Tekit, the chemical compositions of samples are more varied, including 50 per cent or more of the samples within Group 2, Group 5 or unassigned compositions, in addition to the Group 1 dolospar samples that may be present.

In summarising, due to the lack of direct evidence for production, indirect methods were applied to infer the movement of pots within the north-central area. The data are not clear as to whether pots were moved between the north-central sites, but there are a few instances in which the same fabric and chemistry present at two sites could be interpreted as resource sharing, local production, or exchange.

Arguments were presented for the production of vessels with dolospar fabrics being related to Mayapán or an area near this centre. From this location, Mama and Yacman dolospar vessels may have been distributed to minor north-central sites. However, based on the samples, dolospar vessels did not reach Tepich (the northernmost site in this study) and Teabo (the south-easternmost site).

When compared to the wide variety of petrographic and chemical compositions found in vessels from the minor centres, the Mayapán vessels sampled in this study present rather homogenous petrographic and chemical compositions. Mayapán ceramics do not reflect the variety of compositions found at the minor centres, and it may have received few ceramics from other north-central sites.

10.3.2 Long-distance Movement of Pots

In two instances, there is overlap between fabrics found at north-central sites and those found at Cobá, Chac Mool, and Culubá. One is related to the Payil vessels found at Mayapán and Chac Mool. At both sites, the fabric has Calcite-Quartz-B/R-Fine (or CX-QZ-B/R-f), with fine-grained calcite and quartz, within chemical Group 4. The correspondence of petrographic and chemical compositions between these two sites indicates that these Payil samples have a common origin. The chemical analysis (section 7.3.4) that includes clays collected by Cecil (2012) traced the origins of these samples to around Laguna de On, Belize. Therefore, Payil from Chac Mool and Mayapán is foreign to both of these sites.

Sherds classified as Payil may have had other fabrics and probably other origins, given that two sherds were classified as Payil but did not present the fabric described earlier (#349, #135). With only two samples, this issue, however, could not be resolved in this study. It is doubtful that a misclassification of Payil occurred because utilitarian samples in this study (Mama, Yacman, Navula) do not present a compacted and fine-grained texture. The second instance of fabrics shared between north-central and eastern sites is the sparite fabric, Sparite-W/B-Coarse/med (CS-W/B-cm), which is found across the north-central sites and also at Cobá in the east (Table 9-18). As discussed in section 9.1, sascab is the most likely source of the inclusions in this fabric, composed as it is of finely crystalline calcite. It is associated with utilitarian slipped pottery (Mama, Cancun, and Xcanchakan) and Navula, but not with the striated Yacman. It is possible that red-slipped Mama vessels were transported to Cobá. However, the fabric mentioned above is inhomogeneous from a petrographic perspective, and the samples from Cobá may be local. This means that the sascab tradition for the construction of red-slipped pottery was also present at Cobá, while it was not observed at Culubá.

Interestingly, while no overlap was observed in the fabrics of utilitarian ceramics from Chac Mool and Culubá with those of north-central sites, the red-slipped Cancun samples from Cobá present characteristics from both the north-central and eastern areas: sascab sparite fabrics (north-central) and coarse crystalline fabrics (eastern). It appears then that red-slipped utilitarian ceramics from Cobá were manufactured following both the sascab traditions of the north-central sites and the coarse crystalline calcite traditions of the eastern sites.

Four Yacman samples and one red-slipped Mama found at Teabo, Tipikal, Tekit, and Telchaquillo have an unusual fabric, Calcite-Grog-B/R-c (MX-GROG-B/R-c), containing fine to medium calcite crystals, argillaceous inclusions with crystals, and grog (more details in sections 7.3.5 and 8.6.5). This is an unusual fabric because grog has not been reported in Late Postclassic ceramics from these sites. The Euclidian distance points to Santa Rita Corozal, Belize (Figure 7-7) as the most common match in the MURR database, a site at which not only grog (Cecil 2012, p.18-20) but also crystalline fabrics are more common. It is, therefore, possible that imports from Belize include not only elite Payil items but also pottery that appears to be utilitarian, such as unslipped, striated vessels. If this was the case, a key issue is whether the vessels or the contents of the vessels were intended to be transported.

Summarising the findings of this section, the Payil samples found at Mayapán and Chac Mool are foreign to these sites, originating in northern Belize, likely around Laguna de On. The Payil samples share the same chemical characteristics as pottery samples and clays from around Laguna de On (Pierce and Glascock 2015).

In addition to the Payil samples, five samples exhibit an unusual fabric containing grog and single calcite crystals; fabric more consistent with the eastern coast than with the inland areas. Furthermore, the Euclidian distance matches Santa Rita Corozal, Belize. Grog has not been reported during the Late Postclassic of northern Yucatán, but it is common at this site (Cecil 2012).

10.4 Evaluating the Hypotheses

Hypotheses to guide this study were outlined in section 1.3 and described in section 5.3. In this section, the hypotheses are revisited and evaluated in terms of the analytical results from this study.

10.4.1 Hypothesis A: Unslipped jars found at the different sites were locally produced at minor centres and at Mayapán

The results as discussed in this and previous chapters support Hypothesis A. In retrospect, however, this hypothesis is not appropriate to cover the complexity of the interpretations given to the arrangements in the production of unslipped vessels. Navula and Yacman present different orientations in their production. While the plain Navula was most likely locally produced and consumed, the results for Yacman point to a dual path. One arrangement is characterised by several producers and zones of procurement, with some potters producing Yacman with coarsely crystalline temper and others with grey micrite tempers (see section 10.2.1). A second arrangement involved potters producing Yacman *ollas* with fabrics with dolospar sascab (section 10.2.2); potters likely located in the vicinity of Mayapán (dolospar fabric), and with pottery exported to the rest of the north-central sites.

10.4.2 Hypothesis B: In the north-central area, red-slipped pots (Mama) were produced at one production locality that supplied the rest of the north-central centres

The results do not support Hypothesis B. The fabrics found in the Mama samples from the north-central sites are varied and suggest several producers and zones of procurement. Similarly to the production of Navula and Yacman, there was local production of Mama vessels at different north-central centres, with potters preferentially using sascab temper (mainly micrite and sparite). On the other hand, Mama (and Yacman) samples with dolospar compositions, found at several north-central centres, are very homogeneous and appear to have been sourced and produced in one homogeneous geographical area.

10.4.3 Hypothesis C: Payil, the fine-grained red-slipped ceramic type commonly found at eastern coastal sites, when found at north-central sites, was locally made

The results do not support Hypothesis C. The chemical and petrographic results point to a foreign location, Laguna de On specifically, as the source of Payil samples found at Mayapán (and Chac Mool).

10.5 Summary

The associations found in this chapter show that at least two technological traditions were present during the Late Postclassic, or even earlier: the sascab red-slipped vessel tradition and the translucent-rock cooking-pot tradition. These traditions are correlated to the type of vessels. Similar traditions have been documented in ethnographic work in northern Yucatán during modern times (Thompson 1958). Thompson identified different technological paths related to the function of the vessel: sascab for non-cooking vessels (that always include slipped vessels) and translucent rocks for cooking vessels.

Production of utilitarian vessels occurred at many places including Mayapán. At the minor centres, the data are not specific enough to determine whether the find spot corresponds to the location of production. In contrast, the evidence indicates that the pottery found at Mayapán was produced at or near this site. The patterns inferred suggest that complex arrangements existed for the production of red-slipped Mama and striated Yacman jars that point to a dual path.

It was proposed that Mama and Yacman vessels with dolospar fabrics were taken from their zone of production in the Mayapán area to the secondary north-central sites. This proposition is based on a body of evidence including distributional patterns. The distributional patterns, mainly of Yacman and Navula having dolospar fabrics and chemical composition within Group 1, cannot be explained unless Mama, Yacman, and Navula with dolospar are produced in a chemically homogeneous production zone and are not local to the north-central centres.

It was also proposed that Mayapán received few ceramics from other north-central sites. This observation is inferred from the comparison of the variety of chemical and petrographic compositions at the lesser sites with the almost homogeneous chemical compositions of the utilitarian ceramics from Mayapán.

Connections with the east coast are demonstrated by the Payil samples found at Mayapán that are not local but an import from northern Belize. Furthermore, four Yacman samples and one red-slipped Mama found at Teabo, Tipikal, Tekit, and Telchaquillo have an unusual crystalline and grog-tempered fabric (Calcite-Grog-B/R-c) that points to Santa Rita Corozal, Belize as the most common match in the MURR database. Santa Rita Corozal is a site at which grog and a similar fabric are common.

If it is true that Yacman and Mama vessels produced with dolospar inclusions and found at the minor centres were imported, it is apparent that at the smaller centres there was a clear correlation between the vessel types and understandings of how they should be made. At the minor centres, non-dolospar Yacman *ollas* mostly contain macro-crystalline calcite or grey micrite. This understanding has continued until current times. Although some unslipped samples from Mayapán contain macro-crystalline calcite or dolomite, sascab was added to the bulk of all kinds of vessels sampled from this site.

At the end of this chapter, the hypotheses proposed at the start of this book were evaluated based on the results obtained. The hypotheses address key issues in Late Postclassic archaeology and history, and at the same time ensure that the methods selected and the data gathered in the process directly deal with the research questions (section 1.3). In the final chapter, the research questions are debated, and conclusions are drawn.

Chapter 11

DISCUSSION AND CONCLUSIONS

Until the completion of this study, a detailed chemical and petrographic analysis of ceramics from Late Postclassic contexts from several north-central and eastern sites in northern Yucatán had not been conducted, and the variability of raw materials and processing techniques had not been assessed. Current ceramic analyses for this period have been largely based on macroscopic analysis or at the site level, which has resulted in an underestimation of the range of variability of the pottery fabric classes. Work on discriminating between locally, centrally, or foreign-made ceramics has been limited because of a lack of basic research on ceramic manufacture and variability in ceramic fabrics. With this study, crucial basic information for proposing ideas, not only about ceramics but also the organisation and socio-economic interactions during the northern Late Postclassic, is now available.

In this chapter, first, the findings of this research are discussed and summarised. The findings are presented according to the research questions that this study set up to investigate. This chapter evaluates how the results emerging from this study fit current models for ceramic production, distribution, and exchange for the northern Late Postclassic. The chapter also presents the conclusions drawn from this study and suggests future research to build upon the results of this study.

11.1 Discussion of Findings: Research Question 1

Are there sufficiently varied patterns in raw material selection and ceramic attributes or characteristics to allow the characterisation of pottery fabrics into fabric classes and distinct technological classes?

Distinct technological classes exist, shown in the ceramic properties and attributes observed in hand specimens and thin sections that allowed the characterisation of the samples into fabric classes, including those dominated by micro and finely crystalline calcite (sascab), finely crystalline dolospar (sascab), and macro-crystalline or bioclast inclusions. Definition of these classes is complemented with the groups found in the chemical analysis.

11.2 Discussion of Findings: Research Question 2

What might such technological patterns tell us about how the different classes of pottery (such as with different surface finish or forms) were made?

This study showed that, in the north-central area, the types of vessels (red-slipped vessels, unslipped cooking pots, and plain unslipped vessels) are correlated with the raw materials selected and techniques applied and that raw materials were not selected at random. Two technological patterns for the construction of north-central jars and cajetes were observed.

One technological pattern is observed in samples from Mayapán. At this site, dolospar sascab was found in all types (red-slipped Mama and unslipped Navula and Yacman) of vessels sampled. Section 8.3 describes the four dolospar fabrics identified and the associations to the types of vessels. Briefly, two of these fabrics are medium grained and found mostly in Mama samples and some Navula. Hardness differentiates these two fabrics. Two coarse dolospar fabrics are mostly found in Yacman and Navula. One of the coarse fabrics refires to a light reddish brown and is associated with Yacman samples. All dolospar fabrics have a composition within one chemical group, Group 1.

A second pattern is observed at the secondary north-central sites, at which a purposeful selection of raw materials according to the type of vessels occurred. At these sites (as discussed in depth in sections 9.1 to 9.3, and section 10.1), sascab of varied compositions and sources (dolospar, micrite, sparite) was found in Mama and some Navula vessels, while dolospar sascab, crystalline calcite, or gray micrite were found in Yacman vessels. A variety of fabrics was found in Navula vessels, some (but not all) overlapping Mama and Yacman fabrics. These correlations depend on the geographical region. At Culubá and Chac Mool, different patterns were found (see section 9.5 for in-depth results).

Raw materials analysis and experimentation allowed that inferences could be drawn about techniques applied in the construction of the vessels. Clay deposits are sparse, in particular, close to Mayapán. Red clayey soils are present in the area, but they were not selected by Late Postclassic potters (even though, apparently, they were used in one particular item, a candle shield). Marls (plastic sascab) may be common and may have been used for pottery-making. Experimentation with marls showed that some are suitable for pottery making. However, they require a labourious processing that may have included grinding and/or sifting to remove the very coarse (>1 mm) rock fragments observed in the marls and that are common in the sascab marls but absent in the pottery. These are not the only techniques that could have been used. Levigation is a viable option. However, it is not supported by ethnographic work.

A limited surface-feature analysis of a few vessels found a variety of forming techniques. Moulds may have been used for the rounded bottom of two cajetes and one bowl found at Mayapán. Two large jars (found in a small site four km from Mayapán), most likely, were built in sections using a coiling technique. One of the jars was apparently started using the open base technique, which has been documented ethnographically (Thompson 1958).

North-central Yucatán potters dealt with the challenges that working with carbonate materials generate during firing. XRD of one Mayapán marl matching the chemical composition of most Mayapán pottery found montmorillonite/kaolinite. Firing experiments determined that this marl from Mayapán is suitable for pottery making, as well as a marl from the Mama road area. Potters had gained practical knowledge about the interrelation between particle size, crystal size, firing time, and firing temperature of the carbonate rocks and clays of the area. Potters had to play with these interrelations, allowing enough firing time to transform clays or marls into ceramics while keeping the temperature low enough to avoid calcination and destruction of lustre in slipped vessels. That they had mastered the techniques is demonstrated by slipped vessels that survived these firings with few defects (such as popping of carbonate inclusions) and lustre that can be observed even today.

11.3 Discussion of Findings: Research Question 3

Is the observed homogenisation of style and macroscopic composition maintained through different levels of analysis, e.g., microscopic composition and chemical analysis? Does the observed homogenisation of style and macroscopic composition reflect a shared technological tradition? For instance, a technological tradition defined by a shared understanding of how red-slipped Mama vessels should be made.

This study has demonstrated that many fabric classes exist at different locations within the study region and that, for the most part, the observed macroscopic homogeneity throughout large areas of Yucatán during the Late Postclassic (Masson 2001; Pendergast 1985, p.240; West 2002, p.178) is not reflected in a homogeneous chemical or petrographic composition across sites. The homogeneity of style refers to a reduced palate (compared to previous periods) of ceramic types that are found from site to site. In addition, the same ceramic types with similar vessel forms are found from site to site throughout Yucatán. Sherds from the north-central area cannot be distinguished between sites.

More than one level of analysis is involved in the observed homogeneity. Two macroscopic levels are involved: one is related to the form and decorative attributes, and the other, to the macroscopic composition and technological choices (such as the appearance of the paste). At the laboratory level, there are petrographic (technological) and chemical levels of analyses.

This study showed that within the north-central area, the homogenisation of style observed in red-slipped Mama vessels reflects an overarching sub-regional tradition: the sascab, red-slipped tradition. It is identified by formal and decorative (such as surface finish) similarities subsumed in a common ceramic group and matched by uniformity in macroscopic technological choices (such as the use of sascab) and techniques (medium-grained temper, light colour paste). However, laboratory analyses demonstrated that red-slipped ceramics found through northern Yucatán are not mineralogically or chemically homogeneous. The chemical and petrographic variability indicate that several potters' groups were producing Mama vessels at different locations following an overarching tradition.

A second tradition was identified in a group of the striated, unslipped Yacman jars, mainly, from secondary north-central sites, guided by a sub-regional Yucatecan technological tradition dating back at least to the Terminal Classic (Smith 1971a, p.191; 1971b, p.169 – 173) and that extends into the present. In this tradition, translucent rocks (*hi'*) are added to the clay for the construction of cooking pots. The chemical and internal petrographic variability of the fabrics within this tradition indicates that various potters' groups were involved, most likely at different locations within the north-central area. In sum, this study showed that technological traditions exist in the study region for the construction of red-slipped vessels and some of the striated unslipped ollas. It also showed that the notion of homogeneity in pottery composition between and within the two study areas, north-central and eastern, is not supported by the results.

11.4 Discussion of Findings: Research Question 4

What might such technological patterns and traditions say about the organisation of pottery production, such as the number and location of potters groups or associations within and between geographical areas or specific sites, and its social significance?

The main findings and interpretations concerning the organisation of production at the main centre, Mayapán, include:

(a) Mayapán (or an adjacent centre) was a major raw material procurement and production zone. The raw materials for most of the jars and cajetes sampled in this study and found at Mayapán (section 10.2.2) were most likely sourced within the crater's basin (chemical composition within Group 1), probably near Mayapán.

The bulk of the raw materials were collected from sources that are chemically and mineralogically homogeneous, with few exceptions. For instance, a group of Yacman samples has a coarse dolospar fabric and composition within Group 1 but refired to a light-reddish brown (instead of the usual white) colour, which may indicate differences in the iron to carbonate proportions of the clay sources.

(b) At Mayapán or in its vicinity, potters produced a significant portion of the vessels sampled from this site with a uniform technology characterised by the use of sascab as temper. Most of the fragments of sascab in the Mayapán samples (over 70 per cent) are composed of clusters of dolomite crystals (dolospar). Other fabrics, mainly a micritic sascab fabric, are present in Mayapán samples to a lesser degree.

(c) Mayapán potters added sascab to all the types of vessels in this study: slipped (Mama) and unslipped (Navula) jars and cajetes as well as striated, unslipped jars (Yacman). This lack of discrimination in the selection of raw materials represents a break with Terminal Classic and north-central traditions. Instead, Mayapán vessels are discriminated only by grain size into red-slipped Mama vessels having, mostly, medium-grained sascab and the striated Yacman *ollas* presenting coarse sascab. Apparently, most Navula vessels found at Mayapán were also produced with dolospar fabrics but without a grain-size pattern.

(d) Given that the use of the translucent rock *hi'* is a long-lasting inland tradition (as mentioned earlier in this and previous chapters), Mayapán potters differed significantly from potters in the Puuc area, Chichén Itza (Smith 1971a, p.191; 1971b, p.169 – 173) and minor north-central sites in that the type of temper in their vessels was not correlated to the types of vessels built (e.g., slipped jars or cajetes versus unslipped vessels).

(e) Mayapán potters had cultural understandings of how pots should be made that differ from potters from the north-central area, and Mayapán poters may have originated in other areas. Smith (1971a) thought that Mayapán pottery appeared at Mayapán fully developed and that it had developed someplace else. Objects, such as ceramic vessels, represent a cultural body of knowledge that is refined over generations, and, usually, this knowledge is not learned by trial-and-error experimentation within one lifetime (Jordan 2015, p.12).

Pottery of uniform technological characteristics from Mayapán production zone could have been produced under different organisations of production. From section 2.2, current research at Mayapán has found direct evidence for production in domestic contexts of a variety of non-ceramic items (Masson and Peraza Lope 2014). The items include products related to chert, weaving, obsidian, and marine shells (2014, p.281). The types of debris indicate, mainly, late-stage production (Masson and Peraza Lope 2014, p. 294), while artefact densities indicate part-time production. Households with anomalously high artefact densities including ceramics are deemed workshops with production for surplus. Many of these identified crafting houses were multi-crafting households.

Although it can be proposed that production centralisation, mass production, or elite involvement in enforcing uniformity account for the observed uniformity of ceramic composition at Mayapán, the production at this centre of the non-ceramic items mentioned above does not indicate supervised production (Masson and Peraza Lope 2014). Furthermore, the ethnographic record in the Maya areas describes various instances of production of a considerable amount of rather uniform pottery within independent households. Reina and Hill (1978, p.29-41) described one such case from the town of Chinautla in the highlands of Guatemala. Pottery is produced within the town households. The town produces the same eight forms that include *ollas* and *tinajas* (for liquids). Clay is collected from the same mountain area. Around 10 per cent of pumice is used as temper by all potters for all forms. The same clay preparation and manufacturing steps are followed by all potters for all forms. This uniformity follows a *costumbre* that dictates what can and cannot be done in pottery production as well as other aspects of life. As long as *costumbre* is maintained, the pottery is of good quality. It is not up to the individual to interfere with *costumbre*.

This example shows that considerable amounts of uniform pottery can be produced in this way, in domestic contexts, without the need for a central authority or concerted mass production strategy. At Chinautla, Reina and Hill (1978, p.29-41) report that each potter can produce 20 medium-size vessels and 3 or 4 large vessels per week. At the community of Amatenango, highlands of Guatemala, Nash (1966) describes a similar production arrangement of part-time, household production guided by *costumbre*. Of the 280 households, only three did not make pottery.

In an organisation such as this, with production following the norms imposed by *costumbre* (Reina and Hill 1978) or tradition, or a perceived need for uniformity, it is not impossible that the ceramic output accumulated over two or more centuries would appear as very standardised and mass-produced to archaeologists. It is probable, then, that the production of utilitarian pottery followed a path similar to non-ceramic items described earlier, and pottery was, one, produced at Mayapán, and second, within households in a domestic context, of which many were multi-crafting households.

Regarding the north-central minor centres, the main findings and inferences drawn are:

(a) Ceramic production was dispersed. Pottery fabrics from several potters' groups were identified at the different sites (see tables 9-3, 9-11, 9-12 for suggested groups of potters). The locations of production are not known with certainty, mainly because the chemical groups are not linked to any specific location.

(b) At the north-central sites, a correlation between types of vessels and technological choices was found. In-depth discussions of this topic are found in section 9.1.5 for red-slipped Mama potters' groups, section 9.2.3 for plain Navula, and in 9.3.4 for the striated open-mouth Yacman jars.

It is possible that with a larger sampling, more ceramic types would be found correlated to different technological choices. For instance, the Mama variety is always slipped on the exterior wall. However, there is another variety of the same ceramic type and group that is not slipped on the exterior (Unslipped-exterior variety), but it is slipped on the interior wall. It is possible that the Unslipped-exterior variety follows a technological pattern according to unslipped vessels that is Yacman-like, Navula-like, or different from these.

(c) That at the minor north-central sites there is a correlation between pottery fabrics and the types of vessels (different types of materials and processing were used for different vessels) rises the possibility that the work of the potters was divided by raw materials or type of vessels. This type of arrangement predominated in northern Yucatán during modern times (Thompson 1958). What is more, Thompson found that in most communities these divisions extended to the community level, with most communities tending to produce only some types of vessels.

11.5 Discussion of Findings: Research Question 5

What might such patterns and organisation of pottery production tell us about the distribution of utilitarian ceramics and the types of exchanges that may have taken place? In particular, were ceramics produced at each centre? Alternatively, was production centralised and, if so, where?

The main findings and interpretations discussed concerning the mechanisms for ceramic distribution and exchange include:

(a) Vessels produced at Mayapán were consumed at Mayapán and, in addition, transported to the minor centres. This interpretation is based on an array of evidence discussed in section 10.2.2 and 10.3.1.

Of the studied utilitarian types of vessels (Mama, Yacman, and Navula), Mama and Yacman vessels having Mayapán style fabric (dolospar sascab within chemical Group 1) were found, in addition to Mayapán, at Tecoh, Telchaquillo, Tekit, Mama, and Tipikal. The plain Navula jars or cajetes with this fabric apparently remained close to the Mayapán/Telchaquillo area. They were not found outside Mayapán's vicinity.

(b) Vessels with Mayapán fabric style have a restricted sub-regional spatial distribution limited to centres located one day and a maximum of two days' march. Furthermore, the Chac Mool or Culubá vessels sampled do not display north-central style fabrics. At Cobá, a mixed pattern north-central/eastern technology was found but apparently of local production.

(c) Mayapán was not a major consumer of external jars and cajetes (or substances transported in them). Apparently, very few utilitarian vessels (or jars and cajetes) produced in the minor communities were transported to Mayapán (discussed in section 10.3.1).

The movement of vessels from Mayapán to the minor centres represents a close political, social, or economic relationship among them. The nature of this relationship is not clear from the results of this study. This relationship diminishes with distance travelled: none of the vessels sampled from the northernmost site (Tepich located 26 km from Mayapán), or the south easternmost site (Teabo which is 33 km from Mayapán), or the three eastern sites (Cobá, Chac Mool, Culubá) have dolospar fabrics similar to those found at Mayapán.

In cases such as this study, in which direct evidence for ceramic production has not been found, more than one interpretation usually can be given to patterns found. The proposed mechanisms by which the red-slipped and striated vessels moved out of Mayapán is one such case. The rest of this section describes exchange and other mechanisms that may account for the observed distribution of vessels and evaluates how the results emerging from this study fit these mechanisms.

11.5.1 The Case for Market Transactions

Hirth (1998, p.453-454) has summarised indirect approaches that have been used to identify market exchange (discussed in section 2.3.4). Based on these approaches, there are indirect indications of market exchange involving the ceramics found at north-central sites. From a contextual view, a marketplace may have existed based on the assumption that a large city requires a marketplace to complement its supply. Furthermore, the scale of surplus vessels produced that reached areas outside Mayapán may have exceeded that expected to meet reciprocal obligations. For instance, of 22 thin sections analysed from Tipikal, 13 Mama and one Yacman sample have Mayapán fabric style (dolospar).

The indirect evidence gathered in this investigation can be interpreted as market exchange in which some of the ceramics (Mama and Yacman) from Mayapán, found at smaller north-central sites, had been traded for goods from the surrounding areas and centres; granting that these exchanges or market transactions do not necessarily involve the existence of a marketplace (Hirth 1998, p.453), a place with a market.

Only a few of the samples found at Mayapán have fabrics similar in composition to samples from the other communities. It is likely that Mayapán production zone produced most of the jars and cajetes consumed at this centre. Mayapán did not need to supplement its production of these items, which can explain the observation that the vessels from the minor centres are rarely found at Mayapán.

That, on the other hand, red-slipped and striated vessels with Mayapán pottery fabrics are found at the minor centres can be explained by these communities trading with Mayapán. What is more, potentially, large amounts of Mayapán pottery were moved to the surrounding centres. For instance, of 22 thin sections analysed from Tipikal, 13 Mama and Yacman samples have a dolospar fabric (while no plain Navula samples from Tipikal have this fabric). As presented in section 2.1.3, door-to-door trade, itinerant merchants, sub-regional marketplaces, or other mechanisms may have been present.

Given that the north-central communities were also producing red-slipped and striated vessels apparently similar to those from Mayapán, the reasons for the demand for Mayapán vessels needs to be examined. Under the assumption of market exchange, three reasonable explanations for the high proportion (in the samples) of vessels with Mayapán fabrics found in the communities are given next.

It is possible that vessels from Mayapán were not the object of trade with these communities and that the contents of the vessels were more important. However, in a market exchange scenario in which the contents of the vessels might have been traded, it is puzzling that more ceramics from the minor centres were not found at Mayapán.

Alternatively, Mayapán vessels could have been desirable due to their religious, ritual, social, or political association with Mayapán, the major religious and political centre in the northern peninsula. Culber (1962, p.157) described a situation that may have parallels with Late Postclassic northern Yucatán. In San Gregorio, during the Early Postclassic, plain coarse pottery was produced by several communities, while finer utilitarian vessels were imported from major centres. Modern parallels are, for instance, Amatenango (Culbert 1962, p.149) and Chanal, Maya Highland (Deal 1998, p.69).

Lastly, given the observed correlation between pottery fabrics and vessels at the secondary north-central communities, a hypothesis to consider (and a topic for testing in future research) is one of a technological division of potters by raw materials and types of vessels; a division that may have extended to the community level. Community specialisation was observed by Thompson (1958) during his ethnographic work in the pottery towns of northern Yucatán. A division by community during the Late Postclassic remains a hypothesis given that the data in this research, as regard to the minor communities, gives only very coarse indications as to where the vessels were manufactured or the type of vessels that might have been manufactured at each centre. If a community division of potters by raw materials and vessels (as found by Thompson) was also true during the Late Postclassic, each pottery-making community did not produce the whole range of vessel types (as found ethnographically) and needed to supplement their pottery inventory. Market exchange of goods for pottery among the different minor communities and with Mayapán could supplement local production and explain the distribution of ceramics implied by the results.

In summary, a market exchange context can explain the distribution of red-slipped vessels and striated *ollas* having the same fabrics that are dominant in Mayapán vessels and found at the minor north-central centres in this study. However, the sphere of distribution of pottery with Mayapán fabrics has a limited spatial scale and does not represent a widespread regional exchange.

11.5.2 The Case for Gifting

Pottery was an integral part of the many fiestas and ceremonies of Yucatecan Mayas (see section 2.3). In this section, the distribution of Mayapán pottery is discussed in terms of reciprocal exchanges.

Reciprocal exchange represents an effort to re-enforce social relationships by gifting or feasting (Renfrew 1975, p.4), or to respond to social or political collective reciprocal obligations (Nancy Farriss 1992; Fernández Tejedo 1990, p.34; Stanton and Gallareta 2001) as discussed in section 1.1.1.4. These exchange mechanisms played an important role in Yucatecan Maya society and have been well documented in several accounts, including Bishop Landa's Relations at contact time (Landa 1986) and ethnographic work (Redfield and Villa Rojas 1967, p.127–159) during modern times.

As discussed in section 2.3.4, low volume of commodity movement, a small sphere of distribution, and heterogeneity of the sources characterise most reciprocal exchange (Hirth 1998, p.455). Concerning to the volume of pottery moved, the number of vessels transported from Mayapán could have been significant. Regarding the sphere of distribution, the extent of the geographical distribution of vessels from Mayapán was limited to one to two days' march. Lastly, concerning to the last of Hirth's variables, the sources are not heterogeneous. Only red-slipped vessels and Yacman *ollas*, all showing a rather homogeneous raw materials source, were taken to the minor centres. Considering the potentially high volume of transported pottery, reciprocity does not explain the observed movement of vessels to the smaller centres.

11.5.3 The Case for Rituals and Ceremonies

Section 2.3.4 shows the important role of ceramic vessels in Maya rites and ceremonies, as described for instance in the town of Chan Kom (Redfield and Villa Rojas 1967, p.193) or by Landa (1986). Pottery was produced for rites and ceremonies in which, most likely, large amounts of pottery were used, damaged, or broken.

Pottery could have been produced for rituals or festivities at the main centre and taken afterward to the communities. One ethnographic example provided by Reina and Hills (1978) at San José Petén (described in section 2.3) exemplifies pottery being transported from the main town to the minor centres. Cooking pots, usually of dispersed production, once a year were produced in the main town for cooking during the ceremony of the skull in which large amounts of food were consumed. After that, these cooking pots were transported from the main town to the minor centres. This scenario would explain the distribution found in the communities of red-slipped vessels and *ollas* having the same fabrics as the main centre, Mayapán. However, a better assessment of the proportions that Mama and Yacman with Mayapán fabric represent of the assemblage of the different minor sites is necessary to assess whether a ritual context is a viable scenario. Of note, a situation like the one at San José Petén blurs the lines separating definitions of utilitarian from ceremonial vessels. It is possible that a portion of the production of vessels that have been traditionally considered as 'utilitarian' was produced for 'ceremonial' functions.

The distribution of Mayapán pottery found in this study can also be explained using a more complex scenario proposed by Friedel (1981; or see section 2.3) in which market exchange and ritual activities are intertwined. From section 2.3, people from the centre and other communities were coerced or invited to elite-organised festivals and pilgrimages. The influx of people promoted the production of a variety of items including ceramics. It also promoted the existence of a market, and some people came ready to trade, while the rulers increased their wealth by imposing taxes to market transactions. The results of this study better fit a version of this model in which not all the communities but Mayapán and, perhaps, seat-of-the-katun towns (such as Teabo) were the focal point of peregrinations. For instance, Mayapán was the centre of Maya veneration for Cuculcán (Masson and Peraza Lope 2014, p. 88). In this scenario, vessels were produced in anticipation of these occasions to be used during the fiestas or ceremonies and to be traded (or to serve as containers) at the market, making their way to the communities. Furthermore, the ritual/market model proposed by Friedel (1981) can be linked to the scenarios of market exchange and ritual activity described earlier in this section.

11.5.4 The Case for Tribute

Mayapán was at the centre of the tributary system. Barrera Rubio (1984) has proposed that tribute was concentrated at the Maya centres and that the centres monopolised important raw materials, jobs, and manufacturers. The results of this study show that there was no significant movement of vessels from other centres to Mayapán. If a portion of tribute was paid with vessels or transported in vessels to Mayapán, it is not apparent from the results. Nevertheless, based on ethnohistorical accounts and the current state of research, it is likely that tribute and taxes were among the main mechanisms to maintain and increase the position of power of the rulers.

That red-slipped vessels and striated *ollas* from the main centre were transported to the communities can be explained by the tribute redistribution mechanism in which products from Mayapán were taken to the smaller centres. Redistribution could have been in the form of vessels, or goods transported in the vessels. However, there are no references to this type movement within the ethnohistorical record and, apparently, little or no goods were redistributed back to the people (Fernández Tejedo 1990; Barrera Rubio 1984).

On the other hand, redistribution could have taken the form of feasting organised by the community principals. The whole community could have participated or at least benefited. Feasting at Mayapán and the rest of north-central centres required large amounts of vessels. Vessels could have been produced at one centre and moved to others in a significant amount. The material remains left by this scenario are indistinguishable from the ritual scenario described in the previous section.

In summarising, market exchange and distribution related to rituals rather than reciprocal relationships and tribute, better explain the presence of Mayapán vessels at the minor centres, given the potential volume that could have been transported and the homogeneity of the source. In the next section, current models explaining Late Postclassic ceramic production arrangements are evaluated in light of the findings of this study and inferences discussed in this chapter.

11.6 Discussion of Findings: Research Question 6

How do the production, distribution, and exchange patterns described in the previous sections compare with current models for the organisation of Late Postclassic ceramic production, distribution, and exchange?

In this section, current models for the production, distribution, and exchange of ceramics (section 2.3) are examined to understand whether they explain the patterns found and the inferences drawn (summarised in sections 11.1 to 11.5) in this study.

11.6.1 The Mercantile Production and Distribution Model

The results of this investigation partially fit the mercantile model (section 2.3.1). The movement of vessels from Mayapán together with the uniformity of Mayapán pottery technology fit within the mercantile model of production and production for the Late Postclassic (section 1.1.1.3), as proposed by Sabloff and Freidel (1975) and Rathje (1975).

As mentioned in the previous section, in isolation, uniformity of ceramic production can result from more than one scenario. Uniformity can be interpreted as mass production under a cost-cutting strategy of standardisation. In the Mercantile model, during the Late Postclassic there was a shift from mere craft production to mass replication (Rathje 1975, p.441), in particular as a strategy to cut production costs.

On the other hand, the results do not support the notion of a widespread regional exchange of utilitarian ceramics throughout northern Yucatán from Mayapán, or from any other north-central sites, as contemplated in the mercantile view of the Postclassic (Rathje 1975; Sabloff and Freidel 1975). The distribution of the technological fabrics as found at Mayapán does not extend beyond one or two days' march. It was not found in the vessels sampled from Tepich and Teabo, nor was the fabric at any of the three eastern sites. In sum, the results support a more limited sub-regional version of the mercantile model, in which distribution of utilitarian pottery may have occurred from Mayapán to sites located within a limited spatial scale.

11.6.2 The Tributary Mode of Production Model

This model (section 2.3.3) was discussed as part of the answer to research question #5. Tribute is an important component of Late Postclassic socio-economy. However, the spatial distribution of the ceramic vessels sampled in this research does not fit well this model.

11.6.3 The Ritual Mode of Production Model

Concerning Mayapán pottery found at minor centres, one model that can explain the patterns observed is the production of ceramics for rituals and other ceremonies and fiestas (section 2.3.4). In section 11.5, three scenarios were presented that can explain the patterns observed in the data. In one, large amounts of red-slipped (Mama) and striated *ollas* (Yacman), with relatively homogeneous composition and macroscopic characteristics, were produced for rituals at the main centre and taken after that to the communities, as discussed in section 11.5. In the second scenario, large amounts of vessels were produced to be used during the fiestas and rituals linked to festivals and peregrinations, mainly, to Mayapán. In addition, Mayapán potters were producing vessels to be traded or used as containers at the market promoted by the influx of people; vessels that made their way to the communities.

11.6.4 The Calendrically Shifting Production Location Model

This model (section 2.3.5) has political and ritual components. In this model, elite ceramic items were manufactured at different locations according to a ritual calendar (based on the Native chronicles [Edmonson 1979]). These locations were 13 principal centres, which for a *katun* period (20 years), would be the religious capital (seat or host) and entitled to tribute.

The archaeological record left by paid tribute to the seat of the *may* or *katun* when the payment involves pottery, probably as containers (for honey, for instance), is not that different from the archaeological fingerprint of market exchange. It is possible that the movement of red-slipped vessels and striated *ollas* is the result of tribute sent in ceramic containers, or that the vessels function as containers for items paid by Mayapán to the acting religious capital. However, most minor sites in this study probably did not qualify as host of the *katun*, according to the list of towns provided by Edmonson (1979). Teabo, nevertheless, is on that list.

11.6.5 The Dichotomous Ceramic Production Model

Under this model (section 2.3.2), the main centres are consumers in a decentralised market economy, and the production of utilitarian ceramics is dispersed and away from the main centres.

Mayapán as a substantial producer of all types of utilitarian pottery (evidence provided in section 10.2.2 points to production at or near Mayapán) not only for local consumption but also for the movement to other sites runs against a view of main centres as consumers (Fry 1980; Rands and Bishop 1980). Production within the main centre is consistent with Smyth and Dore's (1992) proposition of ceramic production in the non-domestic area of Sayil (Puuc area, Yucatán) during the Terminal Classic. It is also consistent with current findings about the production of, mainly, non-ceramic items at Mayapán (Masson and Peraza Lope 2014). On the other hand, Mayapán received few ceramics from other sites.

A production pattern in which Mayapán received few ceramics from other sites and was a major ceramic producer does not fit the dichotomous production model as discussed earlier (see Fry 1980 or section 2.3). This model expects that the main producers of subsistence products were located on the periphery of Mayapán, while this main centre was a consumer of pottery.

11.6.6 Summary of Comparison between Findings and Current Ceramic Production Models

In sum, two of the models discussed explain better than the rest the results and interpretations of this study. One is a scaled-down version of the mercantile model. Instead of the regional and inter-regional scope of exchange as in the originally proposed model (section 2.3.1), ceramics (or their contents) from Mayapán were exchanged and distributed to sites located relatively close to this centre. The results can equally be explained by a ritual mode of production. Two scenarios were given involving ritual activity that fit the results of this study.

11.7 Conclusions

A variety of organisations of ceramic production and several mechanisms for the exchange and distribution of vessels most likely co-existed during the Late Postclassic. The organising principles of Late Postclassic Maya society did not rest upon one but several mechanisms and institutions (Masson and Peraza Lope 2014, Graham 2011). The goal is to identify the many ways in which ceramic production was organised and understand better the relative contribution of each mechanism to Late Postclassic socio-economy.

11.7.1 Ceramic Production at Mayapán

This study presented evidence supporting the notion of Mayapán as a major producer of pottery. The production zone was located at or near Mayapán and produced the bulk of the pottery found at this main centre. This zone of production is associated with a mineralogically uniform area of raw materials procurement having a Group 1 chemical composition. For instance, after excluding Payil samples, 92 per cent of the chemically analysed and grouped samples from Mayapán have a Group 1 composition. Of the Group 1 samples, one type of fabric accounts for most of the vessels sampled from Mayapán, the dolospar fabrics.

Based on current research, it is likely that, despite the uniformity of composition, ceramic production at Mayapán was not centralised or supervised. Although the general uniformity of composition in the slipped and unslipped vessels from Mayapán can be interpreted as centralised ceramic production (with some peripheral ceramic production), the identified production areas for non-ceramic products occur in a domestic environment and do not present indications of supervised production (Masson et al. 2016, p.232). It is then likely that pottery production was not different. The exceptions are religious artefacts, such as censers and copper bells (Masson et al. 2016, p.232) for which production is linked to elite contexts.

An alternative explanation considers the involvement of the elite or other coercive or influential forces in promoting uniformity without direct supervision. Maya ethnographic sources show that whole communities can be involved in the household level production of uniform ceramics. Elite involvement could have been subtle, encouraging or expecting uniform-looking pottery. Uniformity on the pottery fabrics and stylistic features of the different vessel parallels uniformity in other material remains, and current Mayapán researchers have considered the idea of promoted uniformity. Masson and Peraza Lope (2014, p.550) have observed, for instance, that at Mayapán, one house plan dominates. The uniformity in the construction of outlying buildings located by the major city gates is another example (Masson and Peraza Lope 2014). These instances can be interpreted as a unified government programme in which Mayapán state may have been promoting a symbolism of political unity (Masson and Peraza Lope 2014, p.550). Lastly, *costumbre* or traditions is an influential force able of maintaining uniformity. These traditions pertain to the people who populated Mayapán. As described earlier in this document, the ethnographic record of the Maya shows the production of considerable amounts of rather uniform pottery within independent households.

One tenet of this current (Sánchez) research is that ceramic technology is not disconnected from the broader context of society and that pottery fabrics may have been the manifestation of Late Postclassic traditions, social understandings, or social processes. Therefore, artefacts are fossilised evidence for past traditions or ways of acting and seeing the world (Wobst 2000). They are evidence not only of how things were made but also of the actions taken to make the world conform to a mental schema not necessarily related with ethnicity or identity (Wobst 2000, p.41). A material record presenting those similarities could be interpreted as a situation in which there is a problem keeping the group together, and people's understanding of ethnicity or identity needed to be propped up by constant displaying of material (Wobst 2000:47).

The data suggest that Mayapán potters may have organised their work differently from the secondary centres. In the majority of samples from Mayapán, an association of temper and the type of vessels in this study (slipped and unslipped jars and cajetes) was not found.

11.7.2 Ceramic Production in the North-central Communities

Ceramic production in the north-central communities was dispersed, with many producers or production zones involved in the production of slipped and unslipped vessels. That ceramic vessels from the north-central communities present a correlation between pottery fabrics and the types of vessels (different types of materials and processing were used for the different vessels) makes a division in which potters either produced slipped vessels or cooking pots a likely scenario. Future research involving more ceramic types may uncover finer correlations and divisions in the production of vessels.

11.7.3 Distribution and Exchange Mechanisms

Market exchange and production for rituals better explain the distribution throughout the sites of the different fabrics. That the characteristic Mayapán pottery fabrics are found at the minor centres can be explained by people from the communities trading some of their local goods for Mayapán vessels. Considering that red-slipped vessels and *ollas* were also produced by some of the north-central communities, Mayapán vessels may have been preferred because of their association with this site and what they symbolized. It is also possible that the vessels function as containers for desired traded items.

Market exchange is not the only explanation that fits the potentially large amounts of Mayapán pottery found at minor centres. Two scenarios of production of ceramics for rituals and other ceremonies (see section 2.3.4 and 11.5) have been proposed to explain the patterns observed in the data. Given the periodical nature of many ceremonies and other rituals (Landa 1986), production for ritual also followed a repetitive and periodical schedule, and, potentially, large amounts of people from many communities were involved. As stated by Wells (2006), production for rituals could have been an engine for economic growth.

Mayapán vessels, however, did not have a place in regional markets. They were not found at Culubá, Cobá, and Chac Mool. They were not found at Tepich nor Teabo.

11.7.4 Technology and Traditions

The results of this study provide new insights into the technology involved in the manufacture of Late Postclassic ceramics. This study shows in the minor north-central communities the purposeful selection of raw materials and their processing. Potters followed traditions with a significant time depth that can be traced back to, at least, the Terminal Classic. Petrography has been crucial for the detection of cultural divisions reflected in the data.

A lack of congruence is observed between the Terminal-Classic to Postclassic changes and the ceramic technology found at the minor centres. Deep changes were observed in ceramic style and forms and architecture. However, the Late Postclassic potters who made the striated *ollas* in the minor north-central communities shared with potters of centuries past from Uxmal and Chichén Itza not only a common understanding of how *ollas* should be made but also cultural categories discriminating raw materials for slipped and cooking pots. Despite the collapse of Uxmal and Chichén Itza, some technological traditions and cultural categories present at these centres continued in the different north-central communities.

Mayapán, on the other hand, represents a break from this long-lived differentiation in production technology lacking the discrimination associated with the raw materials for slipped versus cooking pots. Mayapán potters may have originated outside of the northern-central area of Yucatán. It is likely that potters at north-central sites emulated red-slipped Mayapán style pottery while keeping their technological traditions.

At sites further to the east (Cobá, Chac Mool, and Culubá), and similar to Mayapán, a correlation between temper and the utilitarian type of vessels in this study was not observed. At Chac Mool and Culubá, all utilitarian vessels represented by the samples including Navula, Yacman, Mama, Cancun, and Xcanchakan were produced with a coarse macro-crystalline and sascab (micrite) tradition, likely from several production zones. The main division in ceramic composition observed at Chac Mool and Culubá is between Payil, which is fine-grained, and coarsely grained utilitarian ceramics.

11.7.5 Site Integration

The patterns discussed here indicate that complex relations and spheres of political, economic, reciprocal, or religious influence created patterns in the material record. Although these spheres are difficult to discern, a pattern of related centres is appearing.

A series of sites, located at one day march or two at the most, comprises the inner-most sphere of influence of Mayapán. Mayapán red-slipped and cooking vessels were transported within a limited geographical area that encompassed some but not all the north-central sites in this study. No long-distance movement of Mayapán vessels was observed. As suggested by the findings and the distribution of ceramics with Mayapán fabrics, the sites mentioned may correspond to an area mainly integrated by ritual, or sub-regional trade activities, or both.

It is possible that the distribution of Mayapán fabrics reflects a political area. The studied north-central centres were under Mayapán jurisdiction (Roys 1957) and for some of them, the Cocom was the head lineage during Mayapán times (Roys 1957). Although the people who introduced the Tases red-slipped ceramics may have had their roots in the eastern or southern areas of the peninsula (chapter 3), local north-central potters emulated the new ceramic forms and decorative style while producing the new ceramics using their traditional raw materials cultural categories.

Outside the most inner sphere of Mayapán influence, other centres (represented by Tepich and Teabo) were part of a second sphere in which potters were not different from the other north-central centres within the inner sphere, but apparently Tepich and Teabo did not participate with the other communities in their networks of trade or religious rituals. Potters from Tepich and Teabo were not different in that they emulated the macroscopic attributes of Mayapán pottery while selecting and processing the raw materials for their vessels following their traditional cultural categories. Potters from the north-central centres shared technological traditions, potentially, representing a unified cultural block, while some of these centres apparently did not participate in the same networks of trade or rituals.

Centres further to the east, such as Culubá and Cobá, represent a third type of sphere of Mayapán's influence. The ceramics from Mayapán and other north-central sites and the ceramics from the sites further to the east have many stylistic and formal commonalities. On the other hand, the selection and processing of raw materials differ markedly between these two areas. In addition, the results of this study indicate that there was no a significant movement of vessels from Mayapán to areas further to the east.

This does not imply that the east coast was disconnected from the central area. Connections with the east coast are demonstrated by the Payil samples found at Mayapán that are not local but an import from northern Belize. Furthermore, four Yacman samples and one red-slipped sherd found at Teabo, Tipikal, Tekit, and Telchaquillo have an unusual crystalline and grog-tempered fabric (Calcite-Grog-B/R-c) that points to Santa Rita Corozal, Belize (Figure 7-7) as the most common match in the MURR database. Santa Rita Corozal is a site at which grog and a similar fabric are common (Cecil 2012, p.18-20).

Storey and Storey (2016, p.116) have remarked that, when a crisis is a slow moving one, societies have time to adapt and that it is not uncommon for them to rebound after a catastrophe. Given time, a society can not only adapt but also regenerate the old structures into new ones. A new society based on the older one emerges, in a geographical location that is not necessarily the same. Furthermore, the cultural manifestation is not necessarily the same either. After the protracted collapse of the Classic centres, a somewhat different society emerged for which Mayapán is the type site.

The new Postclassic Maya world was not exclusive to northern Yucatán. The ceramic style, the international style, the effigy censer tradition, the world view and ancestor veneration, and the calendrical rituals joined the Maya from Yucatán, Guatemala, and Belize in a common Late Postclassic worldview (Storey and Storey 2016; Masson et al. 2006, p.193) that was not there before.

This study identified the same Late Postclassic ceramic technological traditions in modern Yucatán. According to Farriss (1984, p.395), that the Maya have been able of preserving their traditions, social order, and culture is owed to an agrarian way of life that is adaptable to demands of new rulers and a Yucatán region that has attracted little attention for settlers. The new (and at the same time very old) Late Postclassic order continued up to the present in Yucatán communities and the traditions of the Maya potter.

This study showed that a detailed multi-site scientific analysis of pottery and raw materials, using a combination of analytical methods complemented with experimentation and ethnographic research has advanced our understanding of Late Postclassic ceramic production, organisation, and exchange. It has also provided information with which to propose new hypotheses and new research directions. More important, it has advanced and informed current views about the socio-economic environment during the Late Postclassic.

11.8 Future Directions

This study raised questions about the ceramic technology and the social and economic environment during the Late Postclassic. The information gathered allowed to address the questions raised. However, it is clear from the discussions and results presented in the final chapters that this current study is a stepping-stone to progress in our understanding of the socio, political, and economic nature of the Late Postclassic and that many issues directly related to the research questions remain. Refinements to the research questions and new inquiries have emerged based on the findings of this study.

Some of the new inquiries and refinements to this project are presented below together with a proposed approach to address them. However, it will become clear from the approaches proposed that they are based on the comparison of the variability in ceramic technology either between different contexts within Mayapán or between Mayapán contexts and other sites. A detailed scientific study of the technological and compositional variability of Mayapán ceramic assemblage is needed. The ceramic assemblages from commoners, ceremonial, and elite contexts need to be compared. From each of these contexts, an expanded selection of forms needs to be analysed, and the same forms studied (when possible) in all contexts. The study of Mayapán ceramic technological variability would address basic (fundamental) questions about Mayapán technology. For that reason, the research questions for such study are not very different from this (Sánchez) study's research questions (section 1.4). The research methodology is not different either.

The rest of this section proposes future research aiming to advance our knowledge of the socio-economic and political structures of the Late Postclassic. For instance, this study found that many of the utilitarian vessels sampled from Mayapán show uniformity of technology, composition, and macroscopic appearance. Uniformity has also been observed in other aspects of Late Postclassic Maya society (Masson and Peraza Lope 2014). Were Mayapán rulers promoting uniformity of style? One way to investigate this question involves a detailed study of ceramic production and distribution patterns. This section deals with this and other emerging or long-standing questions. It proposes work addressing these questions and presents the implications that the proposed work might have to advance our understanding of the socio-economic and political structure of the Late Postclassic.

Question I: What was the role of the ruling classes in craft production? Although current research at Mayapán (Masson and Peraza Lope 2014) indicates little involvement of the ruling classes in non-pottery craft production, the observed uniformity of composition of Mayapán pottery warrants a closer look at this uniformity. From the perspective of ceramic production, one way to investigate Question I is by comparing the ceramic production patterns of the utilitarian pottery from Mayapán resulting from this current study with the results of future research examining ceremonial items from various locations within Mayapán. Of special interest is the attached workshop for the production of ceremonial items, described earlier (section 2.2.2), and whether the same paste was used for ceremonial and utilitarian vessels. The implications of the proposed research may shed light on the possible location of utilitarian ceramics production and the extent of elite control over ceramic production.

The proposed research will be enhanced if the technological comparisons extend to neighbouring centres. For example, if indications exist that ceremonial items were produced locally and at the peripheral centres (in addition to being produced at Mayapán), one inference that can be drawn is that elite exerted little centralised control over the production of ceremonial items.

Question II: Were Mayapán rulers promoting uniformity of style? At Mayapán, the uniformity of pottery composition and macroscopic appearance of the pottery fabrics characterised in this study may have been the manifestation of Late Postclassic traditions or social processes. As mentioned in this study, uniformity over extended geographical areas is also reflected in stylistic features and forms.

Future research similar to the current study but including a wider range of types of vessels can provide a finer-grained assessment of the technological variability within Mayapán. The proposed study may provide associations between technological characteristics and contextual information (such as the type of building, or location within Mayapán) to better assess the patterns that might exist linking ceramic production with other aspects of society.

Question III: Did the production of utilitarian ceramics follow the same path than the production of other crafts (such as shell beads, or lithic weapons) as described by Masson and Peraza Lope (2014) at Mayapán? Masson and Peraza Lope (2014) are finding direct evidence for craft production within Mayapán urban area, within the households; however, direct evidence of ceramic production is limited to ceremonial items. Although it is reasonable to propose, based on these findings for non-pottery crafts, that utilitarian ceramics were manufactured within Mayapán, the possibility that they were not produced within this centre cannot be completely ruled out.

Despite the progress in knowledge that the current (Sánchez) study represents, the precise location of production of utilitarian ceramics found at Mayapán is not known with certainty. Therefore, future research should aim to a much finer-grained determination of the place of raw materials collection and manufacture. To this end, future research can build upon the findings presented in the current study and focus on the investigation of vessels from Mayapán and centres in its vicinity. A multi-site research can be undertaken in a manner similar to the approach taken in the current study. The proposed study would differ from the current in that the studied sites are Mayapán and sites within around five km from Mayapán. This distance is based on the survey by Delgado Kú et al. (2014, p.27) of sites located in Mayapán's periphery.

In addition, the composition and technological characteristics of censers, figurines and other vessels from the ceremonial ceramics workshop, identified by Masson and Peraza Lope (2014) at Mayapán, can be compared to the characteristics of censers and other types of vessels (including the samples in the current study and other types of vessels) from other Mayapán structures. The results of the proposed research can potentially point to the locations of production for Mayapán pottery and contribute to a better understanding of the organisation of pottery production and the economic environment during the Late Postclassic.

Question IV: What is the chaine opératoire or manufacturing process of the different vessels in Mayapán assemblage? Through detailed scientific analysis and archaeological experimentation, future research can reconstruct the production processes of a wider variety of types of vessels from Mayapán. Of the utilitarian ceramics, slipped and unslipped jars and cajetes, as well as the unslipped striated ollas, comprise the bulk of the vessels sampled from Mayapán for the current study. Although these vessels are a significant fraction of the total Mayapán assemblage, many other types of vessels are found at this centre. It is expected that with a more extensive sampling more production patterns and more *chaine opératoires* will emerge. Research like this would allow refining the ceramic production patterns resulting from this current study and identifying new patterns.

Question V: Did Mayapán red-slipped (Tases) ceramics develop from local traditions? The same question can be asked for unslipped Mayapán ceramics. This study showed that long-term technological traditions were important factors driving the choices made by potters. The chaine opératoire of a given type of vessel sampled from several periods can provide insights as to the changes or permanence of technological characteristics through time, a macro-tradition (Aimers 2013, p.2). It was said earlier that the history of Mayapán is associated with the Postclassic and that the red-slipped ceramics from this site do not appear to develop from previous ceramics from northern Yucatán. On the other hand, many of the minor north-central centres show associations with several periods. From the compositional and technological analysis of a given type of vessel, such as red-slipped jars or cajetes or striated, unslipped ollas, from the Late Classic, Terminal Classic, and Postclassic, a timeline of changes in technological characteristics can emerge.

Question VI: Where did Mayapán ceramics come from? This long-standing question can be approached with a regional and inter-regional study examining ceramic production patterns. However, based on the findings of the current study (also see Cecil 2012), it is not expected that red-slipped vessels, for instance, display a common fabric or chemical characteristics across distant sites. On the other hand, surface analysis of forming techniques in this study (see section 9.1.6 with analysis of a large jar from a minor north-central site) points to a forming method described in ethnographic data from San José Petén, Guatemala (Reina and Hill 1978), coinciding with modern northern Yucatán forming techniques (Thompson 1958). These data show that forming techniques are culturally ingrained and that they have been preserved by generations of potters despite migrations over a large area. Research can attempt to track over a large area the Late Postclassic red-slipped traditions using technological analyses in which the examination of forming techniques take a preeminent role.

Question VII: What was the impetus behind the movement of pots from Mayapán to the minor centres? Based on the data gathered in the current research, market exchange or ritual activity are likely scenarios explaining the movement of pottery from Mayapán to north-central sites. However, future research with larger number of samples and more variety of vessels from Mayapán and other locations (Tipikal and Tepich for instance) may clarify further the nature of this connexion. Future analyses should consider including censers and other ceremonial items from Mayapán and minor centres. Their movement or lack of it may give indications as to the forces behind the movement of vessels from Mayapán to the adjacent centres.

Question VIII: What do all this mean from the perspective of the type:variety system used at Mayapán and in northern Yucatán in general? In Maya studies, a ware identifies a ceramic assemblage in which paste and surface characteristics are constant (Smith 1971a). The ware concept as applied by Smith (1971a) at Mayapán benefits from a high macroscopic homogeneity of the paste. At Mayapán, wares are then based mainly on stylistic characteristics. At a regional level (Northern Yucatán), macroscopic attributes of the paste divide inland wares with affinities to Mayapán from the eastern wares. However, this current research has shown that, for example, the observed homogeneity of paste displayed by Mayapan Red Ware sherds (Mama in this study) is not maintained at different levels of analyses, such as macroscopic, petrographic, and chemical analyses. Mayapán Red Ware (Mama) is associated with multiple fabric classes. The same can be said about Mayapán Unslipped Ware samples in this study (Navula and Yacman). Furthermore, some of the fabric classes are common to Mayapán Red and Mayapán Unslipped Wares. It is apparent that the complexity resulting from incorporating the fabric classes found in this study into the current type:variety system as used in northern Yucatán, either at the top or the bottom of the hierarchy, would deem useless the classification. A new direction that research may take is to explore concepts such as ceramic systems and macro-traditions (Aimers 2013, Aimers and Graham 2013) to probe their usefulness in describing social divisions, meanings, and interactions associated with the stylistic and technological groups.

BIBLIOGRAPHY

Adams, A.E., and MacKenzie, W.S., 2001. *A Colour Atlas of Carbonate Sediments and Rocks under the Microscope*. London: Manson Publishing.

Aimers, J.J., 2013. Introduction. In: Aimers, J. J., (ed.) *Ancient Maya Pottery: Classification, Analysis, and Interpretation*. University Press Florida, 1-10.

Aimers, J.J., Graham, E., 2013. Type-Variety on Trial: Experiments in Classification and Meaning Using Ceramic Assemblages from Lamanai, Belice. In: Aimers, J. J., (ed.) *Ancient Maya Pottery: Classification, Analysis, and Interpretation*. University Press Florida, 91-106.

Andrews, E.W., IV, 1975. Progress report on the 1960-1964 field seasons, National Geographic Society-Tulane University Dzibilchaltun Program. *Archaeological Investigations of the Yucatan Peninsula*. New Orleans: Middle American Research Institute Tulane University, 31, 23–68.

Andrews, A.P., Andrews, E.W., and Robles Castellanos, F., 2003. The Northern Maya Collapse and its Aftermath. *Ancient Mesoamerica*, 14, 151–156.

Andrews, A.P., and Robles Castellano, F., 1985. Chichen Itza and Coba: An Itza-Maya Standoff in Early Postclassic Yucatan. *In*: Chase, A.F. and Rice, P.M., (eds.) *The Lowland Maya Postclassic*. Austin: University of Texas Press, 62–72.

Andrews, E.W. and Sabloff, J.A., 1985. Classic to Postclassic: A Summary Discussion. *In*: Late Lowland Maya Civilization. Albuquerque: University of New Mexico Press, 433–456.

Arnold, D.B, 1971. Ethnomineralogy of Ticul, Yucatán Potters: Emics and Ethic. American Antiquity, 36 (1), 20–30.

Arnold, D.B, 2005a. Maya Blue and Palygorskite: A Second Possible pre-Columbian Source. *Ancient Mesoamerica*, 16, 51–62.

Arnold, D.B., 2005b. Linking Society with the Compositional Analysis of Pottery: A Model from Comparative Ethnography. *In*: Livingstone Smith, A., Bosquet, D., and Martineau, R., (eds.) *Pottery Manufacturing Processes: Reconstitution and Interpretation*. Oxford: BAR Publishing, 15–21.

Arnold, D.B., 2008. *Social Change and the Evolution of Ceramic Production and Distribution in a Maya Community*. Boulder, Colorado: University Press.

Arnold, D.E., and Bohor, B.F., 1977. An Ancient Clay Mine at Yo' K'at, Yucatán. *American Antiquity*, 42(4), 575–582.

Arnold, D.E., Bohor, B.F., Neff, H., Feinman, G.M., Williams, P.R., Dussubieux, L., and Bishop, R., 2012. The first direct evidence of pre-Columbian sources of palygorskite for Maya Blue. *Journal of Archaeological Science*, 39 (7), 2252–2260.

Arnold, D.E., Neff, H., and Bishop, R.L. 1991. Compositional Analysis and "Sources" of Pottery: An Ethnoarcheological Approach. *American Anthropologist*, 93 (1), 70–90.

Arnold, D.E., Neff, H.N., Bishop, R.L., and Glascock. M.D., 1999. Testing interpretive assumptions of neutron activation analysis: contemporary pottery in Yucatán, 1964-1994. *In*: Chilton, E., (ed.) *Material meanings: critical approaches to the analysis of style*, Salt Lake City: University of Utah Press, 61–84.

Arnold D.E., Neff, H., and Glascock, M.D., 2000. Testing assumptions of neutron activation analysis: communities, workshops, and paste preparation in Yucatán, Mexico. *Archaeometry*, 42(2): 301–316.

Arnold, D.E., Neff, H., Glascock, M.D., and Speakman, R.J., 2007. Sourcing the palygorskite used in Maya blue: a pilot study comparing the results of INAA and LA-ICP-MS. *Latin American Antiquity*, 18(1):44–58.

Barrera Rubio, A., 1984. Consideraciones sobre el modo de producción Asiático entre los Mayas. *In*: Barrera Rubio, P., (ed.) *El Modo de Produccion Tributario en Mesoamerica*, Escuela de Ciencias Antropológicas, México: Universidad de Yucatán, 203–375.

Bautista, F., and Zinck, J.A., 2010. Construction of a Yucatec Maya soil classification and comparison with the WRB framework. *Journal of Ethnobiology and Ethnomedicine*, 2010, 6–7.

Baxter, M.J., 1994. *Exploratory Multivariate Analysis in Archaeology*. Edinburgh University Press.

Beck, M.E., 2006. Linking finished ceramics to raw materials: oxidized colour groups for lowland desert clays. *KIVA: The Journal of Southwestern Archaeology and History*, 72(1), 93–118.

Bishop, R.L., Rands, R.L., and Holley, G.R., 1982. Ceramic compositional analysis in archaeological perspective. *Advances in Archaeological Method and Theory*, 5, 275–330.

Brainerd, G., 1958. *The Archaeological Ceramics of Yucatán*. Archaeological Records, Vol. 19, Berkeley: University of California.

Britton, H.T.S., Gregg, S.J., and Winsor, G.W., 1952a. The calcination of dolomite. Part I. —The kinetics of the thermal decomposition of calcite and of magnesite. *Transactions of the Faraday Society*, 1952, **48,** 63–69.

Britton, H.T.S., Gregg, S.J., and Winsor, G.W., 1952b. The calcination of dolomite. Part II.—The thermal decomposition of dolomite. Transactions of the Faraday Society, 1952, ***48,*** 70–75.

Bronitsky, G. and Hamer R., 1986. Experiments in ceramic technology: The effects of various tempering materials on impact and thermal-shock resistance. American Antiquity, 51(1), 89–101.

Brown, C.T., 1999a. Mayapán Society and Ancient Social Organization. Dissertation (PhD), Tulane University.

Brown, C.T., 2017. Yacman Striated. (Photograph). Maya Archaeology, Mayapán Ceramic Photos. Available from: https://sites.google.com/site/mayaarchaeology/Home/mayapan-ceramic-photos.

Brumfiel, E. and Earle, T., 1987. (eds.) Specialization, Exchange, and Complex Societies. Cambridge University Press, 1–9.

Cabadas-Baez, H., Solleiro-Rebolledo, E., Dedov, S., Pi, T. and Alcalá, J.R., 2010. The complex genesis of red soils in Península de Yucatán, Mexico: mineralogical, micromorphological, and geochemical proxies. *Eurasian Soil Science*, 43(13), 1439–1457.

Canuto, M.A. and Fash, W.L., Jr., 2004. The Blind Spot: Where the Elite and Non-Elites Meet. *In*: Golden, C.W. and Borgstede, G., (eds.), *Continuities and Changes in Maya Archaeology: Perspectives at the Millennium,* London: Routledge, 51–76.

Cecil, L.G., 2012. *Inter-Regional Socioeconomic and Sociopolitical Relations of the Postclassic Maya of North-Central Yucatán, México, Belize, and Guatemala*. NSF Final Report.

Chang, K.C., 1975. Ancient Trade as Economics or as Ecology. *In*: Sabloff, J. and Lamberg-Karlovsky, C.C., (eds), *Ancient Civilization and Trade*. Albuquerque: University of New Mexico Press, 211–224.

Chapman, A.M., 1959. *Puertos de Intercambio in Mesoamérica Prehispánica*. Instituto Nacional de Antropología e Historia, México.

Chase, A.F., 2004. Polities, Politics, and Social Dynamics: "Contextualizing" the Archaeology of the Belize Valley and Caracol. *In* Garber, J., (ed). *The Archaeology of the Belize Valley: Half a Century Later*. Gainesville: University Press of Florida, 320–334.

Chase, A.F., and Chase, D.Z., 1985. Introduction. *In*: Chase, A., and Rice, P. (eds), *The Lowland Maya Postclassic*. University of Texas Press, Austin, 1-8.

Chung, H., Barba, L., and Varela Torrecilla, C., 1995. Inferencias sobre distintivos sociales en Chichén Itzá a través del estudio de la cerámica Pizarra Chichén (Chichén Slate Ware). *In*: Varela Torrecilla, C., Bonor Villarejo, J.L. and Fernández Marquínez, M.Y. (eds), *Religión y sociedad en el área maya*. Sociedad Española de Estudios Maya.

Código de Mendoza, MS. Arch. Seden A.1, Fol. 62r, 113D, 64 Bodleian Libraries, University of Oxford, Available from: http://treasures.bodleian.ox.ac.uk/treasures/codex-mendoza/.

Cruz Alvarado, W.A., 2012. Distribución y Cronología Cerámica de Sitios Arqueológicos del Centro-Sur de Yucatán. Unpublished Thesis. Universidad Autónoma de Yucatán.

Culbert, T. P., 1962. *The Ceramic Sequence of the Central Highlands of Chiapas, Mexico.* Dissertation (PhD), The University of Chicago.

Dahlin, B.H., and Ardens, T., 2002. Modes of Exchange and Regional Pattern: Chunchucmil, Yucatán. *In*: Masson, M., and Freidel, D.A., (eds), *Ancient Maya Political Economies.* Walnut Creek, CA: Altamira Press, 249–284.

De Pablo Galán, L., 1996. Palygorskite in Eocene-Oligocene lagoonal environment, Yucatán, Mexico. *Revista Mexicana de Ciencias Geológicas.* 13(1), Universidad Nacional Autónoma de México, Instituto de Geología, México, D.F., 94-103.

Deal, M., 1998. *Pottery Ethnoarchaeology of the Central Maya Highland.* Salt Lake City: University of Utah Press.

Delgado Kú, P.C., Escamilla Ojeda, B.C, Peraza Lope, C., and Masson, M., 2014. Asentamiento y arquitectura de los sitios periféricos asociados a Mayapán, Yucatán. Primera Mesa Redonda del Mayab, Arquitectura y Sociedad entre los Mayas.

Deutchman, H.L. 1980. Chemical Evidence of Ceramics Exchange on Black Mesa. *In*: Fry, R. (ed), *Models and Methods in Regional Exchange.* 1st Ed. Society for American Archaeology, 119–134.

Dobres, M.A., 2010. Archaeologies of technology. *Cambridge Journal of Economics,* 2010, 34, 103–114.

Doran, J.E., and Hodson, F.R., 1975. *Mathematics and Computers in Archaeology.* Edinburgh University Press.

Drennan, R.D., 1996. *Statistics for Archaeologists: A Commonsense Approach.* New York: Plenum Press.

Edmonson, M.S., 1979. Some Postclassic questions about the Classic Maya. *In*: Robertson, M.G. and Jeffers, D.C. (eds), *Tercera mesa redonda de palenque.* 4, Palenque, Chiapas, Mexico: Pre-Columbian Art Research Institute, 9–18.

Farriss, N.M., 1992. *Maya Society under Colonial Rule: The Collective Enterprise of Survival,* Princeton, New Jersey: University Press.

Feathers, J.K., 1989. Effects of temper on strength of ceramics: response to Bronitsky and Hammer. American Antiquity, 54(3), 579–588.

Feinman, G.M., and Garraty, C.P., 2010. Preindustrial markets and marketing – an archaeological perspective. *Annual Review of Anthropology,* 39, 167-191.

Feinman, G., and Nicholas,L., 2010. A Multiscalar Perspective on Market Exchange in the Classic-Period Valley of Oaxaca. *Archaeological approaches to Market Exchange in Ancient Societies.* pp 85-98. University Press Colorado, Boulder, Colorado.

Fernandez Tejedo, I., 1990. *La Comunidad Indígena Maya de Yucatán: Siglos XVI y XVII,* México: Instituto Nacional de Antropologia e Historía.

Foias, A.E., 2004. The Past and Future of Maya Ceramic Studies. In: Golden, G.,and Borgstede, G., (eds), *Continuities and Changes in Maya Archaeology: Perspectives at the Millennium.* New York: Routledge Press, 143-175.

Folk, R.L., 1980. *Petrology of Sedimentary Rocks,* Austin, Texas: Hemphill Publishing Company.

Freidel, D.A., 1981. The political economics of residential dispersion among the lowland Maya. In: Ashmore, W. (Ed.), *Lowland Maya Settlement Patterns.* University of New Mexico Press, Albuquerque, pp. 371–382.

Freidel, D.A., 1985. Terminal Classic Lowland Mata: Successes, Failures, and Aftermath. In *Late Lowland Maya Civilization,* pp 409-430. University of New Mexico Press, Alburquerque.

Frini-Srasra, N., and Srasra, E., 2008. Effect of heating on palygorskite and acid treated palygorskite properties. *Surface Engineering and Applied Electrochemistry, 2008,* 44(1), 43–49.

Fry, R.E., 1980. Models of Exchange for Major Shape Classes of Lowland Maya Pottery. *In*: Fry, R. (ed.), *Models and Methods in Regional Exchange.* Society for American Archaeology. 3–18.

Garrido Euán, M.A., 2003. *Salvamento Arqueológico en la Carretera Kanasín-Oxkutzcab*, Tesis no publicada de licenciatura, Universidad Autónoma de Yucatán, Mérida, Yucatán.

Garza, S., and Kurjack, E., 1980. *Atlas Arqueológico del Estado de Yucatán*, INAH, México.

Gautam, S., 2001. *Earth's Materials: Minerals and rocks*, Prentice-Hall, New Jersey.

Gelbert, A., 2005. Reconnaissance des techniques et des methodes de faconnage par l'analyse des macrotraces: etude ethnoarcheologique dans la vallee du Senegal. *In*: Livingstone Smith, A., Bosquet, D., and Martineau, R., (eds.), *Pottery Manufacturing Processes: Reconstitution and Interpretation*. Oxford: BAR Publishing.

Gilli, A., Hodell, D.A., Kamenov, G.D., and Brenner, M., 2009. Geological and archaeological implications of strontium isotope analysis of exposed bedrock in the Chicxulub crater basin, northwestern Yucatán, Mexico. *Geology*, August, 2009, 37(8), 723–726.

Glascock, M.D., 1992. Characterization of archaeological ceramics at MURR by neutron activation analysis and multivariate statistics. In: Neff, H. (ed), *Chemical Characterization of Ceramic Pastes in Archaeology*. Madison: Prehistory Press, 11–26.

Glascock, M.D., Neff, H., and Vaughn, K.J., 2004. Instrumental neutron activation analysis and multivariate statistics for pottery provenance. *Hyperfine Interactions*. 2004, 154, 95–105.

Gosselain, O.P., 2000. Materializing identities: an African perspective. *Journal of Archaeological Method and Theory*, 7(3), 187.

Graham, E., 2011. *Maya Christians and Their Churches in Sixteenth-Century Belize*. Gainesville: University Press of Florida.

Graham, E., 2012. Control without Controlling. In Politics, History, and Economy at the Classic Maya Site of Motul de San José, Guatemala, edited by A.E. Foias and K.F. Emery, pp. 419-430. University Press of Florida, Gainesville.

Greensmith, J.T., 1989. Petrology of Sedimentary Rocks. 7th Ed., Oxford: University Printing House.

Hammond, N., 1978. Cacao and Cocineros: an overland trade route between the Maya highlands and lowlands In Mesoamerican communication routes and cultural contacts. *In*: Lee, T.A., and Navarrete, C., (eds.), *Paper of the New World Archaeological Foundation*. Provo, UT: Brigham Young University, 19-26.

Hanna, J.A., Graham, E., Pendergast, D.M., Hoggarth, J.A., Lentz, D.L., and Kennett, D.J., 2016. A new radiocarbon sequence from Lamanai, Belize: two Bayesian models from one of Mesoamerica's most enduring sites. Radiocarbon, Vol 58, Nr 4, 2016, p 771–794.

Hare, T.S., and Masson, M.A., 2010. *Pottery Assemblage Variation at Mayapán Residences*, Presented at the April 2010 SAA Meeting, St Louis, MO. USA.

Hildebrand, A.R., Penfield, G.T., Kring, D.A., Pilkington, M., Camargo, A., Jacobsen, S.B., and Boynton, W.V., 1991. Chicxulub crater: A possible Cretaceous/Tertiary boundary impact crater on the Yucatan Peninsula, Mexico. *Geology*, 19, 867–871.

Hirth, K.G., 1998. The distributional approach: a new way to identify exchange in the archaeological record. *Current Anthropology*, 39(4), 451–476.

Hirth, K.G., 2009. Craft Production, Household Diversification, and Domestic Economy in Prehispanic Mesoamerica. *In* In *Housework: Craft Production and Domestic Economy in Ancient Mesoamerica*, edited by Kenneth G. Hirth, Archaeological Papers of the Anthropological American Association, Vol 19, Issue 1, pp-13-32.

Howie, L., 2012. *Ceramic Change and the Maya Collapse: a study of pottery technology, manufacture and consumption at Lamanai, Belize*. Oxford: BAR Publishing.

Howie, L., Aimers, J., and Graham E., 2014. 50 left feet: The manufacture and meaning of effigy censers from Lamanai, Belize. In: M. Martinón-Torres (Ed.), *Craft and science: International perspectives on archaeological ceramics*. Doha, Qatar: Bloomsbury Qatar Foundation http://dx.doi.org/10.5339/uclq.2014.cas.ch5.

INEGI (Instituto Nacional de Estadística, Geografía, e Informática), 1983. *Carta geológica:* scale 1:250 000. Ciudad de México, Secretaria de Programación y Presupuesto.

INEGI (Instituto Nacional de Estadística, Geografía, e Informática), 1984. *Carta Edafológica*, scale 1:250 000. Ciudad de México, Secretaria de Programacıón y Presupuesto.

INEGI (Instituto Nacional de Estadística, Geografía, e Informática), 2005. Available from http://www.beta.inegi.org.mx/app/mapas/, accessed 19-Jan-2018.

INEGI (Instituto Nacional de Estadística, Geografía, e Informática), 2006. *Carta Geológica-Minera Mérida F16-10 Yucatan, Campeche y Quintana Roo,* scale 1:250 000, available from http://mapserver.sgm.gob.mx/Cartas_Online/geologia/116_F16-10_GM.pdf. Ciudad de Mexico, Secretaria de Programacıón y Presupuesto. Available from: https://mapserver.sgm.gob.mx/Cartas_Online/geologia/116_F16-10_GM.pdf.

INEGI (Instituto Nacional de Estadística, Geografía, e Informática), 2007. *Mapa Topografico F16-10*. Scale 1:250 000. Ciudad de México, Secretaria de Programacıón y Presupuesto. Available from http://www.beta.inegi.org.mx/temas/mapas/topografia/, accessed 2-Feb-2018.

INEGI (Instituto Nacional de Estadística, Geografía, e Informática), 2013. *Carta Edafológica Serie II Mérida F16-10*. Scale 1:250 000. Ciudad de México, Secretaria de Programacıón y Presupuesto.

Isphording, W.C., 1984. The clays of Yucatan, Mexico: a contrast in genesis. *In*: Singer, A., and Galan, E. (eds), *Palygorskite-Sepiolite: occurrences and uses*. Amsterdam: Elsevier Science Publishers, 59–73.

Isphording, W.C. and Wilson, E.M., 1974. The relationship of "volcanic ash," sak lu'um, and Palygorskite in Northern Yucatan Maya ceramics. *American Antiquity*, July, 1974, 39(3), 483–488.

Jeffra, C.D., 2015. Experimental approaches to archaeological ceramics: unifying disparate methodologies with the chaîne opératoire. *Archaeological and Anthropological Sciences*. 7, 141–149.

Jordan, P., 2015. *Technology as human social tradition. Cultural transmission among hunter-gatherers. Origins of Human Behaviour and Culture*. University of California Press, Oakland.

Kemp, R.A., 1985. Soil Micromorphology and the Quaternary. Quaternary Research Association Technical Guide No. 2. Cambridge.

Kepecs, S., 1999. The political economy of Chikinchel, Yucatan, Mexico: a diachonic analysis from the Prehispanic era through the age of Spanish administration. Unpublished Ph.D. dissertation. University of Wisconsin-Madison.

Landa, F.D. de, 1986. *Relación de las Cosas de Yucatán*, 13th ed. México: Editorial Porrua.

Lefticariu, M., Perry1, E.C., and Lefticariu, L., 2003. Diagenetic processes in the Cenozoic sedimentary formations associated with the Chicxulub Impact Crater, northwestern Yucatan Peninsula, Mexico. Presented in The Second International Conference on Saltwater Intrusion and Coastal Aquifers—Monitoring, Modeling, and Management. Mérida, Yucatán, México, March 30–April 2, 2003.

Lemonnier, P., (ed.) 1993. Introduction. *In*: *Technological Choices: Transformations in Material Cultures since the Neolithic*. London: Routledge, 1–35.

Littmann, E., 1958. Ancient Mesoamerican mortars, plasters, and stuccos: the composition and origin of sascab. *American Antiquity*, 24(2), 172–176.

Liu, H.B., Chen, T.H., Chang, D.Y., et al., 2013. Effect of rehydration on structure and surface properties of thermally treated palygorskite. *Journal of Colloid and Interface Science*, 2013. 393, 87–91.

Livingstone, A. and Viseyrias, A., 2010. Shaping Kabambian pottery: identification and definition of technical features. *The Open Anthropology Journal*, 2010 (3), 124–141.

Lohmann, S., 2013. Rational choice and political science. *In*: Durlauf, S.N. and Blume, L.E. (eds),. *The New Palgrave Dictionary of Economics* Second Edition, Palgrave Macmillan, 2008. The New Palgrave Dictionary of Economics Online Palgrave Macmillan. 30 March 2017 Available from: <http://www.dictionaryofeconomics.com/article?id=pde2008_R000253>doi:10.1057/9780230226203.1386.

López, R.E., 1975. Geological summary of the Yucatán Península. *The Ocean Basins and Margins: The Gulf of Mexico and the Caribbean*. New York: Plenum Press. (3).

López Varela, S.L., McAnany, P.A., and Perry, K.A., 2001. Ceramics technology at Late Classic K'axob, Belize. *Journal of Field Archaeology*, 28.

Masson, M.A., 2001. Changing patterns of ceramic stylistic diversity in the pre-Hispanic Maya lowlands. *Acta archaeological*, 12(2), 159–188.

Masson, M.A., 2015. *In the realm of Nachan Kan: postclassic Maya archeology at Laguna De On, Belize.* First paperback edition. Boulder, Colorado: University Press of Colorado.

Masson, M.A., Hare, T.S., and Peraza Lope, C., 2006. Postclassic Maya Society Regenerated at Mayapán, In *After Collapse: The Regeneration of Complex Societies,* edited by Glenn M. Schwartz and John J. Nichols, pp. 188–207. University of Arizona Press, Tucson.

Masson, M.A., Hare, T.S., Peraza Lope, C., Escamilla Ojeda, B.C., Paris, E.H., Kohut, B., Russell, B.W. and Alvarado, W.C., 2016. Household craft production in the Pre-hispanic urban setting of Mayapán, Yucatan, Mexico. *Journal of Archaeological Research*, 24(3), 229–274.

Masson, M.A., and Peraza Lope, C., 2014. *Kukulcan's Realm: Urban Life at Ancient Mayapán*, Boulder: University of Colorado.

Masson, M.A., and Peraza Lope, C., 2004. Commoners in Postclassic Maya Society: social versus economic class constructs. *In*: Lohse, J.C. and Valdez, F., Jr (eds.), *Ancient Maya Commoners*. Austin: University of Texas Press, 197–224.

Masson, M.A. and Rosenswig R.M., 2005. Production characteristics of Postclassic Maya pottery from Caye Coco, Northern Belize. *Latin American Antiquity*, 16(40), 355–384.

Matthew, A.J., Woods, A.J., and Oliver, C., 1991. Spots before the eyes: new comparison charts for visual percentage estimation in archaeological material. *In*: Middleton, A.P., and Freestone, I.C. (eds.), *Recent Developments in Ceramic Petrology*. London: British Museum Occasional Paper 81, 265–276.

McKillop, H., 1996. Ancient Maya trading ports and the integration of long distance and regional economies: Wild Cane Cay in South-coastal Belize. *Ancient Mesoamerica*, 7(1), 49–62.

Mcintosh, R.M., Sharp, J.H. and Wilburn, F.W., 1990. The thermal decomposition of dolomite. *Thermochimica Acta.* 165(1990), Amsterdam: Elsevier Science Publishers B.V., 281–296.

Milbrath, S. and Peraza Lope, C. 2003. Revisiting Mayapán: México's last Maya capital. *Ancient Mesoamerica,* 14(1), 1–46.

Milbrath, S., Peraza Lope, C., 2009. Survival and revival of Terminal Classic traditions at Postclassic Mayapán. *Latin American Antiquity*. 20(4):581–606.

Morales Valderrama, C., 2005. La Alfarería de Yucatán: Una Tradición al Finalizar el Siglo XX. *In*: Merino Carrion, B. and Cook, A.G., (Eds) *La Producción Alfarera en el México Antiguo I*. Mexico: Instituto Nacional de Antropología e Historia.

Munsell, 2009. *Soil-colour Charts: with genuine Munsell colour chips*. Grand Rapids: Munsell Colour.

Nesse, W.D., 1986. *Introduction to Optical Mineralogy.* New York: Oxford University Press.

Ochoa Rodríguez, J.M., 2007. Las Esferas Cerámicas Tases del Posclásico en el Norte de la Península de Yucatán (CA. 1100-1550 D.C.). *In*: *La Producción Alfarera en el México Antiguo V.* Mexico: Instituto Nacional de Antropologia e Historia.

Okoshi Harada, T., 2010. La formación de las entidades políticas en las tierras bajas mayas del posclásico tardio: una nueva perspectiva. *Figuras Mayas de la Diversidad.* Mexico: Universidad Nacional Autónoma de Mexico.

Okoshi Harada, T., 2012. Las entidades políticas de las tierras bajas del norte al tiempo de la invasión española. *Revista Digital Universitaria*. 13(12).

Orton, C., Tyers, P., and Vince, A., 2008. *Pottery in Archaeology*. Cambridge University Press.

Peacock, D.P.S., (ed.) 1977. Ceramics in Roman and medieval archaeology. *In*: *Pottery and early commerce: characterization and trade in Roman and later ceramics*. London: University of Southampton Academic Press.

Pendergast, D.M., 1985. Stability through change: Lamanai, Belize, from the ninth to the seventeenth century. *In*: Sabloff, J.A., and Andrews, E.W. (eds), *Late Lowland Maya Civilization: Classic to Postclassic*. Albuquerque: University of New Mexico Press, 223–249.

Peraza Lope, C.A., 1993. *Estudio y secuencia del material cerámico de San Gervasio, Cozumel.* Tésis profesional, Mérida, Yucatán México: Universidad Autónoma de Yucatán.

Peraza Lope, C., and Barrera Rubio, A., 2006. La cerámica arqueológica de Kulubá, Yucatán. Los Maya de Ayer y Hoy. *In*: Rubio, A.B., and Gubler, R., (eds), Mexico: Eugenia Montalvan Proyectos Culturales S.C.P.

Peraza Lope, C., Masson, M. A., Hare, T.S., and Candelario Delgado Kú, P., 2006. The Chronology of Mayapan: New radiocarbon evidence. *Ancient Mesoamerica*, 17(2), 153–175.

Perry, E., Paytan, A., Pedersen, B., and Velazquez-Oliman, G., 2009. Groundwater geochemistry of the Yucatan Peninsula, Mexico: Constraints on stratigraphy and hydrogeology. *Journal of Hydrology*. 367(1), 27–40.

Perry, E., Velazquez-Oliman, G., and Marin, L., 2002. The hydrogeochemistry of the karst aquifer system of the Northern Yucatan Peninsula, Mexico. *International Geology Review*, 44, 191–221.

Philpotts, A.R., 2003. *Petrography of Igneous and Metamorphic Rocks*. Long Grove, IL: Waveland Press.

Pierce, D., and Glascock, M. D., 2015. *Neutron activation analysis of Late Postclassic Mayan ceramics from the northern Yucatan Peninsula, Mexico.* Archaeometry Laboratory, Research Reactor Center, Columbia: University of Missouri, unpublished report. Appendix C, this document.

Piña Chan, R., 1978. Commerce in the Yucatán Península: The conquest and colonial period. In: Lee, T.A., Jr., and Navarrete, C., (eds), *Mesoamerican Communication Routes and Cultural Contacts*. New World Archaeological Foundation. Provo, Utah: Brigham Young University. 37–48.

Pio Pérez, J., 1877. Diccionario de la lengua Maya. Mérida: Imprenta Literaria de Juan F.Molina Solis. Available: https:/books.google.com.

Pollock, H.E.D., 1952. Department of Archaeology. *Carnegie Institute of Washington*. Yearbook 51, 235–43. Washington.

Pope, K.O., Ocampo A.C., Kinsland, G.L., and Smith, R., 1996. Surface expression of the Chicxulub crater. *Geology*, 24(6), 527–530.

Rands, R. and Bishop, R., 1980. Resource procurement zones and patterns of ceramic exchange in the Palenque region, Mexico. *In*: Fry, R., (ed.) *Models and Methods in Regional Exchange*. Society for American Archaeology, 19–46.

Rathje, W.L., 1975. The last tango in Mayapán: a tentative trajectory of production-distribution systems. *In*: Lee, T.A., Jr., and Navarrete, C., (eds.) *Mesoamerican Communication Routes and Cultural Contacts*. Provo, Utah: New World Archaeological Foundation, Brigham Young University, 147–176.

Reents-Budet, D., 1994. *Painting the Maya Universe: Royal Ceramics of the Classic Period*. Durham, NC: Duke University Press, in association with Duke University Museum of Art.

Redfield, R. and Villa Rojas A., 1962. *Chan Kom: A Maya village, a classic study of the basic folk culture in a village in eastern Yucatan*. University of Chicago Press.

Reina, R.E., and Hill, R.M., 1978. *The Traditional Pottery of Guatemala*. Austin: University of Texas Press.

Rendón, S., 1947. Notas dobre la alfarería Indígena de la peninsula de Yucatán. *Revista Mexicana de estudios antropológicos*. 9, 107–123.

Rendón, S., 1948a. *El Jarro Yucateco. Revista Mexicana de Cultura, Epoca.* 2(78), 4 (Sep. 26 revista dominical en El Nacional), Mexico.

Rendón, S., 1948b. *El "kum" yucateco. Revista Mexicana de Cultura, Epoca.* 2(89), 4 (Dic. 5 revista dominical en El Nacional), Mexico.

Renfrew, C., 1975. Trade as action at a distance: Questions of integration and communication. In: Sabloff, J. and Lamberg-Karlovsky, C.C., (eds.), *Ancient Civilization and Trade*. Albuquerque: University of New Mexico Press, 3–59.

Restall, M., 1997. *The Maya World: Yucatec Culture and Society, 1550-1850*, Stanford University Press.

Rice, P.M., 1987a. *Pottery Analysis: A Sourcebook*. University of Chicago Press.

Rice, P.M., 1987b. Economic change in the Lowland Late Classic Period. In: Brumfiel, E.M., and Earle, T.K., (eds.) *Specialization, Exchange, and Complex Societies*. Cambridge: University Press, 76–85.

Rice, P.M., 2009a. On Classic Maya political economies. *Journal of Anthropological Archaeology*, 28, 70–84.

Rice, P.M., 2009b. Late Classic Maya Pottery Production: Review and Synthesis. *Journal of Archaeological Method Theory*, 16, 117–156.

Robles Castellano, J.F., 1990. *La Secuencia Cerámica de la Región de Coba, Quintana Roo*. INAH.

Robles Castellano, J.F., and Andrews, A.P., 1985. A review and synthesis of recent Postclassic archaeology in Northern Yucatán. *In*: Sabloff, J.A., and Andrews, E.W. V, (eds.) *Late Lowland Maya Civilization: Classic to Postclassic*. Albuquerque: University of New Mexico Press, 53–98.

Roux, V., and Courty, M., 1995. Identification of wheel throwing on the basis of ceramic surface feature and microfabrics. *Journal of Archaeological Science*. 1955(2), 17–50.

Roux, V., and Courty, M., 2005. Identifying social identities at a macro-regional level: Chalcolithic ceramics of South Levant as a case study. *In*: Livingstone Smith, A., Bosquet, D., and Martineau, R., (eds.), *Pottery Manufacturing Processes: Reconstitution and Interpretation*. Oxford: BAR Publishing, 201–214.

Roys, R.L., 1957. *The Political Geography of the Yucatán Maya*, Washington, DC: Carnegie Institution of Washington.

Roys, R.L., 1962. *Literary Sources for the History of Mayapán, Yucatán, Mexico*. Washington, DC: Carnegie Institution for Science, 24–86.

Roys, R.L., 1972. *The Indian Background of Colonial Yucatán*. Norman, OK, University of Oklahoma Press, reproduced from the first edition published by Carnegie Institution of Washington in 1943.

Ruck, L., and Brown, C.T., 2015. Quantitative analysis of Munsell color data from archeological ceramics. Journal of Archaeological Science: Reports 3 (2015) 549–557

Rye, O.S., 1981. Pottery *Technology: Principles and Reconstruction*. Washington, DC: Taraxacum.

Sabloff, J., 2007. It depends on how we look at things: New perspectives on the Postclassic period in the Northern Maya Lowland. *Proceedings of the American Philosophical Society*, 151(1), 11–26.

Sabloff, J.A., Freidel, D.A., 1975. A model of a Pre-Columbian trading center. *In*: Sabloff, J., and Lamberg-Karlovsky, C.C., (eds.), *Ancient Civilization and Trade*. Albuquerque: University of New Mexico Press, 469–408.

Sabloff, J.A. and Rathje, W., 1975. *Changing Pre-Columbian Commercial Systems: The 1972-1973 Seasons at Cozumel, Mexico. Monographs of the Peabody Museum, No. 3.* Cambridge: Harvard University.

Sabloff J.A., Smith R.E., 1997. The Importance of Both Analytic and Taxonomic Classification in the Type-Variety System. In: Lyman R.L., O'Brien M.J., Dunnell R.C. (eds) *Americanist Culture History*. Springer, Boston, MA.

Sackett, J., 1990. Style and ethnicity in archaeology: the case for isochrestism. In: Conkey, M., and Hastorf, C., (eds.), The Uses of Archaeology. Cambridge: Cambridge University Press, 32-43.

de Sahagún, B., 1946. Historia general de las cosas de Nueva Esparta. Notas por M. Acosta Saignes, 3 vols. México.

Sánchez del Río, M., Suarez, M., and García-Romero, E., 2009. The occurrence of palygorskyte in the Yucatán peninsula: ethno-historic and archaeological contexts. *Archaeometry*, 51(2), 214-230.

Sánchez Fortoul, C.G., 2009. Ceramics of Mayapán: A Petrographic Study. Master thesis, Boca Raton: Florida Atlantic University.

Sánchez Fortoul, C.G., 2013. Ceramic composition diversity at Mayapán, the last Maya capital. *Open Journal of Archaeometry*, 1(1), 4.

Sanders, W.T., 1960. *Prehistoric ceramics and settlement patterns in Quintana Roo, Mexico.* Contributions to American Anthropology and History, No. 60. Washington, Carnegie Institution of Washington.

Schultz, L., Shepard, A., Blackmon, P., and Starkey, H., 1971. Mixed-layer kaolinite-montmorillonite from the Yucatán Peninsula. *Clays and Clay Minerals,* 19, 137–150.

Shang, C., and Tiessen, H., 2003. Soil Organic C Sequestration and Stabilization in Karstic Soils of Yucatan. *Biogeochemistry*, February 62(2), 177–196.

Shannon, R.W., 2016. *XRD, elemental, and clay analysis of one Late Postclassic pottery sample from Mayapán,* Unpublished report, Appendix C, this document.

Shannon, R.W., 2017. *XRD analysis of a marl from Mayapán*, Unpublished report, Appendix A, this document.

Shennan, S., 1998. *Quantifying Archaeology,* Iowa City: University of Iowa Press.

Shepard, A.O., 1958. Ceramic Technology. *Carnegie Institution of Washington*, Yearbook 57. Washington, D.C.

Shepard, A.O., 1964. Temper Identification: "Technological Sherd-Splitting" or an Unanswered Challenge. *American Antiquity,* April 29(4), 518–520.

Shoval, S., and Beck, P., 2005. Thermo-FTIR spectroscopy analysis as a method of characterizing ancient ceramic technology. *Journal of Thermal Analysis and Calorimetry*, 82(3), 609–616.

Shoval, S., Gaft, M., Beck, P., and Kirsh, Y., 1993. Thermal behaviour of limestone and monocrystalline calcite tempers during firing and their use in ancient vessels. *Journal of Thermal Analysis and Calorimetry*, 40(1), 263–273.

Sillar, B., and Tite, M.S., 2000. The challenge of 'technological choices' for materials science approaches in archaeology. *Archaeometry,* 42, 2–20.

Smith, R.E., 1971a. *The pottery of Mayapán: Including studies of ceramic material from Uxmal, Kabah, and Chichen Itza.* Vol 1. Papers of the Peabody Museum. Vol 66. Cambridge, MA: Harvard University.

Smith, R.E., 1971b. *The pottery of Mayapán: Including studies of ceramic material from Uxmal, Kabah, and Chichen Itza.* Vol 2. Papers of the Peabody Museum. Vol 66. Cambridge, MA: Harvard University.

Smyth, M.P., and Dore, C.D., 1992. Large site archaeological methods at Sayil, Yucatan, Mexico: Investigating community organization at a prehispanic Maya center. *Latin American Antiquity,* 3:3(1), 3–21.

Stanton, T.W., and Gallareta Negrón, T. 2001. Warfare, ceramic economy, and the Itza: A reconsideration of the Itza Polity in ancient Yucatán. *Ancient Mesoamerica*, 12(2), 229–245.

Stark, M.T., 1991. Ceramic production and community specialization: a Kalinga ethnoarchaeological study. World Archaeology, 23(1), 64-78.

Stark, B.L., and Garraty, C.P., 2010. Detecting Marketplace Exchange in Archaeology: A Methodological Review. *Archaeological approaches to Market Exchange in Ancient Societies*. pp 33-60. University Press Colorado, Boulder, Colorado.

Storey, R., Storey, G.R. 2016. Requestioning the Classic Maya Collapse and the fall of the Roman Empire: slow collapse. In: Faulseit, R.K. (ed.) *Beyond Collapse: Archaeological Perspectives on Resilience, Revitalization, and Transformation in Complex Societies.* Carbondale: Southern Illinois University Press, 99-123.

Terrones González, E., 2006. El asentamiento prehispánico de Chac Mool, Quintana Roo. *In*: Márquez, L., Hernández, P., González, E., (eds.) *La poblacián maya costera de Chac Mool: análisis biocultural y dinámica demográfica en el Clásico Terminal y Posclásico* . México: Escuela Nacional de Antropología e Historia, 15–26.

Thompson, J.E., 1930. Ethnology of the Mayas of Southern and central British Honduras. *Publications of the Field Museum of Natural History. Anthropological Series*, 17(2), 27–213.

Thompson, R.H., 1958. Modern Yucatecan Maya Pottery Making. *American Antiquity,* 22(4).

Trindade, M.J., Dias, M.I., Coroado, J., and Rocha, F., 2009. Mineralogical transformations of calcareous rich clays with firing: a comparative study between calcite and dolomite rich clays from Algarve, Portugal. *Applied Clay Science*, 42(3/4), 345–355.

Urrutia-Fucugauchi, J., Morgan, J., Stöffler, D., and Claeys, P., 2004. The Chicxulub Scientific Drilling Project (CSDP). *Meteoritics & Planetary Science,* 39(6), 787–790.

U.S. Geological Survey, 2001. A laboratory manual for X-ray powder diffraction (Open-File Report 01-041). Online file http://pubs.usgs.gov/of/2001/of01-041/htmldocs/clays/seppaly.htm.

Wells, E.C., 2006. Recent trends in theorizing pre-hispanic Mesoamerican economies. *Journal of Archaeological Research*, 14, 265–312.

Weingast, B.R., and Wittman, D., 2008 *The Oxford Handbook of Political Economy*. (ed.). Oxford UP.

Wendrich, W. (ed.), 2012. *Archaeology and Apprenticeship: Body Knowledge, Identity, and Communities of Practice*. Tucson: University of Arizona Press, 1–19.

Wentworth, C.K., 1922. A scale of grade and class terms for clastic sediments. *The Journal of Geology,* July-August 30(5), 377–392.

West, G., 2002. Ceramic exchange in the late Classic and Postclassic Maya Lowlands: A diachronic approach. *In*: Masson, M., and Freidel, D.A., (eds.) *Ancient Maya Political Economies*. Walnut Creek, CA: Altamira Press, 140–196.

Whitbread, I.K., 1991. Image and data processing in ceramic petrology. *In*: Middleton, A.P., and Freestone, I.C. (eds.), *Recent Developments in Ceramic Petrology*. London: British Museum Occasional Paper 81, 369–388.

Whitbread, I.K., 1989. A proposal for the systematic description of thin-sections toward the study of ancient ceramic technology. *In: Archaeometry: Proceedings of the 25th Interrnational Symposium*. Amsterdam: Elsevier, 127–138.

Whitbread, I.K., 1995. Greek Transport Amphorae: A petrological and archaeological study. The British School at Athens, Fitch Laboratory Occasional Paper 4. Exeter, GB: The Short Run Press.

Whitbread, I.K., 2001. Ceramic petrology, clay geochemistry and ceramic production—from technology to the mind of the potter. In: Brothwell, D.R., Pollard, A.M. (eds.) Handbook of Archaeological Sciences. Chichester, New York: John Wiley & Sons, 449–459.

Whitbread, I.K., Ponting, M.J., and Wells, B. 2007. Temporal patterns in ceramic production in the Berbati Valley, Greece. *Journal of Field Archaeology,* 32, 177–193.

Willey, G.R., 1985. The Postclassic of the Maya Lowlands: A preliminary overview. *In*: Sabloff, J.A., and Andrews, E.W. V, (eds.) *Late Lowland Maya Civilization: Classic to Postclassic*. Albuquerque: University of New Mexico Press, 17-52.

Xie, J.J., Chen, T., Xing, B., Liu, H., Xie, Q., Li, H., and Wu, Y., 2016. The thermochemical activity of dolomite occurred in dolomite–palygorskite. *Applied Clay Science,* 119 (2016) 42–48.

APPENDIX A: XRD ANALYSIS OF THE CLAY PORTION OF A MARL FOUND AT MAYAPAN (S#11)

Pittsburgh
Mineral &
Environmental
Technology, Inc.

March 10, 2017

Ms. Carmen G. Sanchez
6172 NW 88th Ave.
Parkland, FL 33067

Dear Ms. Sanchez:

This report summarizes the results of clay speciation analysis of your soil sample. The sample material was received at PMET's laboratory on February 23, 2017 in a labelled plastic bag along with a PMET chain of custody document. A request for analysis and a sample description were received from Ms. Sanchez via email prior to receipt of sample.

The purpose of the analysis was to determine the species of clay present in the soil.

X-ray Diffraction Analysis

X-ray powder diffraction (XRD) analysis was used to determine the clay species.

The as-received sample was added to a large beaker of distilled water and allowed to soak overnight. Over the next 48 hours the material was alternately stirred with a glass rod and treated with an ultrasonic probe. The sample was then submerged in a column of deionized water, treated for five minutes with an ultrasonic probe, then allowed to settle for 72 hours.

The clay fraction was removed from the top of the column of settled fines using a pipette and deposited on glass slides. The glass slides were treated and analyzed by XRD according to USGS Bulletin 1563.

Clay speciation was determined by running separate scans on the dried oriented, glycolated, heated at 400°C, and heated at 550°C material. The XRD parameters for clay analysis are similar to the above, except that the clay patterns are scanned from 2.9°-21.7°. These traces are color coded black for the oriented slide, blue for the glycolated slide, red for the slide heated to 400°C, and dark red for the slide heated to 550°C. The traces are overlain for comparison to the USGS Clay Mineral Identification Flow Diagram.

700 Fifth Avenue
New Brighton, PA 15066
(724) 843-5000
FAX: (724) 843-5353
www.pmet-inc.com

Discussion

The differential clay analysis shows that montmorillonite, a smectite, is the major clay mineral. Montmorillonite is a swelling clay whose crystal spacing expands upon absorbing glycol vapor. There is a minor amount of kaolinite also present.

Ms.Sanchez,please email me if you would like to discuss these results. Thank you for using PMET's laboratory services on this project.

Sincerely,

Randolph W. Shannon
Laboratory Manager

RFA 6758

Table 1

Sample Identification, As-received Weight

PMET I.D.	Sample I.D.	Sample Description	As-received Wt. (g)
6758-1	S#11	marl	56.07

Table 2

Results of Clay Analysis Estimated Amount

Mineral	Atomic Formula	Amount
kaolinite	$Al_2Si_2O_5(OH)_4$	minor
montmorillonite	$Na_{0.3}(AlMg)_2Si_4O_{10}(OH)_2{\cdot}xH_2O$	major

Shannon 2017 p.3

Figure 2

Clay column and dried slid

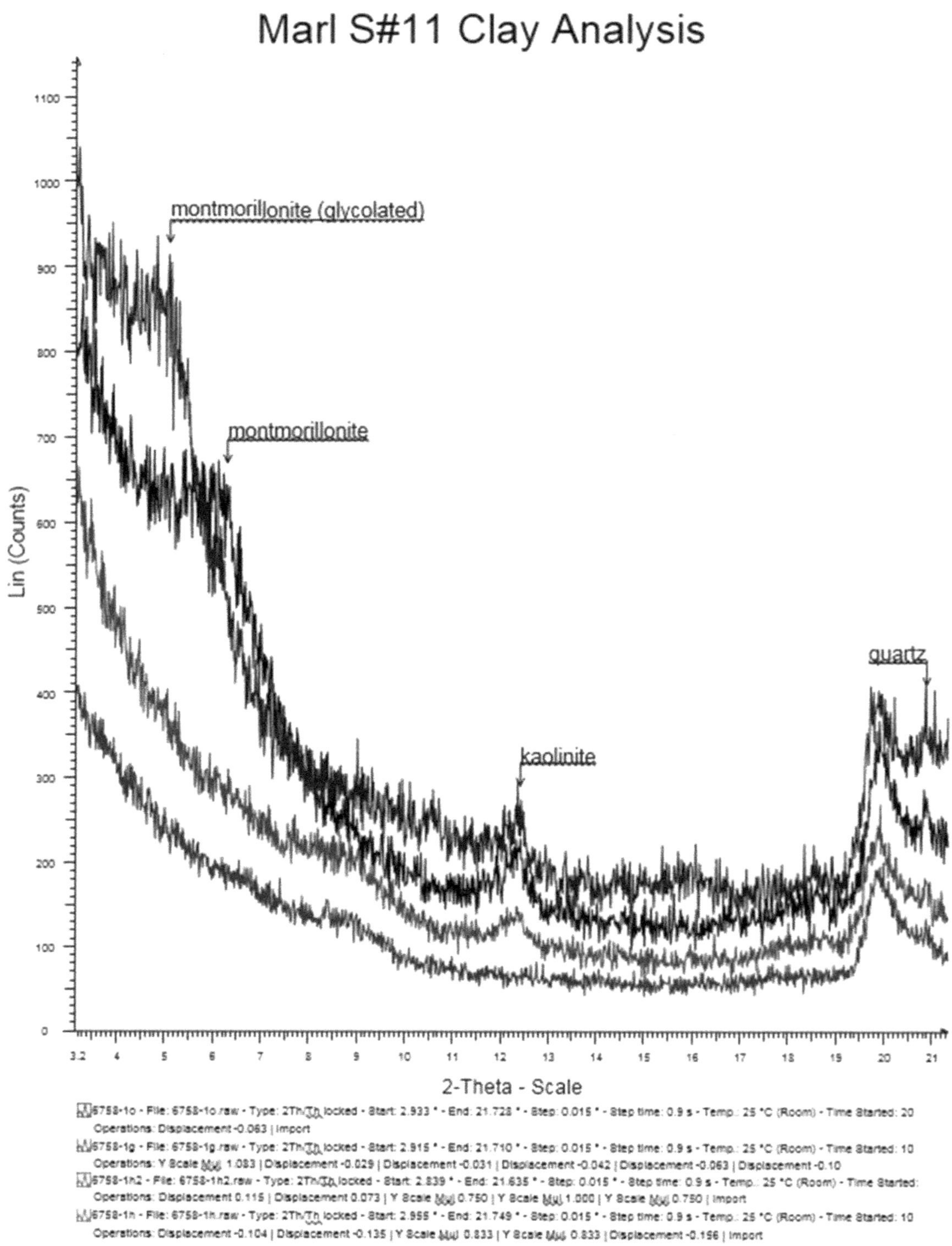

Figure 4

Clay Patterns

Shannon 2017 p.5

APPENDIX B: NEUTRON ACTIVATION ANALYSIS OF LATE POSTCLASSIC MAYAN CERAMICS FROM THE NORTHERN YUCATAN PENINSULA, MEXICO

Archaeometry Laboratory

Neutron Activation Analysis of Late Postclassic Mayan Ceramics from the Northern Yucatan peninsula, Mexico

Report Prepared by:

Daniel Pierce and Michael D. Glascock

Archaeometry Laboratory

Research Reactor Center

University of Missouri

Columbia, MO 65211

Report Prepared For:

Carmen G. Sanchez Fortoul

October 21, 2015

Introduction

This project involves the analysis of 186 pottery samples and 12 raw clay samples from the northern Yucatan peninsula, Mexico. The primary goal of this report is to address compositional variability within the sample and to better understand the consumption and production patterns based on compositional groups. In this report we describe sample preparation, data collection, statistical procedures, address chemical group assignments, and examine possible matches with other ceramic samples previously submitted and recorded in the MURR database to identify possible geographic origin of compositional groups.

This report describes the preparation, analysis, and interpretation of 186 samples of pottery (CGS001-186) and 12 raw clay samples (CGS187-198) from Late Postclassic contexts in the north of the state of Yucatan, Mexico. Specifically, the researcher intends to examine patterns of production and exchange in this largely geologically homogenous region. The sample derives from collections obtained from two case studies and excavated by the Instituto Nacional de Antropologia e Historia (INAH) as part of a salvage project resulting from road expansion. These artifacts and the results presented in the following report is intended to be used for analysis towards completion of Ms. Sanchez's doctoral dissertation. The sample consists of an array of decorative styles, vessel forms, and from multiple sites (Table 1).

Table 1. Late postclassic samples used for the current analysis.

Sample location	Count
Chacmool	12
Coba	13
Coluba	15
Mama	15
Mayapan	39
Teabo	10
Tocah	18
Tekit	15
Telchaquillo	19
Tepich	14
Tipikal	16
***Raw Clays*	12
Total	**198**

The sample was first analyzed in isolation to examine the chemical structure of the dataset. This serves several purposes. First, the chemical data must be analyzed for the purposes of accurately describing the chemical structure of the dataset. Further differences in composition may indicate different production areas even identifying exotic origin at times. Alternatively, while compositional differences may indicate an exotic origin of clays, strong chemical divergences such as enrichment or depletion and different chemical compositions may also be explained by differences in mixing clays and temper into a paste. Strong chemical patterning with regard to the site may also indicate local production and consumption at the site scale when compared to previous studies. Second, many of the statistical techniques employed to compare the current data to previously identified Mesoamerican reference groups have shortcomings that are more readily identified with detailed knowledge of the sample in isolation. Discriminant analysis, for example, has become the standard for ceramic comparison in the Mesoamerica. Discriminant analysis creates new axes that maximize the elemental differences among predefined groups. The current sample may, however, be differentiated from known Mesoamerican reference groups according to elements that are not heavily weighted in the existing discriminant analysis. In this case, one must examine possible group structure through different means (e.g. Mahalanobis calculations, principal component analysis, inspection of bivariate plots, and Euclidian distance searches) and then recalculate the discriminant analysis based on compositional group assignments.

Sample Preparation

Ceramic samples were prepared for INAA using procedures standard at MURR. Fragments of about 1 cm^2 were removed from each sample and abraded using a silicon carbide burr in order to remove glaze, slip, paint, and adhering soil, thereby reducing the risk of measuring contamination. The samples were washed in deionized water and allowed to dry in the laboratory. Once dry, the individual sherds were ground to powder in an agate mortar to homogenize the samples. Archival samples were retained from each sherd (when possible) for future research.

Two analytical samples were prepared from each source specimen. Portions of approximately 150 mg of powder were weighed into clean high-density polyethylene vials used for short irradiations at MURR. At the same time, 200 mg of each sample was weighed into clean high-purity quartz vials used for long irradiations. Individual sample weights were recorded to the nearest 0.01 mg using an analytical balance. Both vials were sealed prior to irradiation. Along with the unknown samples, Standards made from National Institute of Standards and Technology (NIST) certified standard reference materials of SRM-1633a (coal fly ash) and SRM-688 (basalt rock) were similarly prepared, as were quality control samples (e.g., standards treated as unknowns) of SRM-278 (obsidian rock) and Ohio Red Clay (a standard developed for in-house applications).

Irradiation and Gamma-Ray Spectroscopy

Neutron activation analysis of ceramics at MURR, which consists of two irradiations and a total of three gamma counts, constitutes a superset of the procedures used at most other NAA laboratories (Glascock 1992; Neff 1992, 2000). As discussed in detail by Glascock (1992), a short irradiation is carried out through the pneumatic tube irradiation system. Samples in the polyvials are sequentially irradiated, two at a time, for five seconds by a neutron flux of 8×10^{13} n cm^{-2} s^{-1}. The 720-second count yields gamma spectra containing peaks for nine short-lived elements aluminum (Al), barium (Ba), calcium (Ca), dysprosium (Dy), potassium (K), manganese (Mn), sodium (Na), titanium (Ti), and vanadium (V). The samples are encapsulated in quartz vials and are subjected to a 24–hour irradiation at a neutron flux of 5 x 10^{13} n cm^{-2} s^{-1}. This long irradiation is analogous to the single irradiation utilized at most other laboratories. After the long irradiation, samples decay for seven days, and then are counted for 1,800 seconds (the "middle count") on a high-resolution germanium detector coupled to an automatic sample changer. The middle count yields determinations of seven medium half-life elements, namely arsenic (As), lanthanum (La), lutetium (Lu), neodymium (Nd), samarium (Sm), uranium (U), and ytterbium (Yb). After an additional three- or four-week decay, a final count of 8,500 seconds is carried out on each sample. The latter measurement yields the following 17 long half-life elements: cerium (Ce), cobalt (Co), chromium (Cr), cesium (Cs), europium (Eu), iron (Fe), hafnium (Hf), nickel (Ni), rubidium (Rb), antimony (Sb), scandium (Sc), strontium (Sr), tantalum (Ta), terbium (Tb), thorium (Th), zinc (Zn), and zirconium (Zr). The element concentration data from the three measurements are tabulated in parts per million.

Interpreting Chemical Data

The analyses at MURR, described above, produced elemental concentration values for 33 elements in most of the analyzed samples. Nickel was removed from all statistical techniques due to the high number of missing values within the dataset.

Statistical analysis was subsequently carried out on base-10 logarithms of concentrations on the remaining 30 elements. Use of log concentrations rather than raw data compensates for differences in magnitude between the major elements, such as calcium, on one hand and trace elements, such as the rare earth or lanthanide elements (REEs). Transformation to base-10 logarithms also yields a more normal distribution for many trace elements.

The interpretation of compositional data obtained from the analysis of archaeological materials is discussed in detail elsewhere (e.g., Baxter and Buck 2000; Bieber et al. 1976; Bishop and Neff 1989; Glascock 1992; Harbottle 1976; Neff 2000) and will only be summarized here. The main goal of data analysis is to identify distinct and relatively homogeneous groups within the analytical database. Based on the provenance postulate of Weigand *et al.* (1977), different chemical groups may be assumed to represent geographically restricted sources. With pottery, however, chemical composition additionally varies according to the paste recipes that potters employ. A paste recipe reflect the cumulative pottery production steps from the selection of raw materials, preparation of those materials, the mixing of temper and clay, and even the firing of the pottery can affect the final recipe as changes in color and mineral structure can take place. For lithic materials such as obsidian, basalt, and cryptocrystalline silicates (e.g., chert, flint, or jasper), raw material samples are frequently collected from known outcrops or secondary deposits and the compositional data obtained on the samples is used to define the source localities or boundaries. The locations of sources can also be inferred by comparing unknown specimens (i.e., ceramic artifacts) to knowns (i.e., clay samples) or by indirect methods such as the "criterion of abundance" (Bishop *et al.* 1982) or by arguments based on geological and sedimentological characteristics (e.g., Steponaitis *et al.* 1996). The ubiquity of ceramic raw materials usually makes it impossible to sample all potential "sources" intensively enough to create groups of knowns to which unknowns can be compared. Lithic sources tend to be more localized and compositionally homogeneous in the case of obsidian or compositionally heterogeneous as is the case for most cherts.

Compositional groups can be viewed as "centers of mass" in the compositional hyperspace described by the measured elemental data. Groups are characterized by the locations of their centroids and the unique relationships (i.e., correlations) between the elements. Decisions about whether to assign a specimen to a particular compositional group are based on the overall probability that the measured concentrations for the specimen could have been obtained from that group.

Initial hypotheses about source-related subgroups in the compositional data can be derived from non-compositional information (e.g., archaeological context, decorative attributes, etc.) or from application of various pattern-recognition techniques to the multivariate chemical data. Some of the pattern recognition techniques that have been used to investigate archaeological data sets are cluster analysis (CA), principal components analysis (PCA), and discriminant analysis (DA). Each of the techniques has its own advantages and disadvantages which may depend upon the types and quantity of data available for interpretation.

The variables (measured elements) in archaeological and geological data sets are often correlated and frequently large in number. This makes handling and interpreting patterns within the data difficult. Therefore, it is often useful to transform the original variables into a smaller set of uncorrelated variables in order to make data interpretation easier. Of the above-mentioned pattern recognition techniques, PCA is a technique that transforms from the data from the original correlated variables into uncorrelated variables most easily.

PCA creates a new set of reference axes arranged in decreasing order of variance subsumed. The individual PCs are linear combinations of the original variables. The data can be displayed on combinations of the new axes, just as they can be displayed on the original elemental concentration axes. PCA can be used in a pure pattern-recognition mode, i.e., to search for subgroups in an undifferentiated data set, or in a more evaluative mode, i.e., to assess the coherence of hypothetical groups suggested by other criteria. Generally, compositional differences between specimens can be expected to be larger for specimens in different groups than for specimens in the same group, and this implies that groups should be detectable as distinct areas of high point density on plots of the first few components. It is well known that PCA of chemical data is scale dependent, and analyses tend to be dominated by those elements or isotopes for which the concentrations are relatively large. This is yet another reason for the log transformation of the data.

One frequently exploited strength of PCA, discussed by Baxter (1992), Baxter and Buck (2000), and Neff (1994, 2002), is that it can be applied as a simultaneous R- and Q-mode technique, with both variables (elements) and objects (individual analyzed samples) displayed on the same set of principal component reference axes. A plot using the first two principal components as axes is usually the best possible two-dimensional representation of the correlation or variance-covariance structure within the data set. Small angles between the vectors from the origin to variable coordinates indicate strong positive correlation; angles at 90 degrees indicate no correlation; and angles close to 180 degrees indicate strong negative correlation. Likewise, a plot of sample coordinates on these same axes will be the best two-dimensional representation of Euclidean relations among the samples in log-concentration space (if the PCA was based on the variance-covariance matrix) or standardized log-concentration space (if the PCA was based on the correlation matrix). Displaying both objects and variables on the same plot makes it possible to observe the contributions of specific elements to group separation and to the distinctive shapes of the various groups. Such a plot is commonly referred to as a "biplot" in reference to the simultaneous plotting of objects and variables. The variable inter-relationships inferred from a biplot can be verified directly by inspecting bivariate elemental concentration plots. [Note that a bivariate plot of elemental concentrations is not a biplot.]

Whether a group can be discriminated easily from other groups can be evaluated visually in two dimensions or statistically in multiple dimensions. A metric known as the Mahalanobis distance (or generalized distance) makes it possible to describe the separation between groups or between individual samples and groups on multiple dimensions. The Mahalanobis distance of a specimen from a group centroid (Bieber *et al.* 1976, Bishop and Neff 1989) is defined by:

$$D^2_{y,X} = [y - \overline{X}]' I_x [y - \overline{X}]$$

where y is the 1 x m array of logged elemental concentrations for the specimen of interest, X is the n x m data matrix of logged concentrations for the group to which the point is being compared with $\overline{X}$ being it 1 x m centroid, and I_x is the inverse of the m x m variance-covariance matrix of group X. Because Mahalanobis distance takes into account variances and covariances in the multivariate group it is analogous to expressing distance from a univariate mean in standard deviation units. Like standard deviation units, Mahalanobis distances can be converted into probabilities of group membership for individual specimens. For relatively small sample sizes, it is appropriate to base probabilities on Hotelling's T^2, which is the multivariate extension of the univariate Student's t .

When group sizes are small, Mahalanobis distance-based probabilities can fluctuate dramatically depending upon whether or not each specimen is assumed to be a member of the group to which it is being compared. Harbottle (1976) calls this phenomenon "stretchability" in reference to the tendency of an included specimen to stretch the group in the direction of its own location in elemental concentration space. This problem can be circumvented by cross-validation, that is, by removing each specimen from its presumed group before calculating its own probability of membership (Baxter 1994; Leese and Main 1994). This is a conservative approach to group evaluation that may sometimes exclude true group members.

Small sample and group sizes place further constraints on the use of Mahalanobis distance: with more elements than samples, the group variance-covariance matrix is singular thus rendering calculation of I_x (and D^2 itself) impossible. Therefore, the dimensionality of the groups must somehow be reduced. One approach would be to eliminate elements considered irrelevant or redundant. The problem with this approach is that the investigator's preconceptions about which elements should be discriminate may not be valid. It also squanders the main advantage of

\

multielement analysis, namely the capability to measure a large number of elements. An alternative approach is to calculate Mahalanobis distances with the scores on principal components extracted from the variance-covariance or correlation matrix for the complete data set. This approach entails only the assumption, entirely reasonable in light of the above discussion of PCA, that most group-separating differences should be visible on the first several PCs. Unless a data set is extremely complex, containing numerous distinct groups, using enough components to subsume at least 90% of the total variance in the data can be generally assumed to yield Mahalanobis distances that approximate Mahalanobis distances in full elemental concentration space.

Lastly, Mahalanobis distance calculations are also quite useful for handling missing data (Sayre 1975). When many specimens are analyzed for a large number of elements, it is almost certain that a few element concentrations will be missed for some of the specimens. This occurs most frequently when the concentration for an element is near the detection limit. Rather than eliminate the specimen or the element from consideration, it is possible to substitute a missing value by replacing it with a value that minimizes the Mahalanobis distance for the specimen from the group centroid. Thus, those few specimens which are missing a single concentration value can still be used in group calculations.

Summary of Interpretive Techniques

Typically, the approach used to interpret chemical data for pottery involves hierarchical cluster analysis (CA) and principal components analysis (PCA) to establish initial groupings within the sample. Further, to chemically characterize the sample, principal component analyses were useful in identifying which elements are the most significant in creating variation within the sample. After constructing base groups through CA and PCA, bivariate elemental plots were used to refine groups. Next, Mahalanobis distance based probabilities were calculated to assess likelihood of group membership. Initially, raw clay samples were excluded from compositional group construction. Later raw clay samples were integrated to see if they fit any compositional groups. This could indicate a local origin of that group. Finally, the groups were then compared to previously defined groups in the area by Leslie Cecil and the entirety of the MURR database.

To establish chemical groups, we first used a CA to identify clusters of chemically similar samples using 32 of the 33 total recorded elements for each sample. Nickel (Ni) was excluded due to a high frequency of missing values. From this analysis, five rough groups were identified. Following the CA, principle component analyses were used to define the sources of variation between specimens and maximize said variation. Through this, groups were further refined through visual inspection of PC loading plots. However, as shown below, the groups created based on PC loadings have compositional overlap that must be teased out with further analyses. Finally, the intersections of individual elements were projected on bivariate plots to better visually assess group membership. To evaluate these groups, boot strapped Mahalanobis distance calculations then tested each group member against its own group for robustness.

To identify significant correlations and possible geographic source areas for each group, we first considered unpublished NAA data provided by Leslie Cecil from various sites in the region. We

Pierce and Glascock 2015 p.7

then compared the groups to the MURR database to identify the most compositionally similar samples. Finally, using a bivariate plot inspection and Mahalonobis distance calculations, we projected the raw clay samples onto the identified compositional groups to identify local production potential.

Results and Source Assignments

The primary questions addressed here are: 1) What is the chemical group composition of the dataset? 2) Can we identify where each compositional may have originated geographically? 3) Does there seem to be any significant correlations between descriptive variables and assigned compositional groups?

The compositional groups are presented below. All plots below depict all specimens submitted for analysis. The complete results of source assignment for each specimen, can be found in Appendix A.

General Chemical Structure of the Pottery Sample

As an initial step toward interpreting the pottery sample, the dataset was considered in isolation. This allowed perusal of the general structure of the chemical data and identification of important variables for group formation. A PCA was conducted without Ni, (Figure 1). Ni has produced many missing values in the sample because the values were below detection limits

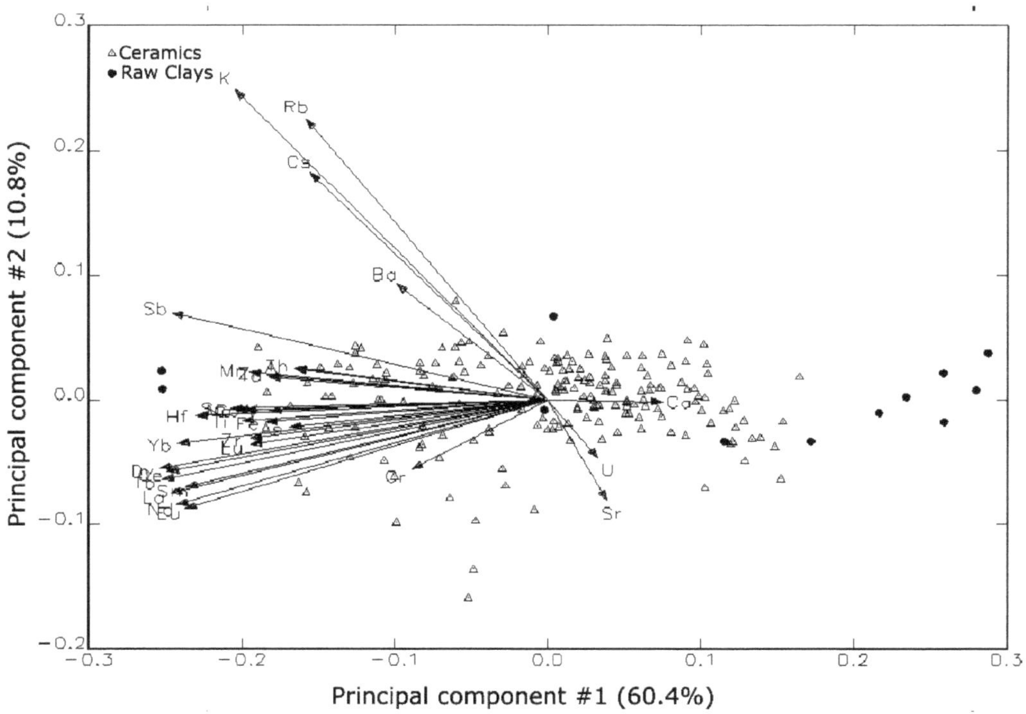

Figure 1. R-Q Mode biplot of the sample on Principal Component 1 and Principal Component 2.

The principal component loadings are presented in Figure 1 through an R-Q mode biplot of the principal component analyses found in Table 2. On PC1 (representing over 60.4 percent of the total sample variation), all elements display negative loadings except U, Ca, and Sr. High values among these elements have generated high PC1 values. It should be noted that cations of the alkaline earth metals (Ca, Sr, and Ba) commonly substitute for one another in the chemical structure of many clays (Ca and Sr in particular). Notably, that all recorded alkali earth metals are positively loaded and all other elements are negative. The first seven principal components alone explain approximately 90.8 percent of the variability in the sample.

Table 2. Elemental Loadings for the pottery sample on Principal Component Axes 1 through 5*.
*Values in **bold** explain the greatest amount of variation within each component. Those in *italics* explain a significant portion of the variation, but less than those in bold.

Variable	Mean	PC1	PC2	PC3	PC4	PC5	PC6	PC7
Al	42879.55	-0.151	0.054	-0.001	-0.054	-0.054	0.119	0.016
As	6.063139	-0.154	-0.047	0.253	*-0.366*	0.004	*-0.591*	0.127
Ba	122.659	-0.090	0.201	0.073	0.197	-0.029	-0.081	0.160
Ca	246083.2	**0.069**	-0.003	-0.057	0.025	0.052	-0.094	-0.083
Ce	28.18013	*-0.227*	-0.124	-0.085	0.082	0.087	0.050	-0.119
Co	2.342636	-0.181	-0.017	**0.308**	0.047	-0.007	-0.022	**0.210**
Cr	48.23178	-0.081	-0.120	**0.597**	-0.088	**0.217**	**0.266**	-0.114
Cs	1.468539	-0.141	**0.394**	0.178	0.051	0.076	0.089	*-0.340*
Dy	1.886598	*-0.231*	-0.118	-0.112	0.120	0.033	-0.093	0.037
Eu	0.400553	-0.216	*-0.188*	0.023	0.177	0.096	-0.017	-0.061
Fe	12607.15	-0.168	-0.038	0.189	-0.083	-0.055	-0.057	**0.173**
Hf	2.609019	-0.210	-0.027	-0.130	-0.103	-0.033	0.171	0.011
K	3127.403	-0.186	**0.537**	-0.021	0.133	-0.018	-0.065	-0.046
La	11.74533	-0.224	-0.161	-0.102	0.179	0.118	0.010	-0.196
Lu	0.181459	-0.177	-0.078	-0.027	0.038	0.021	-0.037	0.056

Mn	86.1236	-0.178	0.048	0.247	**0.353**	-0.227	-0.058	0.501
Na	693.3304	-0.186	-0.018	-0.033	-0.227	-0.693	**0.386**	-0.018
Nd	10.70512	-0.221	*-0.182*	-0.103	0.177	0.135	-0.030	-0.152
Rb	15.46212	-0.144	**0.488**	0.043	0.162	0.056	-0.056	-0.173
Sb	0.60416	-0.224	0.151	*-0.267*	*-0.448*	-0.003	*-0.278*	-0.045
Sc	4.624566	-0.189	-0.014	0.080	-0.063	0.037	0.090	0.033
Sm	2.314391	-0.216	-0.151	-0.100	0.129	0.113	-0.032	-0.107
Sr	301.3819	**0.035**	*-0.174*	0.181	**0.255**	*-0.561*	*-0.370*	*-0.511*
Ta	0.577449	-0.166	0.040	-0.023	-0.104	0.011	0.126	-0.065
Tb	0.316359	*-0.229*	-0.138	-0.114	0.123	0.058	-0.105	-0.004
Th	7.216566	-0.150	0.055	-0.034	-0.076	-0.003	0.110	-0.060
Ti	1446.887	-0.182	-0.035	0.134	-0.056	0.051	0.171	-0.028
U	1.521436	**0.030**	-0.101	**0.279**	-0.123	0.092	0.036	-0.252
V	33.55622	-0.166	-0.018	0.130	-0.315	0.044	-0.097	-0.113
Yb	1.179099	-0.220	-0.075	-0.107	0.079	-0.027	-0.072	0.136
Zn	26.60706	-0.167	0.043	-0.125	-0.073	-0.062	-0.009	0.040
Zr	64.26081	-0.177	-0.066	-0.101	-0.093	-0.022	0.167	0.010
Eigenvalues:		1.206	0.215	0.125	0.098	0.072	0.052	0.045
Total Variation explained:		60.42%	10.75%	6.25%	4.10%	3.61%	2.61%	2.24%

Visual inspection of principal component biplots resulted in blurry and overlapped chemical groups which were used to refine groupings from the Cluster Analysis (Fig 2.) Based on this PCA, it appears that Calcium (Ca), Strontium (Sr), Uranium (U), Antimony (Sb), and Chromium (Cr) explain the greatest amount of chemical variation across the first 5 PC's (85.94% of the variation cumulatively). Other elements have also been found to be useful in visualizing the sample into distinct groups (Figs. 3-6). However, it is apparent that significant chemical overlap between specimens necessitates a confirmation of group membership through other means. Visual inspection alone does not appear sufficient for confident group identification. For this reason, Mahalanobis distance calculations were also used for group final group assignment (Appendix B).

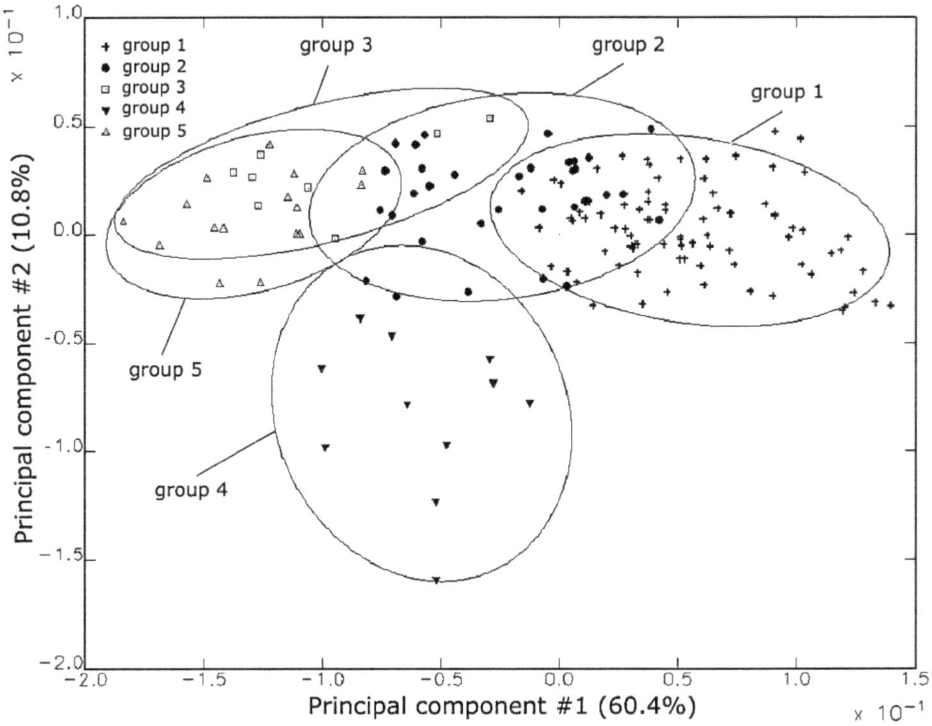

Figure 2. Results of PCA including compositional group assignment

Pierce and Glascock 2015 p.11

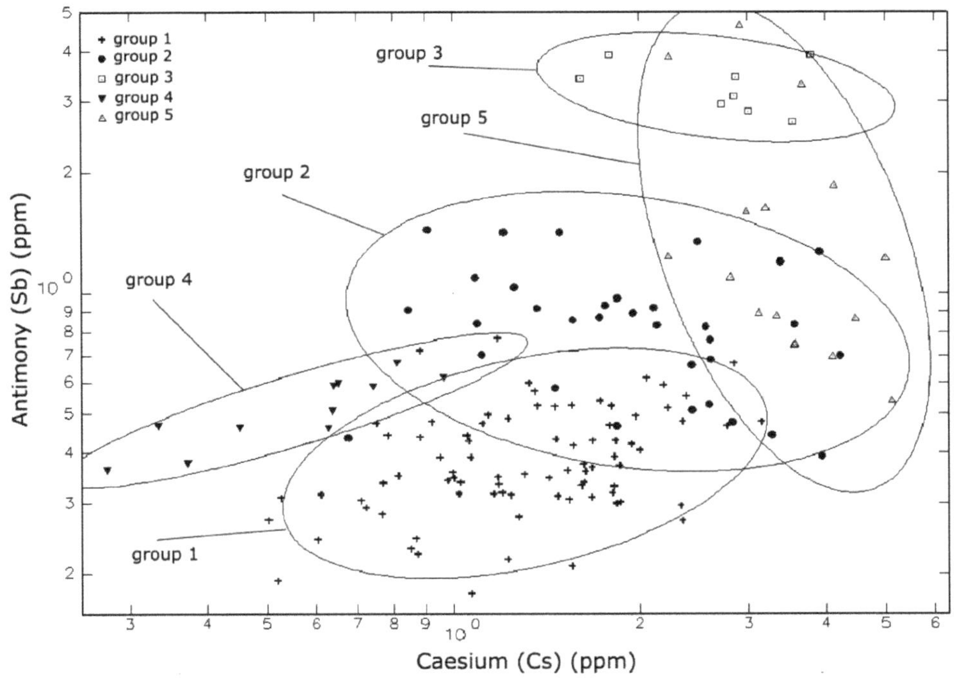

Figure 3. Bi-variate plot of the sample showing the chemical composition of sample on axes of Ce and Sb

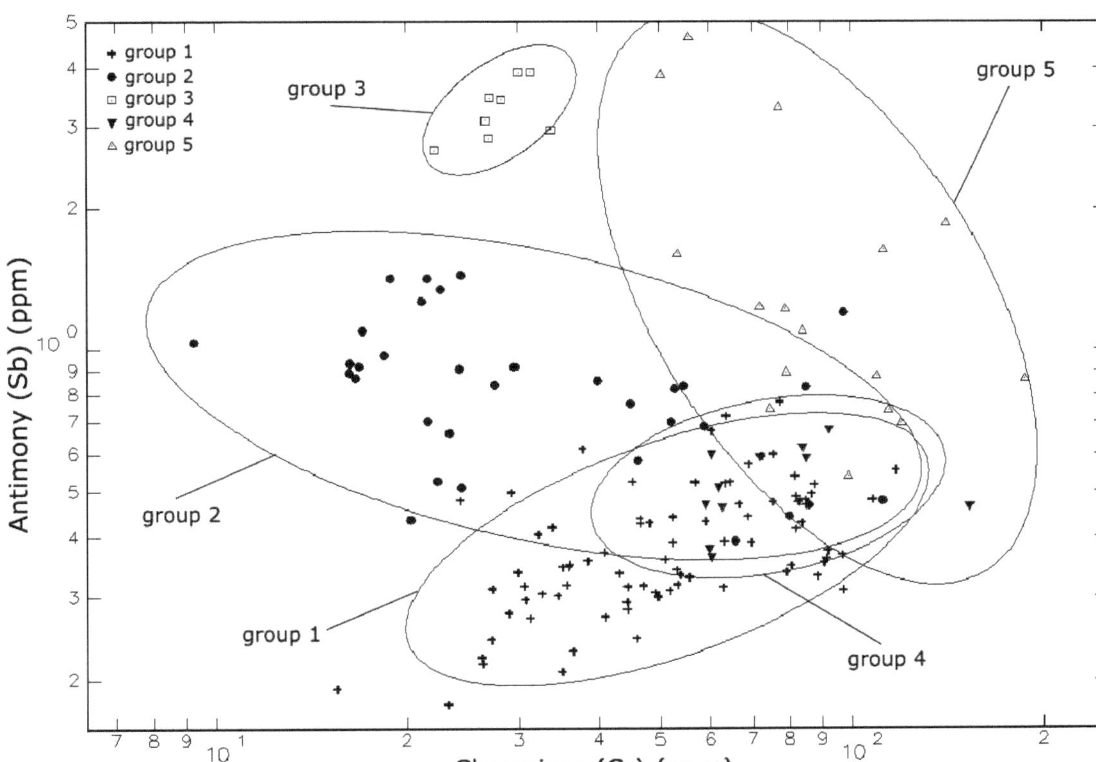

Figure 4. Bi-variate plot of the sample showing the chemical composition of sample on axes of Cr and Sb

Pierce and Glascock 2015 p.12

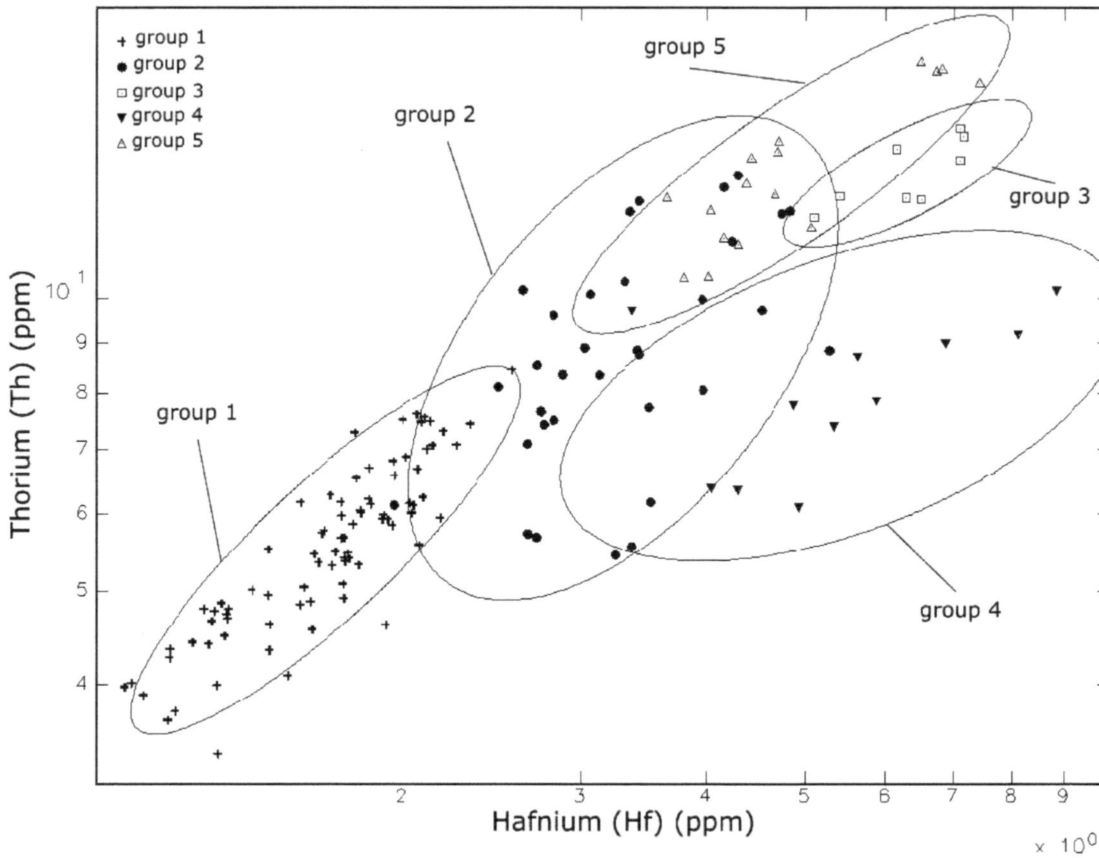

Figure 5. Bi-variate plot of the sample showing the chemical composition of sample on axes of Hf and Th

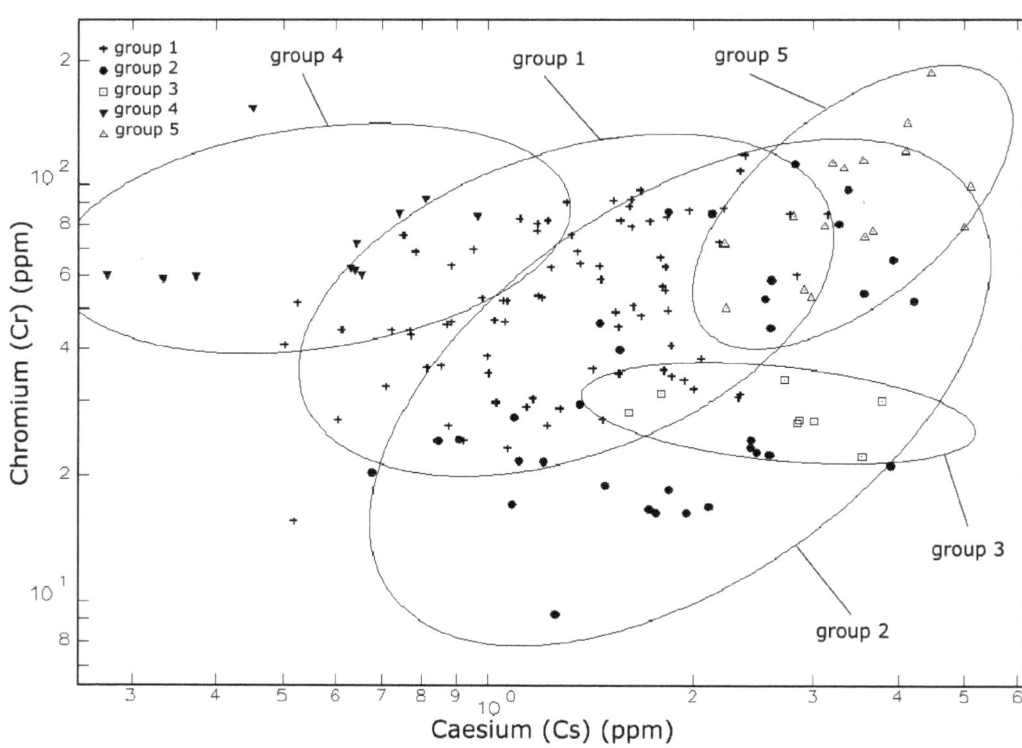

Figure 6. Bi-variate plot of the sample showing the chemical composition of sample on axes of Cs and Cr

Pierce and Glascock 2015 p.13

Group formation based on multiple forms of analysis

Groups were formed within the current sample through multiple analyses as described above and confirmed through the use of Mahalanobis distance calculations. The purpose for this is to provide a consistent, replicable characterization of the pottery sample. After defining chemical groups, a Discriminant Analysis (DA) was conducted using the identified groups (Table 3). This analysis maximizes the variation between groups by maximizing the internal homogeneity as well as external heterogeneity between groups (Fig. 7-8).

Table 3. Canonical Discriminant Analysis of five identified source groups in submitted sample

Element	CD1	CD2	CD3	CD4	
Hf	2.310776	-1.60964	0.673006	0.047925	1.51443
Eu	2.280309	0.334034	1.680123	1.044727	1.083495
Sc	2.102611	0.787027	-1.00333	-1.57547	-0.55927
Sm	1.977309	-0.53578	-1.7251	0.281903	0.753156
Ca	1.400715	0.134875	-0.00039	-1.01388	0.957004
Ta	1.34617	0.351432	0.122469	0.51623	1.186246
Fe	1.193836	-0.89353	0.625309	0.47681	-0.0921
Al	1.19221	0.04924	-1.17942	0.015156	-0.16633
Ce	1.109979	-0.39139	-0.16886	-0.60971	-0.82378
Co	0.951095	-0.1572	0.099531	0.245859	-0.89973
Zr	0.858273	-0.15096	-0.11138	-0.01913	-0.8373
Th	0.830975	0.377674	-0.49008	0.439059	-0.33901
La	0.820045	-0.41863	0.521863	0.060639	-0.47032
Mn	0.794753	0.177012	-0.35652	0.361587	0.58519
Ti	0.762231	0.077456	0.555818	-0.50762	-0.09155
Tb	0.752483	-0.17116	-0.19421	-0.69655	-0.11846
Lu	0.673204	-0.05302	-0.17339	-0.04565	-0.64672
V	0.524624	0.103455	-0.43829	0.24033	0.121129
Sb	0.509273	0.32793	-0.24592	0.109587	-0.28166
Yb	0.504536	0.228579	-0.18346	-0.11965	0.392856
Nd	0.453401	0.439592	-0.06846	0.02698	-0.08317
Dy	0.438237	0.371771	0.064964	0.190015	-0.11624
Cs	0.437149	0.356505	-0.10948	-0.16385	0.158651
Cr	0.432124	-0.2265	0.170409	0.312804	0.09243
Zn	0.41009	-0.04069	0.241641	-0.32881	-0.00356
Na	0.313108	-0.10785	0.282691	-0.01445	-0.07926
Ba	0.276911	0.176914	-0.17929	-0.063	0.09626
U	0.265894	0.128901	0.201191	0.011789	0.116048
As	0.216576	0.024781	0.040395	-0.15359	-0.14515
K	0.176339	-0.10696	0.092491	0.10387	-0.01768
Rb	0.166622	0.125921	-0.07283	0.080756	0.009
Sr	0.141245	-0.04131	0.005535	0.134925	0.002891

Wilk's lambda:	0.001301
Approx. F:	16.04409
p-value:	3E-113

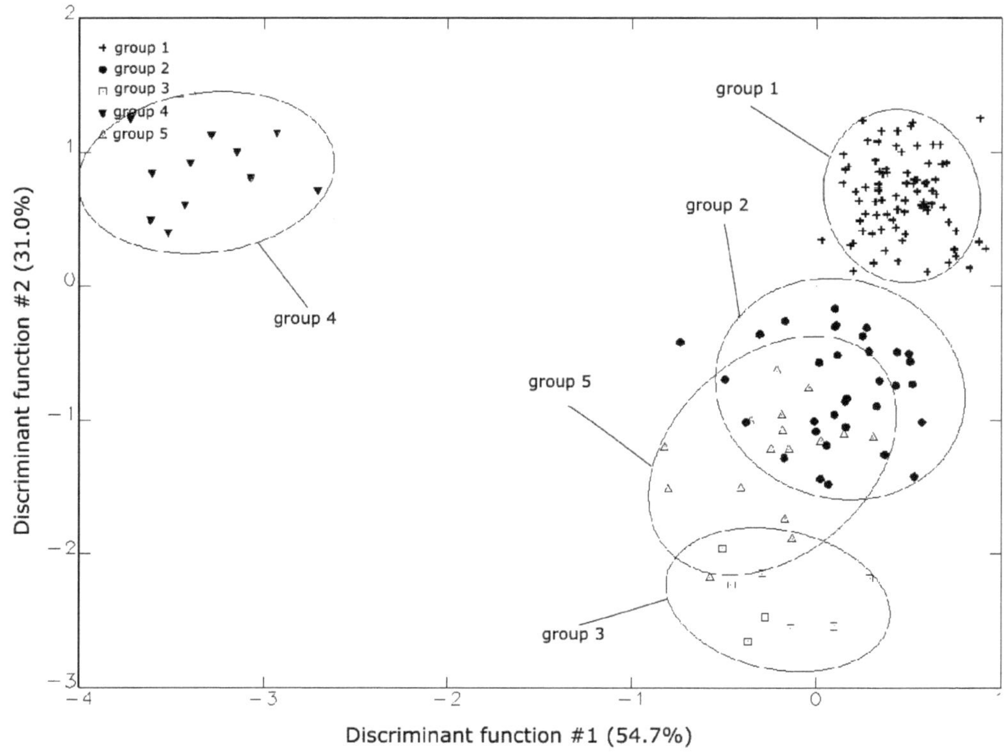

Figure 7. Bi-variate plot of the sample on axes of Discriminant functions 1 and 2.

Pierce and Glascock 2015 p.15

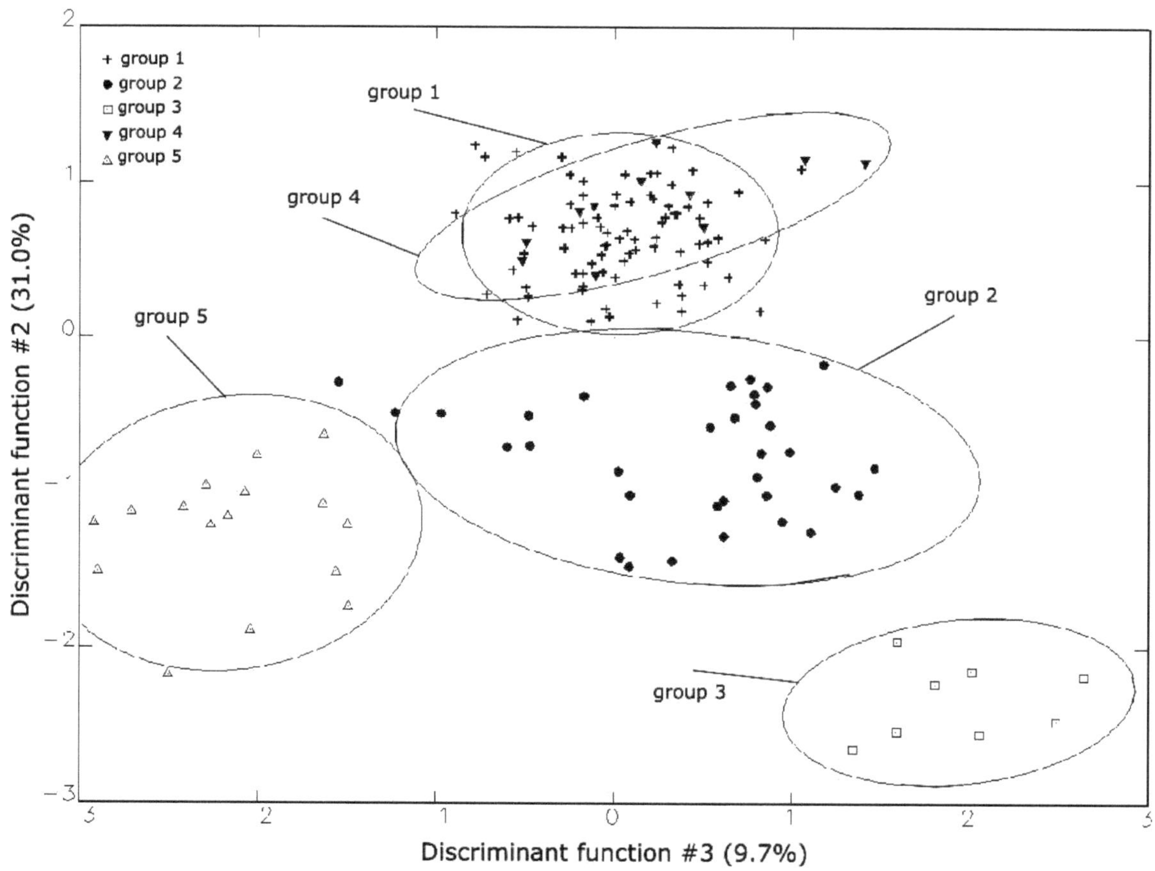

Figure 8. Bi-variate plot of the sample on axes of Discriminant functions 3 and 2.

As indicated here, some overlap occurs between the groups. However, this overlap can be accounted for when viewing the convergence of individual elements. For example, Groups 3 and 5 are greatly overlapped when considering only the PC loadings. But, they are clearly distinguishable based on Chromium (Cr) (see Figs. 4 and 6). Similarly, considering discriminant function 3 (Fig 8) clearly separates the groups. In the same way, groups 1 and 4 are nearly identical in discriminant function 2 and 3 (Fig. 8), but clearly different in Function 1(Fig. 7), PC loadings 1 and 2 (Fig 2), as well as the element Hafnium (Hf), among others (See Figs. 3, and 5). Needless to say, it is important to consider many variables when addressing compositional variability

After constructing source groups based on multiple forms of analysis, we have then a conducted a Mahalanobis distance calculations comparing the current sample to itself. The sample was tested jackknifed, as the likelihood of each specimen being a member of each chemical group was calculated by comparing said specimen to each identified group with that specimen removed. These tests confirmed cohesive group assignment (See Appendix B).

In general, our approach to group formation requires that the groups hold together when presented through multiple statistical techniques and multiple elemental and PCA plots. Since the first two principal components explain over 71% percent of the variability within the sample, we did not rely heavily on the less influential components to formulate compositional groups. It

Pierce and Glascock 2015 p.16

is now believed that the samples do in fact represent separate source groups despite retaining some chemical overlap at times, as these results above demonstrate.

Finally, the combination of Principle Component Analysis, visual inspection of bivariate elemental plots, Mahalanobis Distance Calculations, and Discriminant Analysis provided us with the final group assignment (Appendix A). However, it should be noted that no single type of analysis was in and of itself conclusive in all cases and informed decisions were made considering all analyses. Unassigned specimens were either dissimilar to all groups or may have been chemically consistent with more than one group. However, assignment of over 81.6% of the specimens is sufficiently high, with minimal unassigned specimens. In an attempt to geographically identify these chemical groups, further analysis was then conducted by comparing the collected raw clay samples, and Euclidian distance searches comparing this sample to all other samples from the region in the MURR NAA database.

Geographic identification of source groups

Next, we compared the five identified source groups, with the collected raw clay sample (CGS187-198) (Fig 9). Through visual inspection of the bivariate plot, it appears that only Group 1 is reflected in the raw clay sample. This is identified in a single specimen (CGS193).

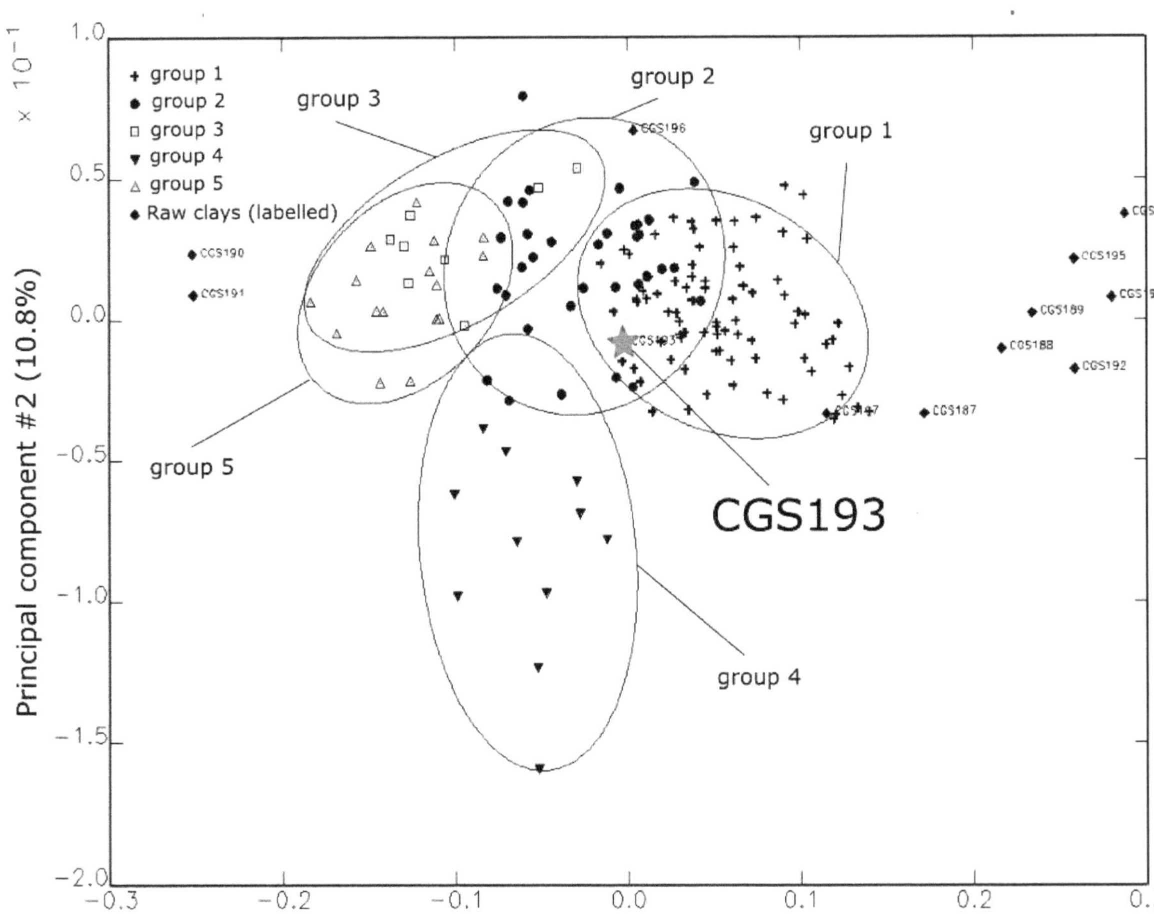

Fig 9. Comparison of Raw clay samples to compositional groups

To confirm the inclusion of clay samples in compositional group 1 rather than 2 as the ambiguity of Fig 9 suggest, a Mahalanobis distance analysis was conducted (Table 4). From these analyses, we can conclude that of the assignable raw clay specimens, compositional Group 1 appears to be the closest match for specimen CGS193. As expected from the above analysis, Mahalanobis distance calculations reveal it is unlikely that other specimens fit in any of these groups. Of note, this specimen was taken from Sascabera Mayapan B. Similarly, Euclidian distance searches suggest that group 1 is also from the site of Mayapan (see below). This provides a strong argument that this is in fact a resource acquisition location and potential production area. However, negative results for other raw clays do not preclude the possibility that they may in fact be source areas. We must remember that NAA is a bulk analytical technique. The data produced from NAA includes not only the composition of the clay, but also the temper, inclusions, etc. Rather, NAA is truly identifying the composition of the entirety of the paste; whereas clay samples reflect only raw clay. For this reason, it is possible that the clay contributing to the pastes may in fact come from the same location as the raw clays provided, and other inclusions may be obscuring the relationship analytically.

Table 4. Mahalanobis distance calculations for raw clay samples based on discriminant functions

MURRID	Alt ID	Group 1	Group 2	Group 3	Group 4	Group 5	Best Group	Source assignment
CGS187	Clay A	0.000	1.861	0.362	0.023	0.005	Group 2	unassigned
CGS188	Clay B	0.000	0.000	0.065	0.009	0.000	Group 3	unassigned
CGS189	Clay C	0.000	0.000	0.033	0.001	0.000	Group 3	unassigned
CGS190	Clay D	0.000	0.000	0.046	0.001	0.017	Group 3	unassigned
CGS191	Clay E	0.000	0.000	0.097	0.001	0.093	Group 3	unassigned
CGS192	Clay F	0.000	0.000	0.024	0.001	0.000	Group 3	unassigned
CGS193	Clay G	77.900	0.366	0.225	0.026	0.084	Group 1	Group 1
CGS194	Clay H	0.000	0.000	0.011	0.000	0.000	Group 3	unassigned
CGS195	Clay I	0.000	0.000	0.045	0.004	0.000	Group 3	unassigned
CGS196	Clay J	0.000	0.000	0.002	0.000	0.000	Group 3	unassigned
CGS197	Clay K	0.073	6.249	1.022	0.016	0.045	Group 2	unassigned
CGS198	Clay L	0.012	0.002	0.121	0.012	0.004	Group 3	unassigned

Finally, to address potential provenance of compositional groups, we have conducted further Euclidian distance searches based on all samples within the region found from the MURR NAA database. This type of analysis is in and of itself not conclusive as it must rely on the assumption of the criterion of abundance. However, it can produce quite compelling results, as we see here. In this analysis, each specimen is compared to every specimen in our database. For this report, we have inquired as to identify the ten closest matches chemically. Though not conclusive in and of itself, the implication is simply that if a high proportion of close matches come from the same area, the identified compositional group may have originated there as well. At times, this can provide useful information and hint at potential provenance. The samples used for comparison are the most proximal geographically found in the database and are generally split into three regions (Fig 10).

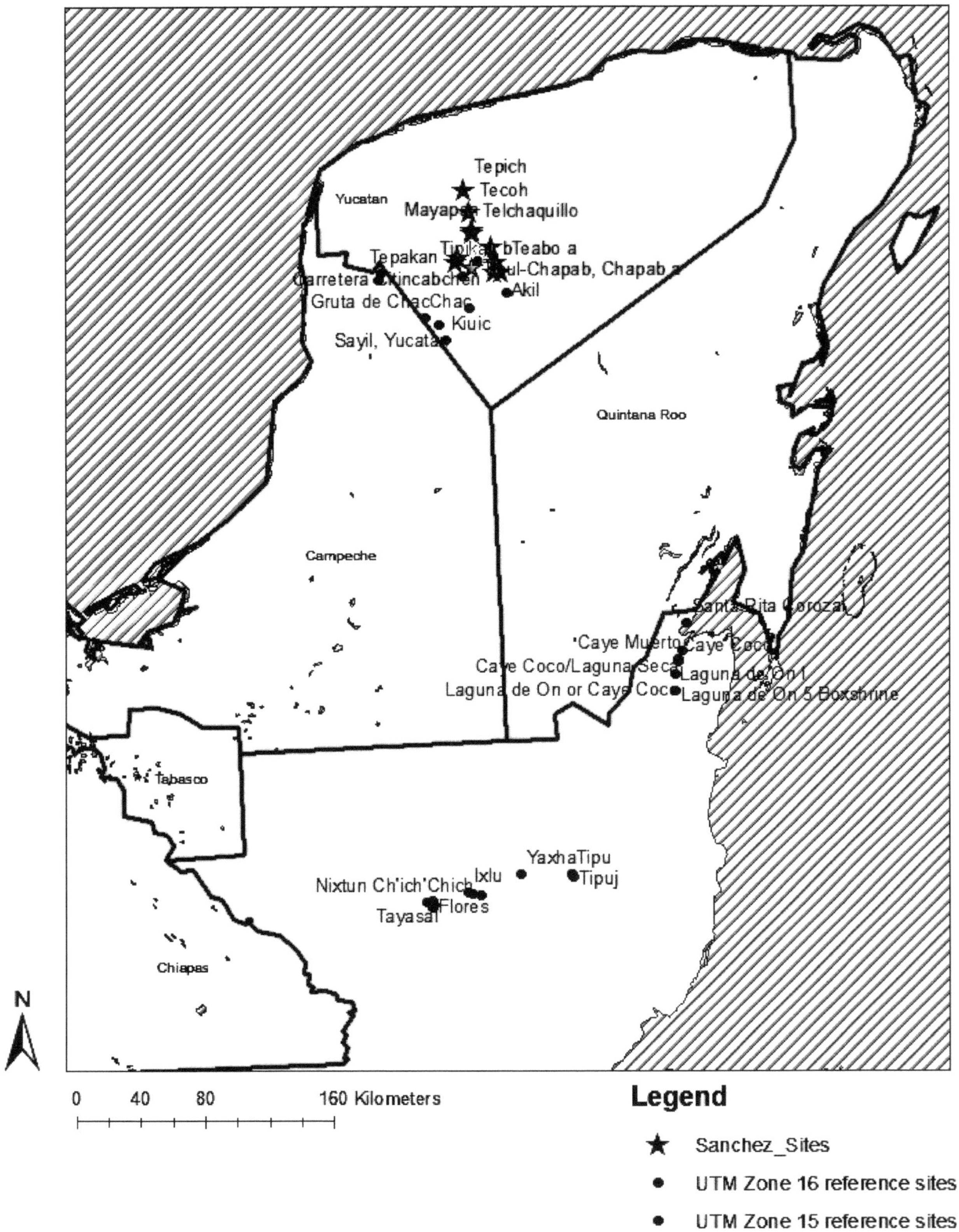

Figure 10. Location of reference samples and current samples.

This analysis did produce some compelling results. Group 1, as alluded to above has a close affinity to Mayapan. Of the 82 specimens identified as belonging to group 1, The most common of the ten closest matches are from Mayapan in 80 (97.6%). This is especially interesting when

Pierce and Glascock 2015 p.19

we consider how rare these samples are in the MURR database. Of the 2071 specimens used for comparison, only 51 (2.5%) were found at Mayapan; a very striking result for such a common match. The majority of the similar matches are from the site of Kiuic. Groups 2 and 3 are highly variable in their closest matches, and little information can be drawn on this account. However, group 4 as well has provided an interesting result. Though small in size (n=11), this group is strongly associated with a completely different area. All 11 specimens are most similar to specimens found at Laguna de On in northern Belize. Again, this is quite compelling when we consider that only 57 of the 2701 (2.8%) reference samples compared are actually from Laguna de On. The other matches that are the most similar are nearly all from this region as well, being from either Santa Rita Corozal or Caye Coco. This is especially compelling when we remember the great distinction of this group observed through discriminant analysis on discriminant function 1 (Fig. 7). Group 5, on the other hand is also compositional similar to samples from this area in many cases, although there is considerably more diversity; while some members of this group are compositionally most similar to specimens from further to the south at sites such as Tipuj.

Final Group Assessment

Unfortunately, the final group designation is often blurry when only individual elements are considered as illustrated in the biplots above. However, some generalizations can be made for each group by using a multi-method approach of group construction.

Group 1 (n=82) is the largest of the five identified compositional groups. It is a relatively homogenous and moderately clustered when considering the PC loadings. Though with some individual elements it is more varied group, it is easily distinguishable from the others when considering certain elements in isolation. Specimens in this group are comparatively low in Hafnium (Hf), Antimony (Sb), for example. Of note, this group is possibly related to raw clay sample CGS193, as shown in Table 4. As discussed above, the specimens of this group are also highly correlated to the Mayapan site, which is the origin of the associated raw clay sample. To further substantiate the validity of this compositional group, 26 of the 33 (79%) sampled sherds collected from Mayapan, are found within this group; a strong case for the legitimacy of utilization of the criterion of abundance. It is therefore possible if not likely that Group 1 represents local production at Mayapan given its large size as seen in the criterion of abundance (Bishop et al. 1982), similarities to other Mayapan ceramics, and raw clay. Also significant is the fact that 12 of 13 compositionally assigned ceramics from Tepich, 12 of 15 Telchaquillo, and 7 of 8 assigned sherds from Mama are members of group 1. Interestingly, of the 27 assigned Cajete form ceramics, 23 of them (85%) are in group 1 despite Group 1 making up only 54% of all assigned sherds.

Group 2 (n=34) is the least homogenous of all groups chemically. Further, in almost all cases, it appears to overlap with another group. However, this can be teased out by considering multiple bivariate plots at once. For example, in many of the biplots, Group 2 overlaps with multiple other groups. However, when we consider discriminant functions, the differences are exaggerated and Group 2 begins to become distinct visually. The internal cohesiveness of this group has then been further substantiated through Mahalonobis testing (see Appendix B). Euclidian distance searches were not able to identify a common geographical match, perhaps a product of the lack of internal homogeneity. There is no single element in which Group 2 is considerably higher or lower than all other groups. In many ways, it is similar to group 1, although it is notably higher in Antimony (Sb) and Hafnium (Hf) than Group 1. When considering descriptive variables and contextual information, it is notable that 5 of the 7 assignable sherds from Teabo are in group 2, yet only 2 of the 33 Mayapan sherds are found in Group 2. It does not appear that any particular vessel shape, form, or decoration is related to this group.

Group 3 (n=8) is the smallest of the groups but is easily distinguishable by numerous elements.

For example, in Figure 4 we see a highly homogenous and tightly clustered distinct group. The distinction is also clear with discriminant analysis and through the use of principal components. In general, this group tends to be higher in Antimony (Sb), Zinc (Zn), and Hafnium (Hf), while trending lower in Chromium (Cr), and Manganese (Mn). Euclidian distance searches were similarly variable in the closest matches, and thus provenance remains unclear. However, a possible point of origin may be the Coluba site, as 7 of the 8 members of this group come from this site.

Group 4 (n=11) is especially compelling given its stark difference observable in Figure 7 based on discriminant function 1. In general, this group appears to be low in Potassium (K), Rubidium (Rb), and Caesium (Cs); while being relatively high in Zircon (Zr), and Neodymium (Nd). As mentioned above, Euclidian distance searches revealed a strong similarity with sherds from Northern Belizean sites (particularly Laguna de On). Notably, 8 of the 9 assigned sherds with Tulum Red decoration, are found in group 4. Similarly interesting, all 4 of the assigned tecomate sherds are members of group 4 as well as all 5 Payil red sherds. This last point is especially significant as Payil ceramics are known to be produced in Belize, far from the research area (Leslie Cecil, personal communication). Can this be evidence of long distance trade? May it be a locally produced imitation? Euclidean distance searches may indicate trade, but perhaps further analysis could confirm this. To explore these issues further, we contacted Dr. Leslie Cecil for assistance. We then compared the current sample to a large dataset provided by her in numerous ways. Cecil has identified 34 of her sherds as Payil (Fig. 11). Of these 34, 18 fit within the currently identified group 4 based on Mahalonobis calculations. Other important descriptive variables appear to be significant as well. Of Cecil's provided data (n=1335), 23 are Navula type ceramics. Of these, all are found at Laguna de On and nearly one third of them fit within the compositional variation of Group 4. This is a reiteration of the previous Euclidian distance search. Finally, we ran another Mahalanobis distance calculation to test this group against all samples. Of the entire reference dataset(both MURR's data and those provided by Cecil), 42 specimens had at least a 10% likelihood of being a member of compositional group 4. Of these 42, twenty eight (67%) of them come from Laguna de On. The significance of this is great given that Laguna de On ceramics make up only 4.3% of the total reference dataset. Given these multiple lines of evidence, we can say with a fair amount of confidence that group 4 does in fact reflect long distance trade from the south, likely near the site of Laguna de On in northern

Belize. Interestingly, despite the small size of this group, 55.6% (5 of 9) of the sherds found at Chacmool in this study are in fact from Group 4; possibly indicating trade between Chacmool and Laguna de On.

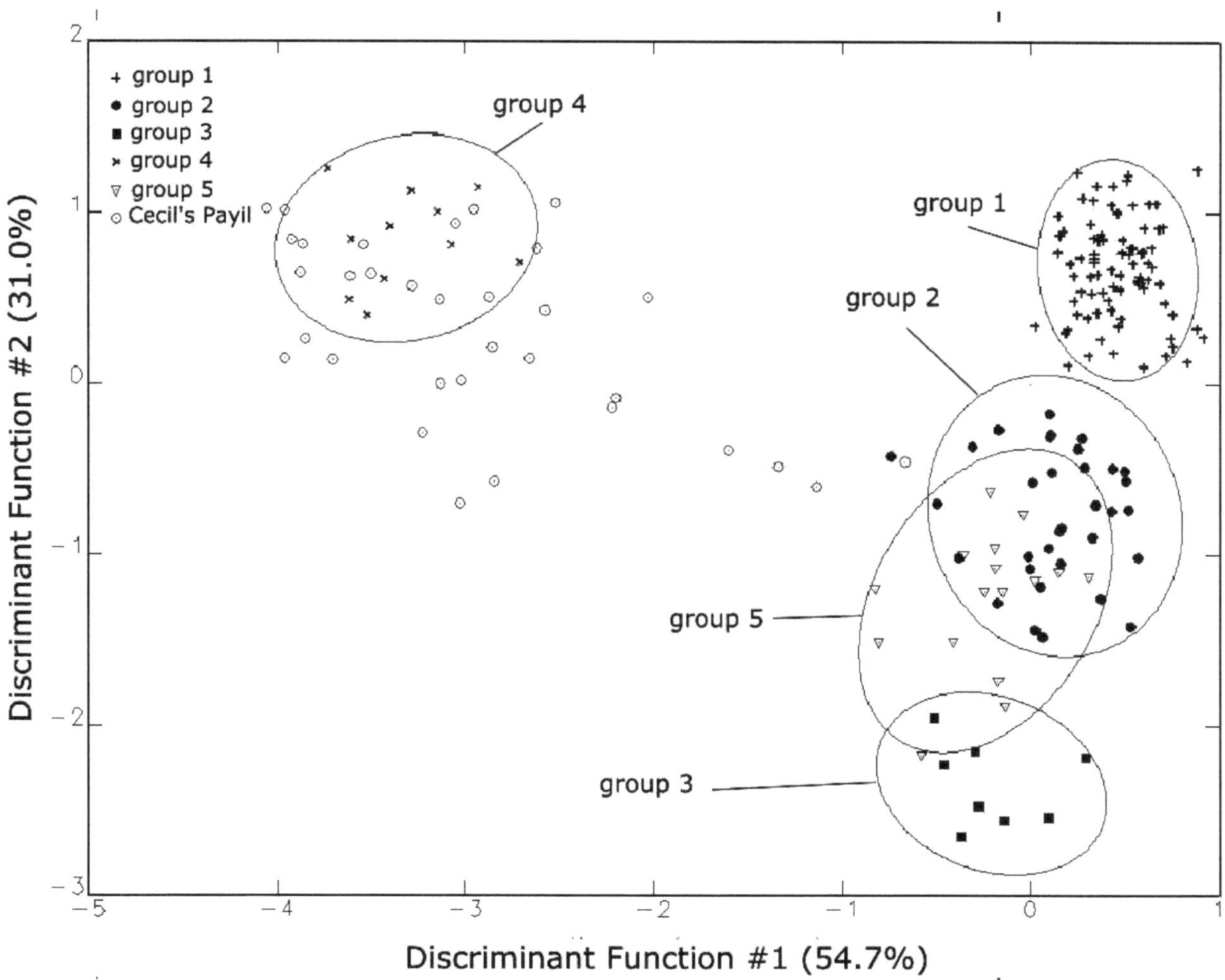

Fig. 11 Comparison with Cecil's Payil samples

Group 5 (n=16) is quite diverse relative to its small size. In many cases it has significant overlap with Group 3. However, as seen in Figures 3 and 5, it can be pulled apart with specific elemental bivariate plots. For instance, group 5 is comparatively high in Caesium (Cs), Cobalt (Co), and Iron (Fe), while being low in Calcium (Ca). Based on Euclidian distance searches, this group is largely similar to specimens from sites in northern Belize, however some specimens are most similar to samples from further south near the Guatemala-Belize border. When considering descriptive variables, it is notable that the remaining Chacmool ceramics not in Group 4, are found in Group 5 (n=4). Also notable, 3 of the 5 (60%) of the Peto Cream ceramics are in group 5.

Finally, the unassigned specimens (n=34) were left unassigned due to either being significantly different from all four chemical groups or being chemically consistent with more than one group.

Pierce and Glascock 2015 p.22

In some occasions, specimens were left unassigned due to conflicting likelihoods based on the different analyses performed. There is little similarity between the unassigned specimens and they likely originate from various source locations as they are wildly variable compositionally and Euclidian distance searches produced a wide array of similar matches.

Conclusions

The above discussions regarding NAA of Late Postclassic Yucatan ceramics has utilized multiple types of analysis to not only define compositional groups within the sample, but to attempt to identify geographically the source location. We feel that we were successful in these goals. However, identifying source location can be a complicated matter. The locations proposed here will require further analysis to confirm. It is advisable to also consider archaeological, ethnographic, geological, and contextual data to better grasp potential source locations geographically.

In this study, we first used Principal Component analysis to identify the compositional structure of the sample. Next we used the PC loadings in combination with various elemental biplots to separate the sample into five clear mutually exclusive source groups; which were then confirmed through Mahalanobis distance calculations.

To get at the potential locations of origin, we have employed various methods. We first compared this sample to provided clay samples and then previously analyzed samples and data from our database and those provided by Dr. Cecil using Euclidian distance searches. It is clear that we have five distinct compositional groups here. While some may reflect local production, as may be indicated with Group 1 at Mayapan, we have also identified strong evidence for long distance trade with group 4 and possibly group 5.

Acknowledgments

We acknowledge Leslie Cecil for her assistance without which many of the results presented here would not be possible. Also, we wish to thank Shilo Bender, and Katie Lane for their roles in preparing the samples for irradiation. This project was supported in part by NSF grant BCS-1415403 to the Archaeometry Laboratory at the University of Missouri Research Reactor. Any forthcoming publication from this report must include this grant number. We also ask that when any publication of this material is underway, please consider the authors of this report for coauthorship.

Appendix A.
Final compositional group

Pages 24 to 29 of this report containing its Appendix A with the final compositional group assigment for the sample, which is present in the original document, have been removed. Their contents were copied into the body of the work in Table 8-18 ("List of samples in this study"). This table contains chemical and petrographic group assignments for the sample.

Appendix B
Mahalonobis results confirming group membership of ceramic sherd samples

Membership probabilities(%) for samples from the group: Group_1
Probability for each sample calculated after removal from original group.

ANID	Group_1	Group_2	Group_3	Group_4	Group_5	Best Group
---------	--------	--------	--------	--------	--------	---------
CGS001	89.554	37.584	0.41	0.018	0.68	Group_1
CGS002	56.385	29.949	0.466	0.043	0.652	Group_1
CGS003	99.567	50.926	1.073	0.114	0.369	Group_1
CGS004	62.69	1.223	0.134	0.004	0.054	Group_1
CGS005	93.731	59.569	2.049	0.341	0.826	Group_1
CGS006	49.255	12.971	0.193	0.005	0.643	Group_1
CGS007	74.655	22.171	0.722	0.026	0.404	Group_1
CGS009	61.151	32.196	0.349	0.008	2.233	Group_1
CGS010	2.691	0.191	0.197	0.002	0.08	Group_1
CGS012	33.495	27.226	3.301	0.182	3.08	Group_1
CGS013	64.334	27.248	0.367	0.007	1.972	Group_1
CGS014	60.219	0.221	0.127	0.003	0.029	Group_1
CGS016	57.426	0.145	0.222	0.024	0.002	Group_1
CGS019	54.726	0.223	0.235	0.026	0.002	Group_1
CGS020	44.719	0.071	0.176	0.022	0.002	Group_1
CGS021	49.85	0.129	0.234	0.026	0.002	Group_1
CGS028	40.149	0.08	0.243	0.026	0.001	Group_1
CGS029	35.062	12.059	1.34	3.292	0.036	Group_1
CGS030	34.762	18.335	1.008	3.244	0.223	Group_1
CGS031	35.414	39.071	3.472	0.231	2.963	Group_2
CGS032	38.605	0.871	0.274	0.016	0.01	Group_1
CGS033	86.039	34.523	0.679	0.041	1.912	Group_1
CGS034	48.799	3.055	1.921	0.179	0.014	Group_1
CGS035	86.724	8.155	1.091	0.313	0.007	Group_1
CGS037	27.483	0.014	0.121	0.005	0.005	Group_1
CGS038	39.112	4.793	3.533	1.04	0.257	Group_1
CGS041	73.37	89.891	3.086	0.302	2.456	Group_2
CGS042	95.135	23.226	0.688	0.154	0.227	Group_1
CGS043	26.491	2.209	0.361	0.03	0.008	Group_1
CGS046	62.059	20.309	3.043	0.471	0.479	Group_1
CGS047	92.614	34.978	1.612	0.294	0.124	Group_1
CGS049	66.995	29.451	2.824	0.328	0.784	Group_1
CGS050	60.27	0.417	0.242	0.014	0.006	Group_1
CGS052	63.126	12.331	0.623	0.096	0.205	Group_1
CGS054	91.258	14.569	1.39	0.707	0.034	Group_1
CGS055	45.634	10.744	0.958	0.475	0.074	Group_1
CGS056	81.99	11.569	2.297	1.37	0.033	Group_1
CGS057	15.202	0.64	0.739	0.802	0.002	Group_1
CGS058	86.274	3.179	0.436	0.039	0.018	Group_1
CGS059	93.981	19.889	0.805	0.114	0.034	Group_1
CGS060	89.96	11.487	1.272	0.436	0.024	Group_1

CGS061	17.158	38.818	0.911	0.585	0.171	Group_2
CGS062	57.607	1.767	0.16	0.012	0.095	Group_1
CGS063	56.27	9.264	0.384	0.102	0.076	Group_1
CGS064	89.365	0.789	0.19	0.006	0.029	Group_1
CGS065	21.733	0.124	0.202	0.008	0.019	Group_1
CGS067	48.758	9.892	0.759	0.054	0.023	Group_1
CGS068	21.133	15.146	2.56	1.111	0.012	Group_1
CGS070	55.096	71.992	4.456	1.639	0.136	Group_2
CGS073	58.844	42.361	0.596	0.043	2.981	Group_1
CGS074	55.281	13.861	0.263	0.013	0.495	Group_1
CGS075	41.267	2.919	0.285	0.019	0.236	Group_1
CGS076	84.761	33.144	1.483	0.265	0.852	Group_1
CGS077	94.652	43.932	1.766	0.175	0.129	Group_1
CGS078	57.969	20.035	4.87	3.185	1.352	Group_1
CGS079	58.716	13.201	0.241	0.018	0.275	Group_1
CGS080	79.638	60.609	3.884	1.092	0.149	Group_1
CGS081	40.693	20.249	7.365	6.846	0.096	Group_1
CGS082	83.133	18.631	2.955	2.09	0.036	Group_1
CGS092	4.183	0.691	0.479	0.109	0.078	Group_1
CGS094	42.276	0.317	0.178	0.009	0.035	Group_1
CGS095	57.359	3.926	0.574	0.047	0.029	Group_1
CGS107	48.253	2.448	0.729	0.026	0.023	Group_1
CGS108	24.926	2.192	0.734	0.015	0.042	Group_1
CGS109	34.7	2.262	0.674	0.017	0.025	Group_1
CGS111	24.579	1.7	0.566	0.009	0.049	Group_1
CGS112	27.873	1.366	0.555	0.012	0.016	Group_1
CGS113	66.303	61.737	1.648	0.039	0.999	Group_1
CGS115	96.915	32.809	1.275	0.076	0.453	Group_1
CGS132	71.482	29.568	0.354	0.014	0.718	Group_1
CGS133	39.71	47.419	0.453	0.05	0.38	Group_2
CGS134	95.646	27.558	3.214	0.968	0.119	Group_1
CGS135	96.823	40.745	3.298	0.68	0.282	Group_1
CGS137	21.673	2.022	0.124	0.002	0.193	Group_1
CGS138	26.471	1.928	0.328	0.013	0.051	Group_1
CGS140	60.915	38.497	2.582	0.276	0.02	Group_1
CGS141	44.352	56.534	2.628	0.365	1.715	Group_2
CGS163	0.958	0.549	0.362	0.119	0.112	Group_1
CGS165	9.213	31.875	1.297	0.055	5.514	Group_2
CGS167	3.742	0.026	0.075	0.002	0.025	Group_1
CGS172	35.858	18.759	0.419	0.028	1.007	Group_1
CGS173	35.697	16.42	0.33	0.013	0.833	Group_1

--------- -------- -------- -------- -------- --------

Pierce and Glascock 2015 p.31

Membership probabilities(%) for samples from the group: Group_2
Probability for each sample calculated after removal from original group

ANID	Group_1	Group_2	Group_3	Group_4	Group_5	Best Group
CGS015	0.002	40.496	0.324	0.004	0.034	Group_2
CGS018	0.007	57.39	0.467	0.007	0.1	Group_2
CGS022	2.546	16.642	2.373	0.232	3.694	Group_2
CGS024	15.240	52.625	3.098	0.539	0.393	Group_2
CGS026	0.000	48.518	0.382	0.008	0.047	Group_2
CGS051	0.418	15.108	0.966	1.512	1.264	Group_2
CGS053	0.454	79.148	2.783	1.181	2.872	Group_2
CGS066	0.776	89.795	0.398	0.075	3.198	Group_2
CGS093	0.000	55.193	2.424	1.049	26.923	Group_2
CGS096	0.014	94.645	0.403	0.111	1.768	Group_2
CGS100	0.171	98.318	0.793	0.057	6.194	Group_2
CGS101	0.020	44.028	1.881	0.774	0.007	Group_2
CGS102	2.180	87.103	0.414	0.015	3.007	Group_2
CGS104	0.000	60.131	1.243	0.918	4.121	Group_2
CGS122	0.000	98.354	0.272	0.086	2.375	Group_2
CGS124	0.000	42.978	0.195	0.021	0.222	Group_2
CGS126	1.306	92.56	1.402	0.027	4.853	Group_2
CGS127	0.025	79.833	1.881	0.069	29.731	Group_2
CGS130	0.000	58.6	0.239	0.104	1.167	Group_2
CGS131	0.000	48.529	0.263	0.401	0.873	Group_2
CGS142	0.000	82.748	0.548	0.217	2.868	Group_2
CGS144	0.000	42.155	0.561	0.044	3.243	Group_2
CGS145	0.000	72.983	0.227	0.036	0.284	Group_2
CGS146	1.143	37.801	1.447	0.03	2.109	Group_2
CGS147	0.000	57.517	0.186	0.042	0.502	Group_2
CGS148	0.000	73.543	0.189	0.05	0.531	Group_2
CGS150	0.000	68.849	0.181	0.042	0.569	Group_2
CGS151	0.000	62.823	0.221	0.074	0.538	Group_2
CGS155	0.000	37.785	0.101	0.039	0.273	Group_2
CGS162	0.000	49.715	0.28	0.068	0.065	Group_2
CGS169	0.000	41.456	0.211	0.038	0.041	Group_2
CGS171	0.007	26	0.898	0.233	18.759	Group_2
CGS174	5.429	74.107	1.964	0.04	16.435	Group_2

Membership probabilities(%) for samples from the group: Group_3
Probability for each sample calculated after removal from original group.

ANID	Group_1	Group_2	Group_3	Group_4	Group_5	Best Group
CGS152	0.000	0.099	91.719	0.079	8.121	Group_3
CGS153	0.000	0.886	92.587	0.097	0.007	Group_3
CGS157	0.000	0.493	93.326	0.038	0.145	Group_3
CGS158	0.000	0.149	92.763	0.153	4.291	Group_3
CGS159	0.000	0.646	92.240	0.212	3.259	Group_3
CGS160	0.000	0.344	92.046	0.098	10.461	Group_3
CGS161	0.000	0.434	99.715	0.095	11.207	Group_3

Pierce and Glascock 2015 p.32

CGS170	0.000	0.425	91.663	0.086	0.179	Group_3

Membership probabilities(%) for samples from the group: Group_4
Probability for each sample calculated after removal from original group.

ANID	Group_1	Group_2	Group_3	Group_4	Group_5	Best Group
CGS083	0.000	0.817	0.618	85.960	0.081	Group_4
CGS084	0.000	0.534	0.969	97.673	0.032	Group_4
CGS085	0.000	0.496	1.277	88.136	0.262	Group_4
CGS086	0.000	3.462	3.919	92.964	0.047	Group_4
CGS089	0.000	0.166	3.724	69.832	0.005	Group_4
CGS105	0.000	0.057	3.197	71.921	0.032	Group_4
CGS175	0.000	0.117	3.100	97.362	0.024	Group_4
CGS176	0.000	0.030	2.499	76.434	0.027	Group_4
CGS177	0.000	0.001	1.966	75.526	0.003	Group_4
CGS179	0.000	0.126	4.126	84.261	0.110	Group_4
CGS180	0.000	0.497	4.339	79.940	0.009	Group_4

Membership probabilities(%) for samples from the group: Group_5
Probability for each sample calculated after removal from original group.

ANID	Group_1	Group_2	Group_3	Group_4	Group_5	Best Group
CGS008	0.151	11.249	0.404	1.500	95.448	Group_5
CGS025	0.045	15.059	0.355	7.143	88.481	Group_5
CGS027	0.000	2.231	0.219	2.869	91.004	Group_5
CGS039	0.000	0.256	0.212	5.235	60.385	Group_5
CGS045	0.017	0.056	0.143	0.781	64.215	Group_5
CGS091	0.063	0.945	0.464	1.089	38.199	Group_5
CGS097	0.258	8.364	0.368	3.275	99.312	Group_5
CGS098	1.065	57.791	1.052	0.857	86.824	Group_5
CGS128	0.075	64.916	2.148	2.266	86.922	Group_5
CGS129	0.016	26.584	0.686	4.236	98.415	Group_5
CGS139	0.000	2.541	0.744	5.975	61.545	Group_5
CGS168	0.001	11.222	0.428	6.511	72.386	Group_5
CGS181	0.000	0.077	.220	0.031	44.018	Group_5
CGS183	0.000	1.447	0.221	6.183	47.466	Group_5
CGS184	0.000	0.092	0.883	0.025	74.928	Group_5
CGS185	0.000	0.253	2.112	0.052	62.709	Group_5

Membership probabilities(%) for samples from the group: Unassigned
Probability for each sample calculated after removal from original group.

ANID	Group_1	Group_2	Group_3	Group_4	Group_5	Best Group
CGS011	29.761	2.033	0.146	0.002	0.185	Group_1
CGS017	0.205	17.335	0.570	0.140	1.279	Group_2
CGS023	12.695	0.031	0.121	0.009	0.004	Group_1
CGS036	20.066	0.011	0.170	0.016	0.001	Group_1

CGS040	0.662	0.440	0.328	1.188	2.833	Group_5
CGS044	4.598	0.118	0.124	0.013	0.008	Group_1
CGS048	4.234	0.003	0.106	0.001	0.003	Group_1
CGS069	21.680	55.672	4.678	6.914	0.442	Group_2
CGS071	2.241	47.304	0.788	0.982	0.808	Group_2
CGS072	1.913	0.029	0.213	0.072	0.003	Group_1
CGS087	0.000	0.325	0.231	16.945	6.754	Group_4
CGS088	0.000	0.562	0.358	19.414	5.995	Group_4
CGS090	0.000	1.123	1.426	0.038	0.006	Group_3
CGS099	0.000	2.682	1.502	0.266	1.830	Group_2
CGS103	0.959	47.432	7.095	3.265	0.092	Group_2
CGS106	0.000	69.128	0.307	0.173	0.811	Group_2
CGS110	30.197	22.678	3.191	3.617	0.075	Group_1
CGS114	0.012	6.974	0.671	0.957	0.191	Group_2
CGS116	0.000	2.313	2.222	26.810	4.238	Group_4
CGS117	0.005	1.145	0.923	0.261	3.402	Group_5
CGS118	14.798	0.008	0.084	0.002	0.005	Group_1
CGS119	23.092	29.562	1.572	0.369	0.787	Group_2
CGS120	0.000	0.990	0.140	0.055	0.361	Group_2
CGS121	0.000	1.011	0.092	0.026	0.111	Group_2
CGS123	4.981	22.169	0.238	0.052	0.073	Group_2
CGS125	0.000	0.226	0.182	2.176	24.035	Group_5
CGS136	2.458	27.074	0.226	0.012	0.881	Group_2
CGS143	3.087	0.082	0.069	0.001	0.039	Group_1
CGS149	0.000	6.274	0.233	0.021	0.166	Group_2
CGS154	0.000	0.684	1.165	0.119	0.633	Group_3
CGS156	0.000	10.626	0.163	0.101	0.730	Group_2
CGS164	2.025	31.858	0.885	0.047	4.954	Group_2
CGS166	0.000	3.140	1.227	0.428	1.176	Group_2
CGS178	0.000	0.101	0.205	0.949	1.856	Group_5
CGS182	0.000	0.117	3.094	0.007	0.032	Group_3
CGS186	0.000	0.004	0.094	0.005	0.384	Group_5
---------	--------	--------	--------	--------	--------	----------

Pierce and Glascock 2015 p.34

References Cited

Baxter, Michael J.
1992 Archaeological uses of the biplot—a neglected technique? In *Computer Applications and Quantitative Methods in Archaeology, 1991*, edited by G. Lock and J. Moffett. BAR International Series S577, 141–148. BAR Publishing, Oxford.

1994 Exploratory Multivariate Analysis in Archaeology. Edinburgh University Press, Edinburgh.

Baxter, M.J. and C.E. Buck
2000 Data Handling and Statistical Analysis. In *Modern Analytical Methods in Art and Archaeology*, edited by E. Ciliberto and G. Spoto, pp. 681-746. John Wiley and Sons,

Bieber, Alan M. Jr., Dorothea W. Brooks, Garman Harbottle, and Edward V. Sayre
1976 Application of multivariate techniques to analytical data on Aegean ceramics. *Archaeometry* 18:59–74.

Bishop, Ronald L. and Hector Neff
1989 Compositional data analysis in archaeology. In *Archaeological Chemistry IV*, edited by R. O. Allen, pp. 576–586. Advances in Chemistry Series 220, American Chemical Society, Washington, D.C.

Bishop, Ronald L., Robert L. Rands, and George R. Holley
1982 Ceramic compositional analysis in archaeological perspective. In *Advances in Archaeological Method and Theory*, vol. 5, pp. 275–330. Academic Press, New York.

Garrett, Donald L.
2004 *Handbook of Lihtium and Natural Calcium Chloride: Their Deposits, Processing, Uses and Properties*. Elsevier Academic Press.

Gibson, Charles
1964 *Aztecs Under Spanish Rule: A History of the Indians of the Valley of Mexico, 1519-1810*. Stanford University Press.

Glascock, Michael D.
1992 Characterization of archaeological ceramics at MURR by neutron activation analysis and multivariate statistics. In *Chemical Characterization of Ceramic Pastes in Archaeology*, edited by H. Neff, pp. 11–26. Prehistory Press, Madison, WI.

Harbottle, Garman
1976 Activation analysis in archaeology. *Radiochemistry* 3:33–72. The Chemical Society, London.

Hillier, Stephen
2003 Clay Mineralogy. In *Encyclopedia of Sediments and Sedimentary Rocks*, edited by Gerard V. Middleton. Springer Publishing.

Laird, David A.
2006 Influence of Layer Charge on Swelling of Smectites. *Applied Clay Science* 34:74-87.

Leese, Morven N. and Peter L. Main
1994 The efficient computation of unbiased Mahalanobis distances and their interpretation in archaeometry. *Archaeometry* 36:307–316.

Neff, Hector
1992 Introduction. In *Chemical Characterization of Ceramic Pastes in Archaeology*, edited by H. Neff, pp. 1–10. Prehistory Press, Madison, WI.

1994 RQ-mode principal components analysis of ceramic compositional data. *Archaeometry* 36:115–130.

2000 Neutron activation analysis for provenance determination in archaeology. In *Modern Analytical Methods in Art and Archaeology*, edited by E. Ciliberto and G. Spoto, pp. 81–134. John Wiley and Sons, Inc., New York.

2002 Quantitative techniques for analyzing ceramic compositional data. In Ceramic_*Source Determination in the Greater Southwest*, edited by D. M. Glowacki and H. Neff. Monograph 44, Cotsen Institute of Archaeology, UCLA, Los Angeles.

Neff, Hector, and Michael D. Glascock
2000 Provenance Analysis of Aztec Period Ceramics From the Basin of Mexico: Final Technical Report. Report written for clients Thomas H. Charlton, Cynthia Otis Charlton, and Deborah L. Nichols. University of Missouri Research Reactor (MURR), Archaeometry Laboratory.

Nichols, Deborah L., Elizabeth M. Brumfiel, Hector Neff, Mary Hodge, Thomas H. Charlton, and Michael D. Glascock
2002 Neutrons, Markets, Cities, and Empires: A 1000-Year Perspective on Ceramic Production and Distribution in the Postclassic Basin of Mexico. *Journal of Anthropological Archaeology*, 21:25-82.

Sayre, Edward V.
1975 Brookhaven Procedures for Statistical Analyses of Multivariate Archaeometric Data. Brookhaven National Laboratory Report BNL-23128. New York.

Sayre, Edward V. and Garman Harbottle
1979 The Analysis by Neutron Activation of Archaeological Ceramics Related to Teotihuacan. Brookhaven National Laboratory, Informal Report C-2250, New York.

Steponaitis, Vincas, M. James Blackman, and Hector Neff
1996 Large-scale compositional patterns in the chemical composition of Mississippian pottery. *American Antiquity* 61:555–572.

Stoner, W. D., M. D. Glascock, and J. R. Ferguson
2008 Instrumental Neutron Activation Analysis of Aztec and Other Figurines from the site of Xaltocan, Valley of Mexico. Report written for client Lisa Overholtzer. University of Missouri research Reactor (MURR), Archaeometry Laboratory.

Williams, H. and R. F. Heizer
1965 Sources of rocks used in Olmec monuments. In *Sources of stones used in prehistoric Mesoamerican sites*, pp. 1-39. University of California Press, Berkeley.

Weigand, Phil C., Garman Harbottle, and Edward V. Sayre
1977 Turquoise sources and source analysis: Mesoamerica and the southwestern U.S.A. In *Exchange Systems in Prehistory*, edited by T. K. Earle and J. E. Ericson, pp. 15–34. Academic Press, New York.

APPENDIX C: BULK XRD AND CLAY ANALYSES OF A NAVULA, MAYAPAN (#256) POTTERY SAMPLE

Pittsburgh
Mineral &
Environmental
Technology, Inc.

November 30, 2016

Ms. Carmen G. Sanchez
6172 NW 88th Ave.
Parkland, FL 33067

Dear Ms. Sanchez:

This report summarizes the results of quantitative mineral phase analysis, clay speciation, and semi-quantitative elemental analysis of your archaeological ceramic. The sample fragments were received at PMET's laboratory on October 24, 2016 in a labelled plastic bag along with a PMET chain of custody document. A request for analysis and a sample description were received from Ms. Sanchez via email prior to receipt of sample.

The purpose of the analysis was to determine the species of clay present in the ceramic, the amount of mineral phases and amorphous material, and the approximate elemental composition of the sample.

X-ray Diffraction Analysis

X-ray powder diffraction (XRD) and Rietveld quantification analyses were used to determine the mineralogical composition of the sample and the clay species.

The as-received fragments were gently disaggregated by hand crushing with a large porcelain mortar and pestle. The crushed material was separated by brushing dry through a 325 mesh (44µ) sieve. The fine fraction was then submerged in a column of deionized water, treated for five minutes with an ultrasonic probe, then allowed to settle for 72 hours.

The clay fraction was removed from the top of the column of settled fines using a pipette and deposited on glass slides. The glass slides were treated and analyzed by XRD according to USGS Bulletin 1563.

The remaining fines in the column were dried and then blended with the original sample. The original sample was then split to obtain analytical aliquots for bulk XRD analysis and for SEM-EDX and IR elemental analysis.

700 Fifth Avenue
New Brighton, PA 15066
(724) 843-5000
FAX: (724) 843-5353
www.pmet-inc.com

Bulk XRD sample preparation included grinding 2 g of sample using the BICO Model VP-1989 mill with a 3.5 inch ring and puck. The pulverized material was spiked with fluorite (CaF_2) on a 90:10 weight basis and mixed using a SPEX Industries Mixer/Mill for 10 minutes. Standard spike intensity was used as a reference to determine the amorphous content of the samples.

Step-scanned XRD data were collected by the Siemens D500 computer-automated diffractometer using Bragg-Brentano geometry. Cu radiation was produced at a power of 45kV and 30 mA. The diffracted beam was collimated by a 0.05° receiving slit. The data was collected in the 2θ range of 4.9°–66.1° with a step size of 0.015° and a dwell time of 1 sec/step using a Sol-X solid state energy-dispersive X-ray detector.

Qualitative analysis of the XRD patterns was performed using proprietary Bruker AXS software Diffrac Plus EVA (v. 7001, 2001) peak search algorithm. The reference database for the crystal pattern search/match is the International Center for Diffraction Data database (ICDD, 2001). A chemical screen using the XRF and ICP data was used to narrow the search.

Quantitative analysis was performed using the whole pattern fitting function of Diffrac Plus Topas R, a proprietary Bruker AXS software (v. 2.0, 2000), which is based on the Rietveld method (Rietveld 1969). The reference database for quantitative analysis of crystal structures is the Inorganic Crystal Structure Database (NIST ICSD, 2010, v.2).

Clay speciation was determined by running separate scans on the dried oriented, glycolated, heated at 400°C, and heated at 550°C. The XRD parameters for clay analysis are similar to the above, except that the clay patterns are scanned from 2.9°-21.7°. These traces are color coded black, blue, red, dark red, respectively, and overlain for comparison to the USGS Clay Mineral Identification Flow Diagram.

Elemental Analysis

Elemental composition was determined using SEM with energy-dispersive x-ray spectroscopy and infrared analysis. Oxygen was not included in the analysis. Carbon and sulfur were analyzed in duplicate using an Eltra induction furnace with infrared detectors. Carbon and sulfur standards were run with errors of 1.36% for C and 2.9%, for S.

The major and minor elements were detected using an Amray 1600 scanning electron microscope with a Tracor energy-dispersive x-ray detector. Elements from Na and higher on the atomic chart were detected. The results were expressed as stoichiometric oxides and normalized along with the carbon and sulfur data to 100%.

Discussion

The only clay mineral present in the sample is a mica species. Its stability under the various conditions of treatment confirms this determination. The presence of a very weak 5- angstrom 002 peak at 8.8° two-theta indicates that the mica species is glauconite

Shannon 2016 p.2

The bulk XRD analysis identified ankerite, which is a solid solution series with Fe and Mg sharing a carbonate ion. The Rietveld refinement indicates that this ankerite contains the maximum amount of iron, as shown in the table of results. This finding may be influenced by other factors, so the Fe:Mg ratio should be considered at the high end of a range and not an absolute.

Ms. Sanchez, please email me if you would like to discuss these results. Thank you for using PMET's laboratory services on this project.

Sincerely,

Randolph W. Shannon

Laboratory Manager

RFA 6710

Shannon 2016 p.3

Table 1

Sample Identification, As-received Weight

PMET I.D.	Sample I.D.	Sample Description	As-received Wt. (g)
6710-1	256	Archaeological ceramics - Mayapan	12.82

Table 2

Results of Bulk XRD Analysis Wt. %

Mineral	Atomic Formula	256
quartz	SiO_2	0.5
mica	$KAl_2(Si_3Al)O_{10}(OH)_2$	5.5
calcite	$CaCO_3$	42.9
aragonite	$CaCO_3$	1.6
ankerite	$Ca(Fe_{0.7}Mg_{0.3})(CO_3)_2$	19.6
amorphous	(clay, hydrocarbons)	29.9

Table 3

Results of Clay Analysis Estimated wt. %

Mineral	Atomic Formula	256
mica	(K,Na)(Fe,Al,Mg)2(Si,Al)4O10(OH)2	99+

Shannon 2016 p.24

Table 4

Results of SEM-EDX & IR Analysis
Approx. Wt. %

Stoichiometric Oxide	256
CO_2	26.8
MgO	6.0
Al_2O_3	7.4
SiO_2	13.9
SO_3	0.5
K_2O	0.3
CaO	40.8
TiO_2	0.5
Fe_2O_3	3.8

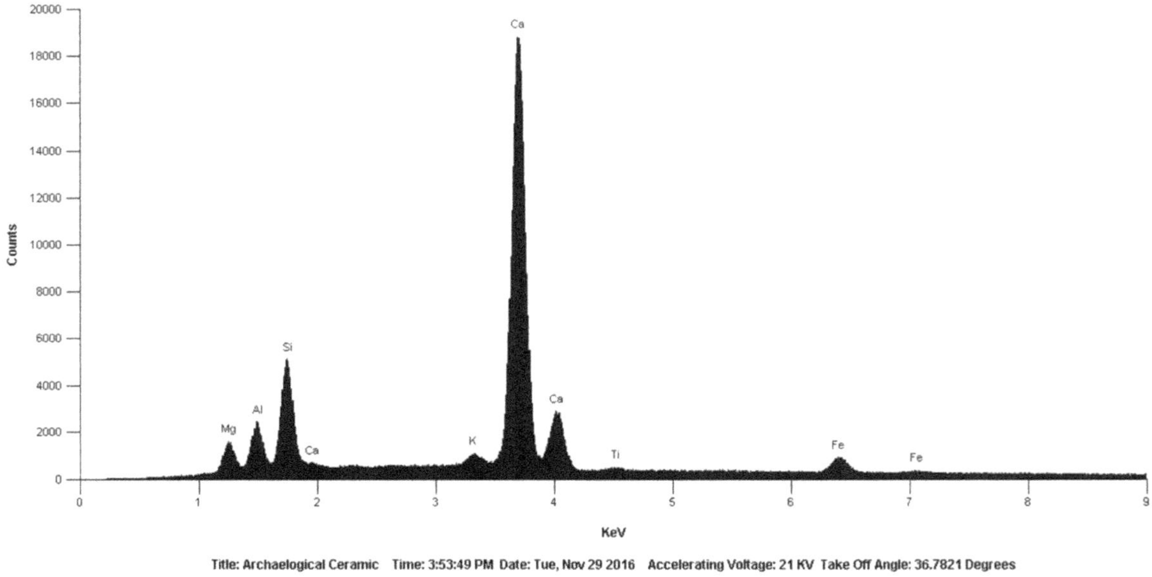

SEM-EDX Spectrum

256

Figure 1

Shannon 2016 p.5

Figure 2

As-received sample fragments

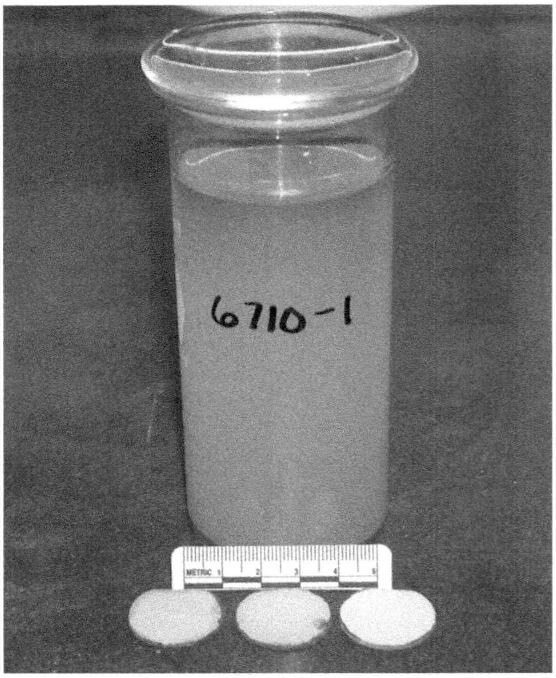

Figure 3 Clay column and dried slides

Shannon 2016 p.7

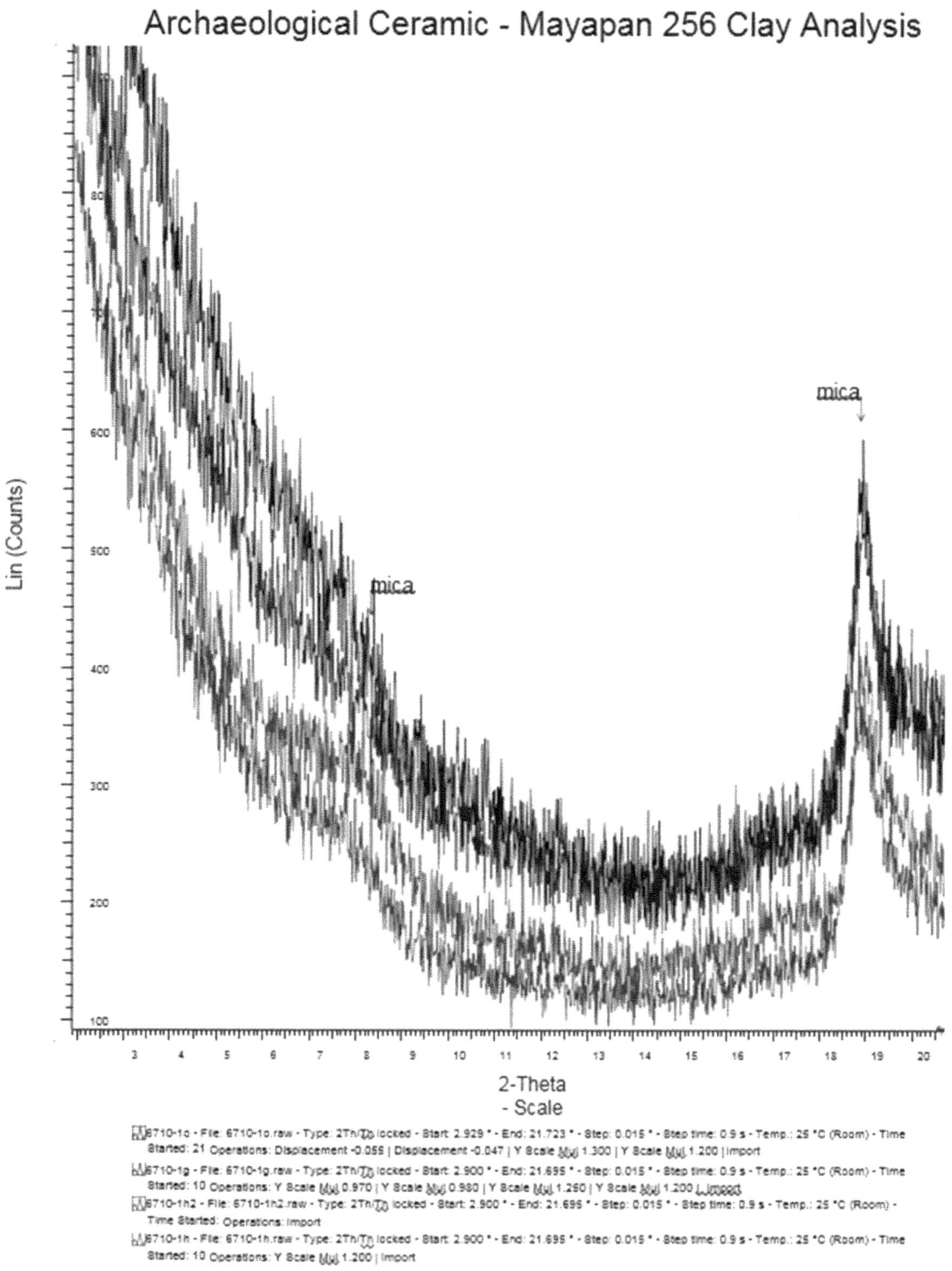

Figure 4 Clay patterns

Shannon 2016 p.8

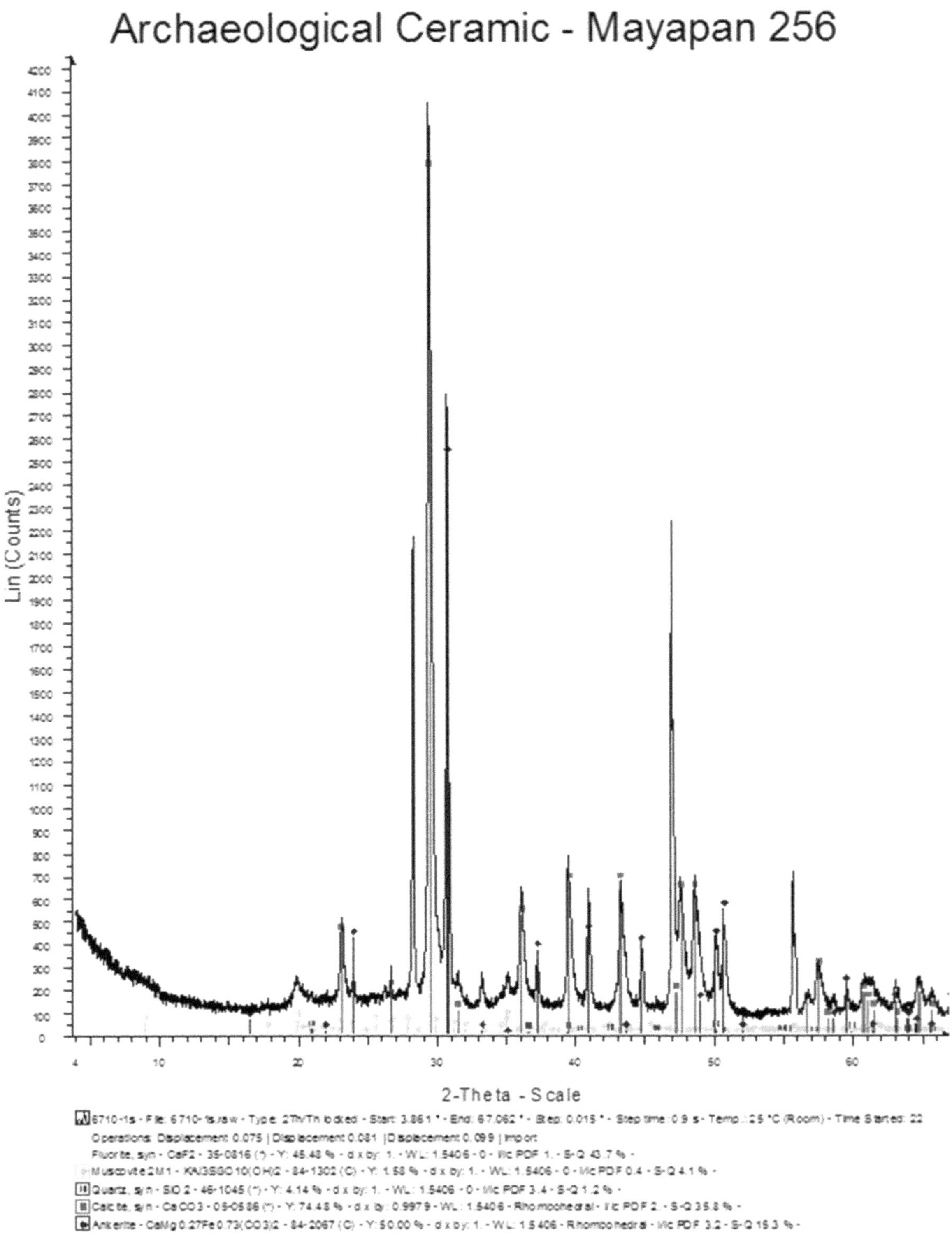

Figure 5 Bulk XRD Scan

Shannon 2016 p.9

APPENDIX D: DETAILED DESCRIPTION OF FABRIC CLASSES IN HAND SPECIMEN

White Micrite Fabric

Micrite-Medium-W (WM-m-W)

Re-fired Color:	Predominantly white, but a few refired to light reddish brown (Table 6.2 for Munsell colors)
Grain size :	Medium
Fired Color:	Varied: light grey (5YR 7/1-7/2, light brownish grey (10YR 8/1), very pale brown (10YR 7/4), pinkish white (7.5YR 8/2)
Hardness:	Soft
Compactness:	Varied: porous to moderately compacted
Rounding:	Varied: subangular to irregular
Fracture:	Varied: hackly, extremely irregular to moderately smooth
Thickness:	Varied: from 5 -12 mm
Inclusions (section 6.1)	
Sorting:	Poor
Composition:	**Common to Few:** MICRITE - white, chalky, opaque fragments, mostly sub-angular to subrounded, medium texture, with particle size ranging from fine to medium. **Very few:** opaque, sub-rounded to irregular grey particles; sparite or micritized remains may also be present
Voids:	mostly moderately compacted, but some samples have long pores or channels.

Sparry Fabrics

Spar-Coarse-W (SP-c-W)

Refired Color	White (Table 6.2 for Munsell colors)
Grain Size:	Coarse
Fired Color:	Mostly light grey (10YR 7/1-7/2), very pale brown (10YR 8/3 -7/3), reddish yellow (5YR 7/8), light red (2.5YR 7/6)
Hardness:	soft; chips with nail
Compactness:	Porous, not compacted, usually very friable
Rounding:	Subangular to subrounded
Fracture:	Hackly, irregular
Thickness:	Varied, 5 – 14.2 mm
Inclusions (section 6.1)	
Over-all Abundance:	Common to frequent
Sorting:	Bimodal/fairly sorted
Composition:	**Common to Frequent**: Clusters of sparry carbonate (calcite or dolomite), subangular to subrounded, with coarse (0.5 – 1.0 mm) mode of particle size ranging from fine to coarse. The size of each spar crystal is difficult to estimate with the binocular. Only crystals larger than 0.04 mm distinguished, but smaller appear to be present. To the naked eye, some particles look whitish and dull and others semi-translucent likely depending on spar crystal size. **Few to Very few**: fine angular individual crystals (likely calcite or dolomite) embedded in the matrix, each around 0.04-0.06 mm **Few to Very rare:** Grey micrite, coarsely to medium grained.
Voids:	Visible channels and pores

Spar-Medium-W (SP-m-W)

Re-fired Color	White (Table 6.2 for Munsell colors)
Grain size (mode):	Medium
Fired Color:	Mostly light grey (10YR 7/1-7/2), very pale brown (10YR 8/3 -7/3), or reddish yellow (5YR 7/8) to a lesser degree
Hardness:	Soft
Compactness:	Moderately compacted, few vughs, rare channels
Rounding:	Subangular to sunrounded
Fracture:	Varied, from moderately smooth to irregular
Thickness:	Wide range, from 5.5 to 11 mm
Inclusions (section 6.1)	
Sorting:	Bimodal/fairly sorted
Average size:	medium
Composition:	**Few:** Mostly opaque, medium grained, subangular to subrounded sparite rock fragments. Many of these bundles are partially coated with micrite. **Few to Very few**: fine angular individual crystals (likely calcite or dolomite) embedded in the matrix, each around 0.04-0.07 mm **Very few to Rare**: Opaque, subangular or subrounded, grey particles. Some look indurated, peloidic.
	Silty matrix
Voids:	Many with long channels and pores

Spar-Coarse-B (SP-c-B)

Refired Color	Light reddish brown (Table 6.2 for Munsell colors)
Grain size:	Coarse
Fired Color:	pinkish grey (7.5 YR 7/2), light reddish brown (2.5YR 7/3), pale brown (10YR 7/3), reddish yellow (5YR 7/8), light red (2.5YR 7/6)
Hardness:	Soft, chipped with nail
Compactness:	High porosity
Rounding:	
Fracture:	Irregular
Thickness:	6 – 10.5 mm
Inclusions (section 6.1)	
Sorting:	Bimodal/fairly sorted
Average size:	Coarse
Composition	Subangular to angular
Composition:	**Dominant to Frequent:** Subangular or angular, coarse ranging from medium to coarse, particles made of spar, probably calcite or dolomite; many present micrite alterations; varied (0.04 mm to .1 mm) crystal size. **Few to Very few**: fine angular individual crystals (likely calcite or dolomite) embedded in the matrix, each around 0.04-0.07 mm **Very rare to Few**: Glassy, rhomboid medium or coarse crystals; opaque subrounded grey micrite; micritized fossils
Voids:	Porous

Dark Particles Fabrics

Dark-Coarse/Medium-W (GM-cm-W)

Fired color:	Very varied: light grey (10YR 7/2), grey (10YR 6/1), light brownish grey (10YR 6/2), very pale brown (10YR 7/3), pale brown (10YR 6/3), light pink (5YR 8/2).
Grain size (mode)	Medium to coarse
Re-fired color :	White (Table 6-2 for Munsell colors)
Hardness:	Soft, chips with nail, and easily scratched w/ # 2.
Compactness:	Not compacted, usually
Rounding:	Rounded to angular
Fracture:	Usually hackly, irregular
Thickness:	Most between 7 and 11mm
Inclusions (section 6.1)	
Sorting:	Fairly sorted
Rounding:	Varied
Composition:	**Common to Few:** Dark grey particles, opaque or sub-opaque, from rounded to angular, medium to very coarse, no internal structure is observed and the individual crystals cannot be seen at the magnification of the binocular microscope. **Few to Very rare:** White micrite, medium to coarse.
Voids:	Deep rounded pores and long channels running on the direction of the wall.

Fabric Dark-coarse-R (GM-cm-R/B)

Fired Color:	Grey (10YR 6/1), light red (2.5YR 7/6) and reddish brown (5YR 6/4)
Grain size:)	Medium to coarse
Re-fired Color:	Red (2.5 YR 5/6) or light reddish brown (Table 6-2 for Munsell colors)
Hardness:	Soft
Compactness:	Porous, not compacted
Rounding:	rounded to angular
Fracture:	Hackly, irregular
Thickness:	6 – 11.5 mm
Inclusions (section 6.1)	
Sorting:	Fairly sorted
Predominant size:	Medium to coarse
Composition:	**Common** – Grey, dark particle, coarse (most frequent grain size), size ranging from coarse to medium, subangular. **Very few to Very Rare** - Sparite - present in a few samples, medium size
Voids:	Varied, from moderately compacted to some with long channels

Single Crystals Fabrics

Crystal-fine-red-refired (XT-f-R/B)

Fired Color:	pink (10R8/3), light grey (10R7/1), light red (10R 7/6), dark reddish grey (2.5 YR 4/1)
Grain size:	Fine
Re-fired Color:	Red and strong pink (Table 6.2 for red refired Munsell colors)
Hardness:	Soft, with some moderately hard but that still are scratched with Mohs # 2..
Compactness:	Compacted to moderately compacted
Rounding:	Angular
Fracture:	Primarily smooth, but one is hackly.
Thickness:	From 5 to 7 mm
Inclusions (section 6.1)	
Overall Abundance:	Veried, from Common to Dominant
Sorting:	Well sorted
Average size:	Fine
Composition of Coarser Portion (>0.0625 mm):	**Dominant to Common***:* under hand lens 10 – 20X, particles are angular, smooth white or light grey. With the binocular more detail is observed: the inclusions are mostly individual crystals (and some crystal aggregates) with size from very fine to fine (mode is fine); glassy, semitranslucent to sub-opaque; rounding is angular with rhomboid shape; color if present: whitish or light grey. **Rare to Very rare***:* medium size (mode), white or grey micrite, with size from fine to coarse
Voids:	very small pores and channels

Crystal-Coarse-R/B (XT-c-R/B)

Fired Color:	Light red (10R 7/6),, light reddish brown (2.5YR 6/3), pink (2.5 YR 8/4)
Grain size:	Coarse
Re-fired Color:	Light reddish brown (2.5YR 6/3,4,6), pink (2.5 YR 8/4), reddish brown (5YR 4/4) and red.
Hardness:	Varied: mostly soft, but some from Xcanchakan from Chac Mool and Cancún from Culubá are moderately hard
Compactness:	Varied
Fracture:	Varied
Thickness:	5.1 – 8.5 mm
Inclusions (section 6.1)	
Sorting:	Good
Composition:	**Frequent to Few** – glassy, semi-translucent to translucent, rhombic or blocky, mostly coarsely grained (mode) single crystals. In some Cancún variety samples from Culubá there are elongated poly-crystalline fragments that still maintain a shell form. **Very Few to Very rare** –at western sites, sparry particles may be present, medium size; grey dark inclusions may be present

APPENDIX E: DETAILED DESCRIPTION OF FABRIC CLASSES IN THIN SECTION

Dolospar Fabric Classes

Dolospar-W-Coarse (DS-W-c)

General Information of DS-W-c Fabric Class:

No. of thin sections: 27
General fabric type : Dolomite spar
Main characteristics: (1) dolospar as main inclusion; (2) turns to white color when refired; (3) coarsely grained.
Photomicrographs: Figures 8-11, 8-12
Ceramic sphere: Tases

Summary of Characteristics in Hand Specimen (Chapter 6)

Hand-specimen Fabric: Spar-coarse-W (SP-c-W)
Hardness : soft
Refired-color: white (Table 6-3)
Texture: coarse
Fired color: 10YR 6/1 Grey; 10YR 7/1 light grey; 7.5 YR 8/2 pinkish white;
10YR 7/2, 6/2 Light grey brown & light brownish grey

DS-W-c in Thin section:

I) Microstructure (see section 8.2.2)

- Voids: High porosity with voids ranging from 5 - 25 per cent of the thin-section area; this is usually a matrix very fragmented by pores and/or long channels that can be up to 1 mm and vughs up to 0.4 mm. Separation voids are observed around some inclusions.
- Packing (particles > silt) : mostly single-spaced.
- Preferred orientation: no visible orientation of voids or particles

IIa) Groundmass: homogeneity and optical characteristic (see section 8.2.2)

- The micromass is largely optically inactive.
- The colors of the micromass are mostly 10YR 7/2 light grey; 5Y 7/2, 6/2 very light olive grey, 5YR 6/4,6 light & reddish yellow.
- Within the individual samples, the micromass is fairly homogeneous in color, degree of optical activity, and distribution of inclusions.
- Across the samples within this fabric class, the fabric is largely homogeneous with the main difference related the frequency of micrite and the degree of micritisation of dolospar grains.

IIb) Groundmass: coarser inclusions (see section 8.2.1)

- The coarse:fine:voids ratio is 30-50 : 25-65: 5-25 with a coarse:fine boundary of 0.0625 mm
- Coarser Inclusions (>0.0625 mm) (frequency ranges refer to the whole thin section, see section 8.2.1).

Frequent to Common: DOLOSPAR – various amounts of dolospar; many with micritic alteration; many with embedded fossilized shells (probably bivalves) similar to the ones floating in the matrix; particle size is from medium to coarse (> 0.25 mm, section 8.2.1 (b)), with a mode of coarse; sub-angular to angular. The dolomite crystals forming the particle are usually euhedral with rhombohedral shape, size: 0.010 – 0.060 mm. Red iron alteration of crystals may be present.
Few to Very rare : GREY or BROWN MICRITE – medium brown-grey opaque or semi-opaque in xp; subrounded to rounded; MICRITE – very coarse to fine; subrounded; whitish, very light brown, or very light grey in xp; sub-translucent..
Rare to Very rare: SKELETAL REMAINS and HOMOGENEOUS CALCITE (CH) – Recognisable skeletal (shells) fossil remains and fragments, likely to be skeletal (made of homogeneous calcite: section 8.1.1); grain size from very coarse to fine; poorly sorted; varied particle shape SPARITE – very finely to finely crystalline sparry calcite, most with particle size from medium to coarse; DOLOMITE CRYSTALS – eroded, subhedral, equant single dolomite crystals with particle size from fine to coarse; CALCITE CRYSTALS – single crystals with size from fine to coarse, with planar boundaries; also mosaics made of medium-crystalline calcite with over all particle size from fine to coarse particle size, many with light brown or greenish mineralisation.
Very rare: QUARTZ – two or three single-crystal grains, usually with particle size from fine (0.0625 – 0.25 mm, or section 6.2) quartz crystals.

IIc) Groundmass: finer inclusions (see section 8.2.1):

The total fine (<0.0625 mm) portion takes 25 to 65% (see c:f:v), and the finer inclusions or silt (0.04 - 0.0625 mm) are relative to the fine portion only.
Frequent to common: mainly single dolomite crystals, finely crystalline, mostly euhedral, rhombohedral

III) Textural Concentration Features

The following two may be present

- Light brown to light-yellow brown in pp, greenish grey in xp, homogeneous with no internal structure, inclusion, or orientation; mostly sub-rounded; lower optical density than micromass; most common size is 0.15 mm.
- Brownish-red to dark-brown-red (xp), more than 5 are present in most thin sections; no internal structure, inclusions or orientation; high optical density; most have clear boundaries; the mode is 0.15 mm with a range from 0.05 to 0.3 mm; usually rounded to subrounded.

Dolospar-W-Medium (DS-W-m)

Fabric class code DS-W-m general information:

No. of thin sections = 28
General fabric type : Dolomite spar
Main characteristics: (1) dolospar as main inclusion; (2) medium-grained; (3) turns to white color when refired
Photomicrographs: Figure 8-13
Ceramic sphere: Tases

Summary of Characteristics in Hand Specimen (Chapter 6):

Hand-specimen Fabric: Spar-medium-W (SP-M-W) and Micrite-Medium-W (WM-m-W)
Hardness : soft
Refired-color: white (see Table 6-3)
Fired color: 10YR 7/1 Light grey; 5YR 8/2 Pinkish white; 2.5YR 7/6 Light red; 10YR 7/3 very pale brown; 10YR 6/2 light brown grey

DS-W-m in Thin section:

I) Microstructure (see section 8.2.2)

- voids: moderate porosity; less porous and more compacted fabric than DS-W-c; very few vughs comprising around 3 - 5per cent of the thin section; usually without voids separating particles from matrix (but some inclusions do have them).
- Packing (particles > silt): spaced to singled
- Preferred orientation: no orientation of vughs or particles were observed

IIa) Groundmass: homogeneity and optical characteristics (see section 8.2.2)

- Homogeneity: color between thin sections fairly homogeneous mostly very pale brown (10 YR 7/3) color. Within one sample, color fairly homogeneous throughout most of the micromass. In some samples, inhomogeneous bands of darker brown by on wall are present.
- Small and few areas of optically faintly active micromass. In pp, faint pleochroism may be observed in areas of lighter color of the micromass. In areas of lighter color, the colors of the micromass in fabrics Dolospar-W-Medium and Dolospar-W-Coarse are similar.
- The colors under xp are 5Y 7/2, 6/2 very light olive grey; 10 YR 6/2 light brownish grey; 10 YR7/3 very pale brown; 10YR 5/4 yellowish brown; 10YR 5/3 brown;

IIb) Groundmass: coarser inclusions (see section 8.2.1):

- This fabric is defined with a wide coarse:fine:voids ratio of 20-45:77-50: 3-5, with a coarse:fine boundary of 0.0625 mm
- Coarser Inclusions (>0.0625 mm)

Common to Few: DOLOSPAR – particle size ranges from fine to coarse, mode is medium; particles are mostly subangular; the individual dolomite crystals forming the particles are usually euhedral with rhombohedral shape; size: 0.010 – 0.060 mm; particles may be altered by micrite; some with embedded fossilised shells (probably bivalves), similar to the ones embedded in the micromass;
Few to Very rare: MICRITE – semi-translucent in xp; white and chalky in hand specimen; SPARITE – medium to coarse, finely crystalline, rounded to subrounded; GREY or BROWN MICRITE – likely, with organic material; usually rounded to subrounded, opaque to semi-opaque in xp.
Rare : SKELETAL REMAINS and HOMOGENEOUS CALCITE – recognisable skeletal remains and homogeneous calcite fragments likely to be skeletal (made of homogeneous calcite, section 8.1.1), sub-translucent in polarised light; SPARITE – coarse, subrounded.
Rare-Very rare: SINGLE CALCITE CRYSTAL — planar crystals usually from fine to coarse; also medium-crystalline calcite mosaics with fine to coarse particle size; some blocky and prolate shape; many with brown alteration; SINGLE DOLOMITE CRYSTALS usually planar boundaries, euhedral to subhedral, medium particle size.
Very rare: QUARTZ – two or three single-crystal grains, usually, fine size (0.0625-0.25 mm, or see section 6.2).

IIc) Groundmass: fine portion (see section 8.2.1):

The fine portion (<0.0625 mm):takes 77-50 per cent of the whole field of view (see c:f:v), and the finer inclusions or silt (0.04 - 0.0625 mm) frequencies are relative to the fine portion only.

Frequent to common silt size single dolomite crystals from 0.03 to 0.0625 mm, euhedral, rhombohedral. Reddish brown iron alteration of crystals may be present

Dolospar-W-Hard (DS-W-mc-h)

Fabric class code DS-W-mc-h general information:

No. of thin sections = 11
General fabric type: Dolomite spar
Main characteristics: (1) dolospar as the main inclusion, also grey micrite ; (2) hard paste; (3) turns to white when re-fired; (4) coarse to medium medium-grained particle size
Photomicrographs: Figure 8-14
Ceramic sphere: Tases

Summary of Characteristics in Hand Specimen (Chapter 6)

Fabrics: Micrite-White-hard (WM-cm-W-h)
Hardness : hard
Refired color: white (see Table 6-3)
Fired color: 10 YR 7/2 light grey; 10 YR 6/2 light brownish grey; 10 YR 6/3 pale brown

DS-W-mc-h in Thin section :

I) Microstructure (see section 8.2.2)

- Moderate to high porosity with few voids, usually vesicular vughs; separation voids are observed around are common
- Packing (particles > silt) : single spacing
- Preferred orientation: no voids or particle orientation observed

IIa) Groundmass: homogeneity and optical characteristic (see section 8.2.2)

- Fired (original) color in xp, pp: yellow brown (10YR 5/4), reddish brown (2.5 YR 5/4), light brownish or light greenish grey (10 YR 7/3)
- The micromass is mostly brown color, with large opaque areas, but bands of light brownish grey, or light greenish grey areas by one wall may be present. Mottled micromass occurs in many areas.
- Micromass largely optically inactive. In pp, faint pleochroism is observed in the micromass in some samples.

IIb) Groundmass: coarser inclusions (see section 8.2.1)

- The coarse:fine:voids ratio is 20-45: 75-50: 5 with a coarse:fine boundary of 0.0625 mm
- Coarser Inclusions (>0.0625 mm) (frequencies are taken of the whole thin section).

Common to Few: DOLOSPAR – particle size ranges from fine to very coarse, mode is varied with coarse and medium modes included within this fabric; subangular; The dolomite crystals forming each particle are usually euhedral with rhombohedral shape, with size from 0.010 to 0.060 mm.
Few: BROWN MICRITE – opaque to semi-opaque (xp), medium to dark brown, rounded to subrounded, also irregularly shaped; no internal structure, without inclusions; particles are tinted with alizarin red; coarse to fine, mode is medium. In areas of lighter micromass, the color of the brown inclusions is very light brown indicating that the brown color is due to organic material. MICRITE – medium to coarse, mode is medium; subrounded to rounded.
Very Few to Very rare: SKELETAL REMAINS and HOMOGENEOUS CALCITE (CH) – Recognisable skeletal remains and fragments likely to be skeletal (made of homogeneous calcite, section 8.1.1); MICRITE – semi-translucent in xp.
Very rare: SINGLE DOLOMITE CRYSTAL — some prismatic and prolate in shape; CALCITE CRYSTALS — single crystals, planar boundaries, with particle size from fine to coarse; medium-crystalline clear calcite mosaics may be present, with various particle size from fine to coarse; QUARTZ – absent in half the samples and, in the rest, from one to five particles may be present.

IIc) Groundmass: finer inclusions (see section 8.2.1)

The fine portion takes 85-50 per cent of the whole field of view (see c:f:v), and the finer inclusions or silt (0.04 - 0.0625 mm) frequencies are relative to the fine portion only.
High percentage of silt made of single dolomite crystals 0.03 to 0.0625 mm (finely crystalline), mostly euhedral, rhombohedral comprising 20- 50 per cent of the fine portion.

III) Textural Concentration Features

Textural concentration features were not observed, but the dark micromass and lumps may be preventing their observation

IV) Amorphous Concentrations Features

Common dark brown opaque fine to medium nodules, most likely organic matter. In addition areas of the micromass show impregnation with same matter.

Finely Crystalline Sparite Fabric Classes

Sparite-W/B-Coarse/Medium (CS-W/B-cm)

Fabric class code CS-W/B-cm general information:
No. of thin sections = 20 General fabric type: finely crystalline sparite Main characteristics: (1) finely crystalline sparite as main inclusion; (2) turn to white or light reddish brown when refired; (3) coarse or medium-grained Photomicrographs: 8-16, 8-17 Ceramic sphere: Tases
Summary of Characteristics in Hand Specimen (Chapter 6):
Hand-specimen fabrics: various (some characterised as single crystals, others as sparite, and others as white and dark micrite). Hardness : soft Refired color: white, light reddish brown. Grain size: medium to coarse Fired color: very light grey, light pinkish grey or light pink
CS-W/B-cm in Thin section:
I) Microstructure (see section 8.2.2)
• Most samples are low to moderately porous presenting in thin section vughs taking around 3 - 7 per cent of the thin section areas, with only a few thin sections containing significant channels. • No preferred orientation of particles or voids was observed.
IIa) Groundmass: homogeneity and optical characteristic (see section 8.2.2)
• The micromass is optically inactive to optically slightly active. • Within individual specimens, the micromass is homogeneous in color and optical activity. • Between samples this is not a homogeneous fabric, with a wide range or porosity and micromass color. The colors found include (pp), light red (7.5 YR 7/6), light.reddish brown (5yr 6/4), dark brown, light greenish grey 7.5YR 7/2 - 7/1, very pale brown (10 YR 8/2).
IIb) Groundmass: coarser inclusions (see section 8.2.1)
• The coarse:fine:voids ratio is around 15-40:82-53:3-7, with the coarse:fine boundary at 0.0625 mm • Coarser Inclusions (>0.0625 mm; percentages are taken of the whole thin section). **Common :** SPARITE (CLEAR) – particles usually subangular, with size that ranges from medium to very coarse with a mode that varies. Particles made of finely crystalline, anhedral, granular calcite crystals; each crystal ranges in size from 0.015 to 0.06 mm (15-60 microns); particles usually subangular, with size that ranges from medium to very coarse with mode that varies. **Common to Very rare:** SKELETAL REMAINS and HOMOGENEOUS CALCITE – recognisable fossil shells and fragments likely to be skeletal (made of homogeneous, section 8.1.1). Particle size ranges from coarse to fine. This type of inclusion is absent in about half the samples. **Few to Very rare:** BROWN or GREY MICRITE – particles are opaque to semi opaque (xp), without any internal structure; fine to coarse grain size; sub-rounded to rounded; with some particles with holes as colander; brown micrite is present in more than half of the samples; MICRITE – subrounded, medium-grained micrite, sub-translucent (xp); micrite is present in half or less of these samples. **Very rare**: SINGLE DOLOMITE CRYSTALS – euhedral and subhedral; SINGLE CALCITE CRYSTALS
IIc) Groundmass: fine portion (see section 8.2.1)
The total fine portion takes 82-58 per cent of the whole field of view (see c:f:v). The finer inclusions or silt (0.04 - 0.0625 mm) frequencies are relative to the fine portion only. Common to rare, usually made of fragments of calcite crystals, or rounded dark micrite, or tcf.
III) Textural Concentration Features (tcf)
The following tcf were found, but they are not common: Brownish-red to dark-brown-red (xp) concretions: present to a frequency count of up to 5 – 10 particles in the thic-section; the particles have no internal structure, inclusions or orientation; high optical density; most with clear boundaries; grain-size mode 0.15 mm with a range from 0.05 mm to 0.3 mm; subrounded and acicular (needle like). Tipikal has an anomalous sherd with around 15 per cent (of thin section area) red concretions. Yellowish Light brown particles present in one third of the samples; homogeneous with no internal structure, inclusion, or orientation; faintly pleochroic (pp); mostly sub-rounded; lower optical density that matrix; mode is 0.15

Micrite Fabric Class

Micrite-W-Medium (CM-W-m)

Fabric CM-W-m General Information:

No. of thin sections = 17
General fabric type: Micrite
Main characteristics: (1) micrite (semitranslucent in xp) as main inclusion; (2) turns to white when refired; (3) medium-grained
Photomicrographs: 8-18
Ceramic sphere: Tases

Summary of Characteristics in Hand Specimen (Chapter 6):

Fabrics: Micrite-Medium-W (WM-m-W)
Hardness : soft
Refired-color: white (see Table 6-3 for refired Munsell colors)
Fired color: varied color including pinkish grey(7.5 R 7/2) to reddish brown 2.5 YR 5/4)

CM-W-m in Thin section:

I) Microstructure (see section 8.2.2)

- Voids: Most samples have voids in the form of vughs (many subrounded) taking 3-8 per cent of thin section, but the range is wider with one sample taking around 15 per cent. Most voids are vughs, with only a few channels. Most micrite grains show diffused grain boundaries.
- The space between coarse particles (>0.0625) is varied. Sorting is varied. The whole range of sizes is present from fine to very coarse, with mode = medium.
- No preferred orientation of particles or voids was observed.

IIa) Groundmass: homogeneity and optical characteristic (see section 8.2.2)

The fabric is characterised by

- The micromass is not homogeneous within thin sections and between samples. The color of the micromass is very varied from reddish brown and medium brown to light grey and a light brown. Some have darker medium brown areas, other samples (#161 for instance) have a mottled appearance with dark or opaque micromass spotted by small speckled areas of optical activity, while in some other instances (#332 and #340), the groundmass is largely homogeneous in color, optical activity, and inclusion distribution.

IIb) Groundmass: coarser inclusions (see section 8.2.1)

The c:f:v varies from 20-35: 57-77 :3-8 with a c:f boundary at 0.0625mm,
Coarser Inclusions (>0.0625 mm) (frequencies are taken of the whole thin section).

Common: MICRITE (semitranslucent in xp) –mostly sub angular to subrounded; fairly sorted, with grain size ranging from fine to coarse but most (mode) are fine to medium.
Few to Very rare : SPARITE (clear) – medium to coarse; rounded, subrounded
Very few to Very rare: BROWN or GREY MICRITE – Semi opaque, mostly brown micrite (xp); fine to coarse grain size; subrounded, rounded; homogeneous appearance. HOMOGENEOUS CALCITE and SKELETAL REMAINS – Recognisable remains such as shells in addition to homogeneous calcite (which likely are skeletal remains) grain size ranging from very fine to coarse.
Very rare: QUARTZ – quartz is a very minor component in about half the samples (three or four particles in the thin section) and absent or one particle in the rest; DISCRETE CALCITE CRYSTALS – fine grained; DOLOMITE CRYSTALS – finely crystalline

IIc) Groundmass: fine portion (see section 8.2.1)

Fine Portion (<0.0625 mm) (frequencies are based on the fine portion only)
Rare to very rare silt size fragments of calcite crystals

III) Textural Concentration Features

Brownish-red to dark-brown-red (xp), with no internal structure, inclusions or orientation; varied optical density; most with clear boundaries; the most frequent size is 0.15 mm with a range from 0.05 mm to 0.3 mm; subrounded and acicular (needle like).

Medium and Coarsely Crystalline Calcite Fabric Classes

Calcite-Micrite-01-Coarse (CX-CM-01-c)

General Information of the CX-CM-01-c Fabric Class:

No. of thin sections: 16
General fabric type: Single calcite crystals
Main characteristics: (1) single calcite crystals (and sparite or micrite) ; (2) moderately compacted; (3) varied refired color; (3) coarsely grained
Photomicrographs: Figures 8-19, 8-20
Ceramic Sphere: Oriental Tases:

Summary of Characteristics in Hand Specimen (Chapter 6):

Hand-specimen fabrics: Crystal-Coarse-R/B (XT-c-R/B)
Hardness : soft
Refired color: red, light reddish brown
Fired color: light red, pinkish grey
Texture: coarse

CX-CS-R/B-c in Thin section:

I) Microstructure (section 8.2.2)

- Voids: Low to moderately porous fabric with vughs taking 1-5 per cent of the whole field of view.
- Packing: single to double spaced
- Preferred orientation: no orientation of particles or voids was observed.

IIa) Groundmass: homogeneity and optical characteristic (definition of terms and scales: this chapter, section 8.2.2)

- Micromass homogeneity within individual specimens: mostly homogeneous color and optical activity. Heterogeneous samples across the group.
- Color xx, xp: medium grey, light reddish brown or brown; optical state: not active micromass to optically slightly active; semi opaque micromass in many

IIb) Groundmass: coarser inclusions (section 8.2.1)

The coarse:fine:voids ratio is 25-40:55-75:1-5 with a coarse:fine boundary of 0.0625 mm

Coarse inclusions (>0.0625 mm) (frequencies of the whole thin section).

Frequent to Common: SINGLE CALCITE CRYSTALS – coarse (mode) particles ranging in size from very fine to coarse; most are angular with planar boundaries, subhedral and some anhedral, many altered with reddish brown alteration .
Few to Very rare: MICRITE – birefringent and semitranslucent (xp), mostly sub rounded micrite particles, usually coarsely grained (mode) but ranging from fine to very coarse; SPARITE – grain size ranging from medium to coarse.
Rare to Very rare: DARK MICRITE – subrounded to rounded, fine to coarse; SKELETAL REMAINS; MEDIUM to COARSELY CRYSTALLINE SPARITE – medium to coarse; red concretions can take at least 2-3 per cent of the thin section vary in size from fine to coarse (Quartz is absent).

IIc) Groundmass: fine portion (section 8.2.1)

Fine Portion (<0.0625 mm) (frequencies are of the fine portion only)
Presence of silt is varied, but usually very few silt, mostly calcite fragments taking 3 to 7 per cent of fine portion

III) Textural Concentration Features

- Brownish-red to reddish brown (xp) concretions are very common, and mostly rounded to sub-rounded, with no internal structure, inclusions or orientation; high optical density; size ranging from fine to coarse with the most frequent size 0.15mm. The brown and the red concretions are two different colors depending most likely on temperature: brown opaque in no oxidation and red in oxidation.
- Yellow concretions: yellow concretions are a mineral alteration of calcite crystal grains that overcomes the grain and becomes what looks as an isolated grain by it-self.

Sparite-m-W/B-Coarse (CSm-W/B-c)

General Information of CSm-W/B-c Fabric Class:

No. of thin sections: 11
General fabric type : single calcite crystals
Main characteristics: 1) Usually few sparite in which each crystal is relatively large in size, or medium-crystalline, sometimes in combination with clear finely crystalline sparite or coarse single calcite crystals (2) soft paste; (4) coarsely grained
Photomicrographs: Figure 8-22
Ceramic sphere: Tases

Summary of Characteristics in Hand Specimen (Chapter 6)

Fabrics : Spar-Coarse-B various (SP-c-B); DarkParticles-Coarse/Medium-W (GM-cm-W)
Hardness : soft
Refired-color: white and light reddish brown
Texture: coarse
Fired color: light red, reddish brown

CSm-W/B-c in Thin section:

I) Microstructure (definition of terms and scales: this chapter, section 8.2.2)

- Voids: very porous (with one exception), with vughs and long channels taking 10 - 15 per cent of the thin section.
- Packing (particles > silt) : double spaced or more.
- Preferred orientation: no orientation of particles or voids was observed.

IIa) Groundmass: homogeneity and optical characteristic (section 8.2.2)

- The micromass color is not homogeneous between thin section, but it is fairly homogeneous within a thin section.
- The colors of the micromass under xp are usually browns: light brown, light reddish brown, and medium brown.

IIb) Groundmass: coarser inclusions (section 8.2.1)

- The coarse:fine:voids ratio is 20-30-: 55-70: 10-15 with coarse:fine boundary of 0.0625 mm
- Coarser Inclusions (>0.0625 mm) (frequencies are taken of the whole thin section).

Common to Few: SPARITE (mediun-crystalline or larger) – coarse (mode) particles made of mosaics of sparite, or sparry calcite, with a coarse mode. The crystals comprising the particle are medium-crystalline or larger, usually between 0.06 mm to 0.18 mm (medium crystalline defined as = 0.0625 – 0.25 mm), contrasting to the finely crystalline sparite in the sparite fabric CS-W/B-cm.
Few to Very rare: MICRITE – micrite (sub-translucent and translucent in xp); and DARK BROWN INCLUSIONS (opaque or sub-opaque in xp).
Very rare: SINGLE CALCITE – medium grained single calcite crystals, usually subangular; SPARITE (finely crystalline) – medium to coarse, subrounded to subangular, particles of finely crystalline sparite.

IIc) Groundmass: fine portion (scales and measurements: this chapter, section 8.2.1):

The total fine portion takes 55-70 per cent of the whole field of view (see c:f:v), and the finer inclusions or silt (0.04 - 0.0625 mm) frequencies are relative to the fine portion only.
Rare to very rare calcite crystal fragments

III) Textural Concentration Features

The following two may be present

- light-yellow brown in pp, homogeneous with no internal structure.
- Brownish-red (xp); no internal structure; with clear boundaries; usually rounded to subrounded.

Calcite-Quartz-B/R-Fine (CX-QZ-R/B-f)

Fabric class code CX-QZ-R/B-f general information:
No. of thin sections: 9 General fabric type: calcite and quartz single crystals Main characteristics: (1) fine grained calcite and quartz; (2) mostly compacted Photomicrographs: 8-23, 8-24 Ceramic Sphere: Eastern Tases (and one classified as Mama)
Summary of Characteristics in Hand Specimen (section 6.2):
Hand-specimen analysis: Crystalline-Fine-R/B (XT-f-R/B) Hardness : mostly soft but two hard samples are included Refired color: red and light reddish brown Porosity: compacted to moderately porous Texture: fine-grained Fired color: grey, dark reddish grey, light red
CX-QZ-R/B-f **in Thin section:**
I) Microstructure (section 8.2.2)
• Few voids, usually vesicular vughs taking around 1-5 per cent of the thin section area • Packing (particles > silt) : single spacing • Preferred orientation: no voids or particle orientation observed
IIa) Groundmass: homogeneity and optical characteristic (section 8.2.2)
(a) The micromass is homogeneous, without crystallites, with colors (pp, xp) brown (7.5 YR 5/4), dark yellowish brown 10YR 4/6, pale brown 2.5Y 8/3
IIb) Groundmass: coarser inclusions (section 8.2.1)
(a) The c:f:v from 20-65: 30-80: 1-5, with a c:f micromass' boundary of 0.0625, (b) Coarser Inclusions (>0.0625 mm) (frequencies are taken of the whole thin section). **Frequent to Common:** SINGLE CALCITE XTALS – discrete particles ranging in size from very fine to coarse, with very fine to fine mode; angular, planar boundaries, anhedral. **Few to Very few:** QUARTZ – discrete particles ranging in size from fine to medium, with fine mode; usually subrounded to rounded, anhedral
IIc) Groundmass: fine portion (section 8.2.1)
The Fine Portion (<0.0625 mm) is 30-80 per cent of total, and the frequency is of the fine portion only. Common mostly calcite fragments
III) Textural Concentration Features
Count of 3 to 10 brownish-red to dark-brown-red (xp), with no internal structure, inclusions or orientation; high optical density; most with clear boundaries; the most frequent size is 0.15 mm with a range from 0.05 mm to 0.3 mm; subrounded and acicular (needle like).

Calcite-Grog-B/R-Coarse (MX-GROG-B/R-c)

Fabric class code MX-grog-B/R-c general information:

No. of thin sections: 6
General fabric type : dark argillaceous
Main characteristics: (1) dark argillaceous particles with embedded calcite or quartz crystal inclusions ; (2) micrite or single crystals of calcite ; (3) red or light-reddish brown refired
Photomicrographs: Figure 8-25, 8-26, 8-27
Ceramic sphere: Tases

Summary of Characteristics in Hand Specimen (section 6.2)

Fabric : Crystal-Coarse-R/B (XT-c-R/B) or DarkParticles-Coarse-R/B (GM-c-R/B)
Hardness : soft
Refired-color: red and light-reddish brown
Texture: coarse
Fired color: grey, dark grey, light brown

MX-GROG-B/R-c in Thin section:

I) Microstructure (section 8.2.2)

- Voids: few to common porosity with large vughs and channels
- Packing (particles > silt) : double spaced
- Preferred orientation: no preferred orientation of voids or particles was observed

IIa) Groundmass: homogeneity and optical characteristic (section 8.2.2)

- The micromass is not homogeneous in color and optical activity. The colors of the micromass are dark colors such as strong brown (7.5YR 4/6), reddish brown 2.5 YR 5/6 and lighter colors such as light brown or light red (2.5 YR 6/6).

IIb) Groundmass: coarser inclusions (section 8.2.1)

- The coarse:fine:voids ratio is 35: 55 – 60:5-10 with a coarse:fine boundary of 0.0625 mm
- Coarser Inclusions (>0.0625 mm) (frequencies are of the whole field of view).

Few to Very few: ARGILLACEOUS PARTICLES with CRYSTALS – angular to subrounded, coarse in mode (but ranging from fine to coarse), high optical density, opaque, dark brown argillaceous particles with silt to very fine inclusions of calcite crystals; these particles, probably, are sherd fragments.
Few to Very few: SINGLE CALCITE CRYSTAL – Fine to coarse (mode medium) mostly with planar boundaries, euhedral calcite crystals
Few to Very rare: MICRITE – fine to very coarse, rounded to subrounded;
Rare: SPARITE
Rare to Very rare: GROG – medium sherd fragments, some with the same fabric than the dark brown argillaceous inclussions; QUARTZ – two or three fine particles in most thin sections. SINGLE DOLOMITE CRYSTALS

IIc) Groundmass: fine portion (section 8.2.1):

Fine Portion (<0.0625 mm) (frequencies are of the fine portion only)
Frequent to rare angular calcite crystals; few angular and sub-angular dark argillaceous particles

III) Textural Concentration Features

Light greenish grey (10Y 7/1) medium to coarse size, rounded particles with ghosts of acicular skeletal remains giving them the appearance of a ball of coquina; less dense than the fabric and of contrasting color and inclusions.

Medium and Coarsely Crystalline Dolomite Fabrics

Dolomite-W/B-Medium with (DX-B/W-m)

Fabric class code DX-B-m general information:
No. of thin sections : 5 General fabric type: single dolomite crystals Main characteristics: (1) fine-to-medium particle size, euhedral dolomite; (2) fine to medium texture; (3) few silt Photomicrographs: Figure 8-28 Ceramic Sphere: Tases
Summary of Characteristics in Hand Specimen (Chapter 6)
Hand-specimen fabrics: samples originally assigned to DarkParticle-Coarse/Medium-W (GM-cm-W) and DarkParticle-coarse-R/B because fine dolomite crystals were not detected during hand-specimen analysis. Hardness : soft Refired color: light reddish brown and white Porosity: low
DX-B/W-m in Thin section:
I) Microstructure (8.2.2)
Mostly compacted sherds with voids, usually in the form of thin channels taking 2 -7 per cent of field of view. The spacing between grains in the coarse portion (> 0.0625mm) is double to open.
IIa) Groundmass: homogeneity and optical characteristic (section 8.2.2)
• Largely homogeneous groundmass • lightly active mostly without crystallites, but in some (#339) some faint crystallites were observed • very pale greenish brown (xp, pp)
IIb) Groundmass: coarser inclusions (section 8.2.1)
This is a medium-grained fabric. with the micromass' boundary at 0.0625, the c:f:v around 25:70:5 **Common to few:** DOLOMITE (single crystals) – fine (mode) particles of rhombohedral single crystals of dolomite, with size ranging from fine to medium, usually 0.1 - 0.35 mm in longest diameter particle size; euhedral to subhedral merging into the clay matrix; colourless in pp light and light grey in xp. These inclusion are well sorted. **Very few:** DARK MICRITE – sub opaque (xp); medium to coarse, with medium mode. **Rare to very rare:** MICRITE – micrite that is birefringent or semi translucent (xp), with wide variety of roundness (rounded, surrounded, and sub angular). CLEAR SPARITE (finely crystalline) – sparite fragments in which the most common size of each crystal is 0.020-0.010 mm; fine (mode) particle size, when present, but the size could range from fine to coarse. SINGLE CALCITE CRYSTALS – very fine to medium size particles, subhedral.
• IIc) Groundmass: fine portion (section 8.2.1):
Fine Portion (<0.0625 mm) (frequencies are of the fine portion only) Very rare silt size inclusions, but some silt-size red concretions are present
III) Textural Concentration Features
Reddish-yellow to olive-green in pp and xp, with no internal structure, inclusion, or orientation; faintly pleochroic with mottled extinction; mostly rounded; higher optical density that matrix; diffuse boundaries; most frequent size is 0.05 mm ranging from 0.05 to .2 mm Brownish-red to dark-brown-red (xp), with no internal structure, inclusions or orientation; frequency of more than 10 particles; high optical density; most with clear boundaries; the most frequent size is 0.04 mm with a range of 30 microns to 0.15 mm; most are rounded, but some subangular and irregular.
IV) Amorphous
Black (pp and xp); very rare; rounded or rhombic probably showing relic shape of dolomite, replaced by most probably, magnetite; most frequent size is 0.08 mm with range 0.06 to 0.1 mm.

Dolomite-Calcite-B-Coarse (DX-CX-B-c)

Fabric class DX-CX-B-c general information:
No. of thin sections = 4 General fabric type: dolomite crystals Main characteristics: coarse (mode) single dolomite crystals Photomicrographs: 8-30, 8-31, 8-32 Ceramic sphere: Tases
Summary of Characteristics in Hand Specimen (section 6.2):
Hand-specimen Fabric: Crystalline-Coarse-R/B (XT-c-R/B) Hardness : soft Refired color: reddish brown, light reddish brown Texture: coarse Fired color: grey, light reddish brown, pale brown
DX-CX-B-c in Thin section :
I) Microstructure (see section 8.2.2)
• Porosity: varied, from high to moderate taking approximately from 10 to 3 per cent • Packing (particles > silt) : doubled spaced • Preferred orientation: no orientation was observed
IIa) Groundmass: homogeneity and optical characteristic (section 8.2.2)
• Homogeneous: the inclusions are fairly homogeneous throughout the thin section. Variations in color were observed within the Tekit samples. • Fired (original) color in xp, pp: light yellowish brown0YR 6/4, very pale brown 10YR 7/4
IIb) Groundmass: coarser inclusions (section 8.2.1)
The c:f:v is 35:55-62:3-10, with the micromass boundary at 0.0625, Coarser Inclusions (>0.0625 mm) (frequency ranges are taken of the whole thin section). **Frequent to Few**: SINGLE DOLOMITE CRYSTALS – coarse (mode) single crystals, with size ranging from fine to coarsely grained; euhedral to anhedral; in Mayapán samples, some are very eroded or altered. **Very few to Very rare:** SINGLE CALCITE CRYSTALS –medium (mode) with size from very fine to medium; subhedral; DARK MICRITE – sub opaque (xp), fine to medium, with mode fine; MICRITE – semitranslucent (xp), with wide variety of roundness (rounded, surrounded, and sub angular).
IIc) Groundmass: finer inclusions (section 8.2.1)
Fine Portion (<0.0625 mm) (frequencies are related to the fine portion only) Very few to very rare silt, no crystallites.

Fabric Classes with Skeletal Remains

Bioclasts-W-Coarse/Medium (CH-W-cm)

Fabric class code CH-W-cm general information:

No. of thin sections = 11
General fabric type: Skeletal remains and fragments of calcite homogeneous
Main characteristics: (1) skeletal remains and homogeneous micrite; (2) white refired color; (3) coarse-grained or medium-grained
Photomicrographs: Figure 8-35
Ceramic sphere: Tases

Summary of Characteristics in Hand Specimen (section 6.2):

Hand-specimen fabric: Micrite-Medium-White (WM-m-W), which is explained by the micritic nature of the fossil remains.
Hardness : soft
Refired color: white

CH-W-cm Thin section:

I) Microstructure (section 8.2.2)

- Porosity: Mostly compacted sherds with voids usually in the form of vughs taking 2 -7 per cent of field of view
- Packing (particles > silt) : double space
- Preferred orientation: no orientation

IIa) Groundmass: homogeneity and optical characteristic (section 8.2.2)

- Largely homogeneous groundmass
- white to v. pale brown 10YR 8/1 to 8/2, very light greenish grey (xp, pp); birefringent, faintly pleochroic matrix

IIb) Groundmass: coarser inclusions (section 8.2.1)

- Coarser Inclusions (>0.0625 mm) (frequencies are related to the whole thin section).
- the 30: 63-68:2-7, with the micromass' boundary 0.0625,

Common to Few: MICRITISED SHELLS and HOMOGENEOUS CALCITE – many recognisable fragments of skeletal remains and of homogeneous calcite (likely fragments of micritised skeletal remains); mostly prolate; many showing zoned extinction, setting them apart from micrite. They range in size from fine to coarse, mode is coarse to medium.
Rare to Very rare: MICRITE – single calcite, coarse, micrite-altered sparry calcite.
Very rare: QUARTZ – present with frequency of 1 or two particles in a thin section.

IIc) Groundmass: fine portion

Fine Portion (<0.0625 mm) (frequencies are of the fine portion only)
Very rare silt

III) Textural Concentration Features
Absent to a count of 3 or four are brownish-red to dark-brown-red (xp), with no internal structure, inclusions or orientation; high optical density; most with clear boundaries.

Bioclast-B-Coarse (CH-B-c)

Fabric class code CH-B-c general information:

No. of thin sections = 9
General fabric type: skeletal remains and homogeneous calcite
Main characteristics: (1) Skeletal remains; (2) light reddish brown refiring color; (3) coarse-grained or medium-grained
Photomicrographs: Picture 8-36
Ceramic sphere: Tases

Summary of Characteristics in Hand Specimen (section 6.2):

Fabrics: various (sparitic, micritic, or single crystals)
Hardness : soft
Refired color: light reddish brown

CH-B-c in Thin section :

I) Microstructure (section 8.2.2)

- porosity : varied, ranging from 5 to around 20 per cent of the thin section, with vughs and long channels
- Packing (particles > silt) : close
- Preferred orientation: no orientation observed

IIa) Groundmass: homogeneity and optical characteristic (section 8.2.2)

- Homogeneous:
  - Fired (original) color in xp, pp: yellowish brown to strong brown (7.5 YR 5/8), brown 10YR 5/3 opaque

IIb) Groundmass: coarser inclusions (section 8.2.1)

- Coarser Inclusions (>0.0625 mm) (frequencies are related to the whole thin section).
- The c:f:v is 15-50 :30-80: 5-20

Frequent to Common: SHELLS and HOMOGENEOUS CALCITE – a mixture of recognisable fragments of skeletal remains and of homogeneous calcite (which are likely fragments of micritised skeletal remains); mostly prolate; many showing zoned extinction, setting them apart from other micrite. They range in size from fine to coarse without a more frequent size.
Very Few to Very rare: DARK MICRITE (sub opaque in xp)

IIc) Groundmass: finer inclusions (section 8.2.1)

Fine Portion (<0.0625 mm) (percentages are of the fine portion only)
Varied silt frequency from very few to common

Dark Micrite Fabric Classes

DarkMicrite-Micrite-Medium (CG-CM-W-m)

Fabric class code general information:

No. of thin sections = 7
General fabric type: dark micrite
Main characteristics: (1) dark micrite,
Photomicrographs:
Ceramic sphere: Tases

Summary of Characteristics in Hand Specimen (section 6.2):

Hand-specimen fabric: GM-cm-W
Hardness : soft
Refired color: white

CG-CM-W-c in Thin section :

I) Microstructure (section 8.2.2)

- Porosity: very low, around 2 per cent
- Packing (particles > silt) : double
- Preferred orientation: no preferred orientation observed

IIa) Groundmass: homogeneity and optical characteristic (section 8.2.2)

- homogeneous : color may vary within a thin section from brownish grey to light grey
  - Fired (original) color in xp, pp: mainly light yellowish or reddish grey, but also brownish grey was observed

IIb) Groundmass: coarser inclusions (section 8.2.1)

- Coarser Inclusions (>0.0625 mm) (frequencies are related to whole thin section).
- The c:f: v is 10-30: 68-88: 2

Common to Few: DARK MICRITE -- dark brown calcareous mud (xp, pp), fine to very coarse, mode is medium
Very few to Very rare: MICRITE may be present; CALCITE HOMOGENEOUS -- Micritised fossil shells, medium to coarsely grained present in most thin sections.
Very rare: QUARTZ – fine grained quartz, with a count of 1 or 2 per thin section when present.

IIc) Groundmass: finer inclusions (section 8.2.1)

Fine Portion (<0.0625 mm) (frequencies are of the fine portion only)
No silt observed in most, rare dolomite crystals in thin section.

III) Textural Concentration Features

Very rare red concretions

www.ingramcontent.com/pod-product-compliance
Ingram Content Group UK Ltd.
Pitfield, Milton Keynes, MK11 3LW, UK
UKHW061846041026
15591UKWH00015B/939

9 781407 316406